ONE HUNDRED RINGS AND COUNTING:
FORESTRY EDUCATION AND FORESTRY IN
TORONTO AND CANADA, 1907–2007

Examining Canada's first Faculty of Forestry at the University of Toronto from its birth in 1907 to its hundredth year anniversary, *One Hundred Rings and Counting* is a detailed account of one of the country's most successful and influential institutions. Although its founding was marked by opposition arising from both the university's uncertainty of the field's importance and the provincial government's concern about how such an institution would affect the government's control over forests, the faculty has produced a disproportionate number of leaders in world of forestry and beyond.

Demonstrating the Faculty of Forestry's longstanding commitment to conservation and environmental stewardship, Mark Kuhlberg depicts its struggles with governments and the public to implement sustainable natural resource practices. Using unexamined archival materials and contextualizing the Faculty within the major educational, social, and political changes of the last century, *One Hundred Rings and Counting* is a solid institutional history that also traces the development of conservationism in Canada.

MARK KUHLBERG is an associate professor in the Department of History at Laurentian University.

Dave Fayle (5T7), who went on to teach at the University of Toronto's Faculty of Forestry (1978–1995), drew these sketches to illustrate the four main sites the Faculty has occupied over its existence. These include: 11 Queen's Park, a building it shared with the Department of Botany from 1908 to 1925 (top left); 35 St. George Street from 1926 to 1958 (top right); 45 St. George Street from 1958 to 1989 (bottom left); and Earth Sciences Centre from 1989 to present (bottom right). (Sketches courtesy of Dave Fayle).

One Hundred Rings and Counting

Forestry Education and Forestry in Toronto and Canada, 1907–2007

Mark Kuhlberg

UNIVERSITY OF TORONTO PRESS
Toronto Buffalo London

© University of Toronto Press Incorporated 2009
Toronto Buffalo London
www.utppublishing.com
Printed in Canada

ISBN 978-0-8020-9685-2

Printed on acid-free, 100% post-consumer recycled paper with
vegetable-based inks

Library and Archives Canada Cataloguing in Publication

Kuhlberg, Mark, 1966–
    One hundred rings and counting : forestry education and
    forestry in Toronto and Canada, 1907–2007 / Mark Kuhlberg.

    Includes bibliographical references and index.
    ISBN 978-0-8020-9685-2 (bound)

    1. University of Toronto. Faculty of Forestry – History.   I. Title.

    SD256.T6K84 2009   634.9071'1713541   C2009-903143-4

University of Toronto Press acknowledges the financial assistance to its
publishing program of the Canada Council for the Arts and the Ontario
Arts Council.

 Canada Council   Conseil des Arts    ONTARIO ARTS COUNCIL
for the Arts   du Canada   CONSEIL DES ARTS DE L'ONTARIO

University of Toronto Press acknowledges the financial support for its
publishing activities of the Government of Canada through the Book
Publishing Industry Development Program (BPIDP).

To my mother

~

May they live in a world of wonderful forests

Contents

Illustrations follow page 148

Acknowledgments

My forestry education began unceremoniously. A desperate sister working for a dubious reforestation company in northwestern Ontario, and urgently in need of help, coaxed me into believing that treeplanting was the life for me. While I was eager for the challenge ('There was life north of cottage country?' I wondered as a typical Torontonian) and the opportunity to generate desperately needed income, my experience was so miserable that I swore I would never return. Nevertheless, that was the summer of 1984, and it turned out to be the first of twenty seasons spent toiling in the wonderful woodlands of northern Ontario and Alberta.

While clearly it turned into far more than a one-shot deal, I had never imagined the insight my time in the bush would give me into our forests. In addition to learning a great deal about the practical side of silviculture (when given the choice between planting black spruce or jack pine on a humid, calm day, choosing the latter will leave me enough blood to donate later in life!), I came to see the myriad difficulties that pervade managing woodlands in Canada. In particular, I became fascinated by the disconnect that exists between the reality and perception of forestry. I arrived at this point because a string of foresters took the time to answer my endless questions about the forests in which they worked. They did so not as company spokespersons or industry representatives, but as folks whose future depended upon prudent stewardship. I was aghast when I learned that foresters were not the evil-doers I had been led to believe they were. Gary McKibbon was the first to teach me this lesson. Over the course of countless walks through a patchwork of cutovers in the Ear Falls/Red Lake area, he explained the trials and tribulations of practising forestry in Ontario. Later, 'Mac' Squires, Bill Smith, and Paul

Poschmann performed the same service when I worked in the 'Freehold' and up the Spruce River Road north of Thunder Bay. Back in Toronto's urban environment, Ken Armson, truly Canada's forestry grandpère, always found the time in his jam-packed calendar to continue my education. To all of these individuals, I owe an immense debt of gratitude.

All historians are equally indebted to countless others. In preparing a manuscript of any length, we are dependent upon archives for our manna, and the archivists can either make or break a project. Fortunately, I have had the privilege of receiving assistance from a gang of wonderful people in this field. For this project, they have included those working at the Archives of Ontario, the Biltmore Company Archives, Library and Archives Canada, the Queen's University Archives, the United States Forest Service (Forest Products Laboratory) Library, and the University of British Columbia Archives.

Those at the University of Toronto Archives deserve special mention. Harold Averill, Barbara Edwards, Marnee Gamble, Loryl McDonald, Lagring Ulanday, and Garron Wells provide truly exemplary service in answering my endless requests for materials. While I would like to say thank you for their effort in that regard, most of all I am grateful that they make the UTA's fourth-floor reading room such a warm and welcoming place to conduct research.

Other individuals also deserve credit for contributing to this effort. In addition to those who agreed to answer my questions about this history and whose names are listed in the back of this book, Susan Watts of UBC's Faculty of Forestry, Michael Roche of the College of Humanities and Social Science of Massey University in Palmerston North, New Zealand, and Dr Mary Ann Fieldes of the Department of Biology at Wilfrid Laurier University, deserve mention for enthusiastically providing me with information in a way that made my job infinitely easier. The two anonymous readers offered valuable and thoughtful feedback, as did Dr David Fayle and Dr Charles Levi. Harold Averill, archivist par excellence at the UTA, read the manuscript with a fine-toothed comb and identified a slew of errors and ways in which it could be improved: thanks – again – Harold!

I would also like to thank the University of Toronto's Faculty of Forestry for supporting this project. It is a rare privilege indeed to be invited to write an institutional history *and* be granted editorial autonomy. There were certainly queries about some of my conclusions and claims, but never the heavy-handedness that so often quashes the life out of such projects. Marilyn Wells, Ian Kennedy, and John McCarron were fantastic at facilitating this project at the faculty. Moreover, the faculty's book committee – Dr David Balsillie, Ken Armson, Dr Sandy Smith, and Dean Tat C. Smith – reviewed chapters and forwarded probing questions about them. Dr Shashi Kant was supportive from the outset and merits special mention. So, too, does Dr Sandy Smith, for her kindness in both words and actions. I must also acknowledge her advice in helping me grapple with the philosophical challenges inherent in enjoying the privilege of working at a job that one loves.

Then there is my family. Carling and Nolan continually remind me of the wonders of life; I hope that we can sustain healthy forests for them. My wife Cindy has been a constant pillar of support even when times were grim. And man, they were grim at times! Nevertheless, she helped buck up my spirits through the dark moments, and I hope that she understands how much I appreciate all that she has done and continues to do for me. Ironically, my sister deserves thanks for having – however unintentionally – cast my fate down this path that has taken me on a wonderful journey so far. And Marce. What can one say about a mother who was the best parent one could have imagined? As a single mom raising two kids with barely the means to do so, she led by example. She has always personified hard work and dedication to family, even when her son barely warranted such consideration. Saying thank you seems so trivial, but it is the best I can do.

Finally, the help I have received from these varied and invaluable sources was critical to any strengths readers may find in this book. Its weaknesses, however, are entirely of my own making.

ONE HUNDRED RINGS AND COUNTING:
FORESTRY EDUCATION AND FORESTRY IN
TORONTO AND CANADA, 1907–2007

Introduction

'The Most Spirited Faculty'

This is not a simple story to tell. Centennial histories are almost always celebratory tomes that, although they chronicle ups and downs, ultimately describe the gradual ascent of a subject whose survival speaks to its overall success. This is, however, hardly the case for the University of Toronto's Faculty of Forestry. Established in 1907 as Canada's first school of its kind, with an enrolment that could be counted almost on one hand, it grew surely and steadily. Although over time the faculty developed strong graduate and research programs, its core was its undergraduate program, whose alumni had earned the degree of bachelor of science in forestry and were credentialed members of the forestry profession. Then, in 1993, the University of Toronto dealt the faculty a mighty blow. The university abruptly decreed that the forestry faculty was to terminate the bachelor of science in forestry program to focus on graduate studies and research, thereby ripping out the faculty's very soul. This begs two questions: how does one explain the evisceration, and what is there to celebrate?

To answer the first question, we must start by explaining the nature of the forestry profession itself and particularly the factors surrounding its origins in Canada. Those in fields such as law, medicine, and dentistry, established faculties long after their practitioners had carved out their niche in society and ensconced themselves among the ranks of its intellectual elite. Even if some of these professions were still in their adolescence, society had at least recognized the utility of the work their accredited members do. As a result, there was no questioning of their privileged and respected standing, and the desirability of training more individuals in these fields. Not so with foresters. When Toronto's Faculty of Forestry was established in 1907, there were only a dozen foresters in Canada, and although there was some support for creating a school to educate more of them, there was little understanding of what silviculture – growing and cultivating trees – entails. Certainly, there was a desire to increase the efficiency with which the woodlands were being harvested, but there was precious little understanding of, or interest in, the complementary 'regenerative' aspect of the forester's job. As a result, both the forestry faculty's students and the true work of foresters lacked social legitimacy at the outset, and it will be seen, for the longest time thereafter. James Douglas 'Jim' Coats (5T2), secretary manager of the Canadian Forestry Association, lamented this situation just after the forestry faculty's fiftieth birthday. In early 1958 Coats urged a fellow alumnus to create a booklet that illustrated to Canada's youth precisely 'where the "Forester" fits into the social scale. He should be referred to in the same breath as doctors, lawyers, dentists and engineers. Many may think of him with lab technicians, rangers, etc. Somehow it should be pointed out that for the most part he is a civilized animal capable of doing executive work.'[1]

A closely related factor that influenced the nature and perception of foresters and the forestry faculty's work over the past century was the singular milieu – both political and intellectual – within which those in this field operated. The legacy of Crown control of the public domain in Canada meant that governments controlled the development of the country's natural resources. The elected officials were thus well positioned to manage our timberlands in a prudent manner. At the same time, since at least the late 1800s, Canadians have collectively professed a profound affection for our woodlands and celebrated the bountiful harvest that they have provided. While logic suggests that these forces ought to have coalesced into an exemplary level of forest stewardship in Canada, quite the opposite is true. Spending on silvicultural measures was and is highly unattractive to politicians, whose terms are five years at the utmost, because the return on the investment is not realized for decades. More importantly, notwithstanding our ostensible reverence for all that 'the woods' have given us, Canadians have paradoxically refused to attach a cost to this windfall and have acted instead as if we have an inherent right to enjoy it. The result has seen elected officials respond to the electorate's relative disinterest in 'managing' the forest by neglecting their fiduciary duty to do so. If it registered at all, it has typically ranked near the bottom of voters' priorities. In *The Politics of Development*, H. Vivian Nelles describes the forest in Ontario prior to 1941 as something 'from which revenue was derived, not something upon which money was to be spent,' and sadly this situation has persisted until very recently.[2]

To be sure, attitudes did change over time, but this did not make the work of foresters and those training them any easier. For most of the twentieth century, the lack of public support for forestry in Canada prevented the Faculty of Forestry's graduates from practising their profession in the field. John Walton (2T4) fittingly captured the essence of this situation in the mid-1920s in a letter to James Herbert White (0T9), the faculty's first graduate and one of its veteran professors: 'I often wonder if Dr Howe [the faculty's dean from 1919 to 1941] was absolutely correct in saying that a forester is a "John the Baptist crying in the wilderness."' Walton added that, in his view, the forester was 'an Israelite in Egypt, trying to make bricks without straw. Possibly the true forester is both.' When, in the late 1960s, society finally began to recognize the need to care for its woodlands ironically the forestry profession became tainted. Those engaged in raising timber crops were now vilified for aiding and abetting an enterprise that was deemed injurious to the environment. The upshot saw forestry emerge as one of Canada's most stigmatized professions.[3]

These and other factors also turned forestry into arguably the country's most ill-defined profession. Individuals engaged in other pursuits, such as medicine or engineering, have seen technology dramatically alter the nature of their work over the past century. One hundred years has not, however, altered their fundamental raison d'être. The doctor still aims to ensure a patient's health, the engineer the bridge's stability. The same cannot be said of forestry. While forestry is defined as 'the science and practice of planting, caring for, and managing forests,' there has rarely been agreement between those inside and those outside the profession regarding what this actually means. Initially, foresters were expected to ignore their training in the field of regenerating the woods and instead focus on extracting timber as cheaply as possible. Then, long before Canadians had awakened to the need to care for their woods, they had begun to insist that many tracts of forest be administered in a manner that paid pre-eminent attention to recreational values. As more time passed, the number of considerations only increased, leaving foresters in a most unenviable position. They could not simply apply their technical knowledge because each of Canada's myriad constituencies had its own definition of what constituted proper forest management. Thus, what the forester ought to be doing in caring for the trees has been a slippery concept that has meant very different things to different persons at different times in different locations.

The Faculty of Forestry also had to contend with major challenges at its host institution, the University of Toronto. Politics shaped developments at the 'provincial university' from the time King's College, its forerunner, was founded in 1827. Elected officials at Queen's Park determined the amount of funding the university received each year, thus, for example, leaving them free to wield major influence over the forestry school. More importantly, the University of Toronto was rarely comfortable with the Faculty of Forestry, and embraced it on only the rarest of occasions. Furthermore, the university spent much of the twentieth century striving to be Canada's leading centre of graduate studies and research work, an evolution that was incongruent with the continued existence of tiny professional schools that taught mere dozens of undergraduates.

At times the Faculty of Forestry's own behaviour made the uphill battle in which it was already engaged that much steeper. It is probably the only academic division at the University of Toronto whose deans fired two of their own colleagues, and within its first few decades of existence no less. While the first termination was justified, the second enshrouded the forestry school in a lingering melancholic haze. More significant, however, was the ultra-conservatism that the faculty displayed in the face of rapidly changing conditions. This intransigence was supported by fervent adherence to tradition and the occasional tinge of hubris that made the forestry faculty feel as though it was both good enough not to change and far more important in Ontario and on the University of Toronto's campus than it actually was. This attitude was clearly illustrated in the late 1980s, when the faculty was advised to alter its name to better reflect forestry's broadening scope. On this occasion, the late Donald David Lockhart (5T0) wrote to Justin Roderick 'Rod' Carrow (6T1), the faculty's dean, to express his unequivocal opposition to this plan of action. 'I'm agin it,' Lockhart declared. He stressed that he was a strong advocate of constructive change, but that 'adding some jazzed-up mumbo jumbo [to the faculty's name] does nothing. The person who dreamed up the proposed changes is probably the same person who wants to modernize the Lord's Prayer. A pox on him or her.'[4]

As a result of these factors, this story is as much about Canada's human ecology as it is about the country's oldest forestry school. In many ways the faculty's establishment and development was a mirror that reflected the temper of our times, particularly in terms of environmental stewardship. The Faculty of Forestry's first hundred years can thus be seen as so arduous because we have collectively had such a difficult time coming to grips with the ever-more pressing need to live in harmony with the planet that supports us.

This may leave the reader wondering how, in fact, the Faculty of Forestry is even still with us, but there is a good explanation for both its continued existence and success. For despite all the doom and gloom, and its external and internal challenges that turned its development into what seemed like a never-ending battle against all odds, the forestry faculty not only survived but prospered. Herein lies the ground for celebration.

The forestry faculty devised creative strategies to endure, and often thrive, within this difficult setting. Whereas the faculty's small size – it was consistently one of the University of Toronto's smallest undergraduate schools – often worked against it when it was confronted by external challenges, it learned to derive strength from being the underdog.

One major way in which the forestry faculty did so was to capitalize on its relatively small class sizes and have its professors teach its students much more than simply 'the three Rs.' From the moment it opened its doors, the faculty's academic staff focused their efforts on producing not merely exceptional foresters but also, and perhaps more importantly, exemplary, well-rounded citizens. The school recognized the many challenges that its graduates would face in confronting political and public opposition in their professional careers. As a result, the guiding forces behind the forestry faculty strove to produce technical experts whose model characters would eventually earn the respect of senior bureaucrats and elected officials alike, and thereby influence forestry policy.

There was also the cohesion among the members of the faculty's student body. Until the recent termination of its undergraduate program, the forestry faculty's students had not only taken practically every class together, but they had also worked and lived side by side during their field camps. The upshot was an intimacy that fostered tight-knit, fiercely proud, and extraordinarily vivacious graduates; to this day many of the forestry faculty's surviving 'classes' gather annually to reminisce. Andrew Lewis Kenneth 'Ken' Switzer (3T4) personified this spirit. After graduating, he acted as secretary for the 'Class of '34.' He not only compiled a newsletter each year updating his cohorts on their confrères' activities, but also organized their annual rendezvous. As they died, he continued to send their widows his annual bulletin. Fittingly, when Switzer passed away in 1999, he was the final member of his class to die. To be sure, this elan was not simply a product of a bygone era. Peter John Johnson (8T9) congratulated incoming 'frosh' in the summer of 1987 on choosing to study forestry at the University of Toronto. 'You can take pride in the fact that you are entering the most "Spirited" faculty in Canada's largest and best university,' he rightfully boasted. 'Last year, the Faculty of

Forestry in association with Nursing won the university-wide Spirit Challenge. This alone indicates the high profile and extensive participation we exhibit despite being one of the smallest faculties in the University ... The Foresters' Club wants to help you discover the entire University and get the most out of it, but we need your participation. We want these next four years to be your most memorable.'[5]

While the school's deans provided paternal support to the 'the Faculty family,' their female assistants were essential to its functionality in terms of providing both motherly advice and the glue to keep the kin united through good times and bad. In succession, Olivia Fernow (the first dean's wife), Grace McAree, Marion Harman, Pat Balme, Marilyn (Candy) Wells, and Amalia Veneziano played this critical role at the Faculty of Forestry. Fittingly, when McAree died in October 1950, the Forestry Alumni Association set up a fund in her memory.[6]

Moreover, the Faculty of Forestry derived immense strength from the nature of its curriculum. Although it did not adapt its undergraduate program as much as it could have in the few decades prior to its demise, its course offerings exposed its students to a wonderful breadth of subjects, something with which Robert Dicks 'Bob' Carman (5T4) was intimately familiar. As one of the forestry faculty's most eminent graduates, he had been deeply involved beginning in the late 1970s in discussions about updating its curriculum. On the eve of being named secretary to the cabinet, Ontario's highest-ranking civil service position, in November 1984, Carman urged the forestry faculty to focus on its core courses, which he believed provided the students with skills that were crucial to rising to upper management positions. 'It is no accident,' he told Dave V. Love, the faculty's acting dean, 'that so many of the Faculty's graduates have attained senior ranks throughout government ... It is not too difficult to extrapolate the lessons of ecology and silviculture to the domains of societal competition and issue resolution and the biology of organization life.'[7] In addition, learning about forestry gave the faculty's students an appreciation for bringing a long-term vision to their work. Whereas other professions conceived of issues in terms ranging from nanoseconds to maybe years at the most, foresters framed their work in terms of decades if not centuries. They had been trained to see the forest *and* the trees.

The upshot of these factors is an irreconcilable conundrum: despite all the forestry faculty's problems, it produced graduates who were, in the main, truly extraordinary. Counted among them are a disproportionately high number of presidents of forest and other companies, ministers and deputy ministers, and ambassadors. Considering it has only generated roughly 2,500 graduates at both the undergraduate and graduate levels, the Faculty of Forestry has been one of the University of Toronto's most successful academic divisions in terms of producing stellar alumni.

There is another important layer to this story. As much as the forestry faculty appears to be a radically different place today than it was at the time of its establishment, its history has been stamped by several consistent themes that have woven their way through its development from year 1 to year 100. Some of these are as pronounced as ever. For example, the faculty has always had strong international connections in terms of its student body, academic staff, and professional involvement. Similarly, nearly all the students it has drawn were attracted by their love of the woods, and many were bona fide 'environmentalists' long before the term came into vogue. That most were prevented from acting on this impulse by forces that have already been described should never be mistaken for evidence of their lack of concern for the woodlands in which they worked. Lloyd Morley Lein (3T7) grew so distraught when he came to grips with this disheartening situation at his first job with industry in northern Ontario, in the late 1930s, that he sought solace from his dean in Toronto. 'I still want to be a forester,' Lein lamented, 'but every day sees me approaching the career of a forest engineer. As long as I stay here with this company the more remote becomes my desire and yet there seems to be no opening for people who wish to plant trees, create forests and parks and otherwise add to the material resources of our country along this line.'[8]

And as much as this story analyses the evolution of Canada's first forestry school since 1907, it also tells numerous other 'histories,' all of which influenced – to varying degrees – the development of the Faculty of Forestry at the University of Toronto. Some of these forces are surprising. Religious differences that marked Canadian society at the turn of the twentieth century, for example, played a major role in determining where the faculty was established. Likewise, social perceptions of what constituted 'women's work' determined that the faculty was a male enclave for over half a century. This story also touches upon urban and provincial history, specifically in terms of how the forestry faculty benefited from the decision, taken by a handful of the country's financial tycoons, to preserve their estates within Ontario's boundaries. Some of the other historical factors that shaped the forestry school's first century were predictable. At the top of this list is society's changing view of the forest and how best to interact with it. Closely related is the history of forestry in Canada in general, and particularly the repeated promises to effect meaningful reforms with little action taken to back them up. Not surprisingly, this is also a tale of university education over the previous century both across the country and in its largest urban centre, and thus it provides valuable insight into both the diffusion of 'higher' education from Canada's metropolises to the country's hinterland and the lean days of budgetary constraints since the 1970s. In addition, this story follows the peaks and valleys that both Canada's and Ontario's economies experienced, especially in their respective forest industries. It also recounts the horrendous toll the two world wars took on our country, particularly on Canada's youth, and it can only be understood within the political environment in which it occurred.

Finally, this tale combines the traditional with the not-so-new approaches in terms of examining events from both the top-down and the bottom-up. At its core, this is certainly an institutional history of a forestry school. It focuses on the Faculty of Forestry at the University of Toronto as both an actor influencing developments and an entity upon which contemporary forces acted. More often than not, this is a tale of the forestry faculty's attempt to navigate through waters, often hazardous, in which it was usually buffeted about by people who simply did not value much of what the faculty was doing. At the same time, this is the story of individuals who either worked at or graduated from the Faculty of Forestry at the University of Toronto. These people, and not the forestry building nor the faculty's curriculum, gave the school its vitality. Through telling a smattering of their stories, we gain insight into the heart and soul of the Faculty of Forestry and its contribution to society at large.

In the end, both the evisceration and the celebration are understandable. Undeniably, the history of the University of Toronto's forestry faculty's first one hundred years is fraught with the struggles it endured and the many setbacks it suffered, and it is sprinkled with but a few boons that came its way. At the other end of the spectrum, there is every reason to exalt its graduates. This is thus the story of an institution the sum of whose parts is greater than the whole.

Chapter 1

'There Is Nothing in It Practically for the Government,' 1894–1907

Canada was a fledgling nation in the late nineteenth century, on the cusp of a series of new beginnings. Although still under Great Britain's wing, the country was about to undergo a period of dramatic development. Over the course of the 'Laurier Era' (1896–1911), Canada would experience the arrival of hundreds of thousands of immigrants from Europe, the construction of two new transcontinental railways, and an unprecedented economic boom. These events made Laurier's prediction of the twentieth century being Canada's seem possible. With Ontario's storehouse of untapped natural resources in its hinterland, the province was well placed to share in the period's growth and optimism about the future.

Yet the young country was not without its problems. All did not share equally in the prosperity, as large numbers of the working class laboured and lived in less than pleasant conditions. More importantly, serious cleavages divided Canadians along racial and religious lines. The country's sizable minority of francophone Catholics, most of whom lived in Quebec, had already learned that, when issues affecting their world rose to the national level, the views of the anglophone Protestant majority would triumph. But even the latter group was not united, as there were many strains of thought under its banner. In an era when one's faith truly mattered, these divisions ensured that battles would be waged among and within religious denominations.

These forces all influenced the founding of the Faculty of Forestry at the University of Toronto, and the specific circumstances surrounding its establishment explain much about its development over its first century. From our colonial heritage and the British North America Act, Ontario inherited the principle of 'Crown' ownership of natural resources. This dictated that politicians would determine who would cut timber, how much was paid for this privilege, and the regulations governing the harvesting. The provincial government would also decide when, where, and how to establish a forestry school because the BNA Act gave the provinces jurisdiction over education.

But creating the new forestry school turned out to be an intensively competitive undertaking. By the early 1900s, three candidates – Queen's University in Kingston, the Ontario Agricultural College (OAC) in Guelph, and the University of Toronto – were aggressively lobbying for the provincial government to place it on their campuses. Logic suggested the foresters ought to be trained at either Queen's or Guelph. The former was located in the heart of eastern Ontario, which had been one of Ontario's first timber production centres and still served as the home base for many of Canada's iconic lumbermen. Queen's also taught many of the 'support' courses that a forestry program would require. Situated in farm country, the OAC (forerunner to the University of Guelph) specialized in teaching courses akin to silviculture, supported a large arboretum, and had begun offering forestry courses. Moreover, those who led the campaigns at Queen's and the OAC to win the forestry school were clearly enamoured of their projects. They truly wanted the new program per se. While the University of Toronto offered the supplementary courses that forestry students would require, it had little else going for it. Nevertheless, its president was determined to 'win' the forestry school because he wanted to deny both Queen's and the OAC – particularly the former – this privilege. There is little evidence that he truly wanted the program for its own sake.

When it came time to select the winning suitor, political expediency, and not logic, drove the Ontario government. The Liberals had ruled the province since Confederation, but by the early 1900s their hold on power was slipping. Those who wished to see the new forestry school land in Toronto sensed the government's vulnerability, and they pounced. No sooner had they slyly communicated to the Grits that the latter's political fortunes hung in the balance over the question of where the foresters would be trained than the University of Toronto was declared the winner in this bitter fight.

Political considerations also determined other key aspects involved in founding the forestry school. In the context of early twentieth-century Canada, public interest in forestry was growing but it still ranked very low on the national and provincial governments' lists of priorities. Neither investing in meaningful timber management nor training a cadre of foresters was a pressing concern for the politicians. The forestry school thus came into existence in Ontario largely because its creation rode the coattails of an overall policy aimed at dramatically improving the quality of education at the University of Toronto. Politics also shaped who would become the Faculty of Forestry's inaugural dean. All interested parties agreed on the ideal candidate – a young, dynamic forester from the Maritimes, but he declined the offer after learning first-hand that the Ontario government was uninterested in implementing bona fide forestry reforms. The position was then offered to Bernhard E. Fernow.

Fernow accepted the post, but again political considerations – this time both at the provincial level and within the University of Toronto – determined that he did so only after being hoodwinked in a most cruel way. In his eyes, taking on this job was an integral part of effecting a veritable revolution in forestry in the province in particular and the country in general. A few of Fernow's professional colleagues undoubtedly shared this vision, but those at Queen's Park and on the University of Toronto's board of governors definitely did not. The politicians saw the forestry school essentially as a nuisance that they would be forced to tolerate, while the university's senior administrators viewed it as a miniscule part of their overall operation deserving of but nominal support.

From the outset, then, the manner in which the 'provincial forestry school' was established at the University of Toronto was striking. In so many ways, it was a harbinger of things to come.

As a profession forestry came to North America relatively late. Its advent is associated with the rise of the conservation movement in both the United States and Canada in the late 1800s. This movement was fuelled by several forces, including a realization that resources were not inexhaustible and a hope that scientific principles could be adopted when exploiting resources to ensure that they were developed with maximum efficiency. This meant that, as far as harvesting trees was concerned, professionals who were trained to raise successive crops of them in perpetuity would be needed. So, too, would administrative bodies and legislation to regulate how the publicly controlled woodlands were managed.

The upshot was a string of landmark developments in the few decades before the turn of the twentieth century, although their import was more symbolic than real. Bernhard Eduard Fernow, a Prussian immigrant about whom much more will be said in the pages that follow, broke the ice in North America when, in the 1870s, he became the continent's first practising professional forester. Over the next few decades, the Ameri-

can and Canadian governments established departments (albeit relatively tiny 'divisions' or 'bureaus') to administer the timber under their
jurisdiction. Fernow, for example, served as chief of the U.S. Division of
Forestry from 1886 to 1898. He left this post to create the continent's
first professional forestry school, at Cornell University. However, these
steps did not translate into sound forest management in either Canada
or the United States. Foresters were few in number, and they exercised
very limited influence over the formation and enforcement of policy.[1]

This same paradigm marked Ontario's early forestry history, whereby landmarks were passed but only nominal tangible progress was realized. The province enacted legislation to protect the forest from fire in
1878, for example, and began training a crew of fire rangers seven years
later, but the conflagrations seemed to devour ever-increasing expanses
of woodlands each season. Likewise, the Ontario government implemented the Forest Reserves Act in 1898, which authorized it to set aside
prime tracts of timber to ensure a future wood supply, but it did little to
manage these areas.[2]

These developments prompted W.L. Goodwin, the director of the
School of Mining at Queen's University, to suggest that his institution
ought to establish a course in forestry. As early as 1894 Goodwin asked
the senate at Queen's to invite Fernow to Kingston to deliver a series of
lectures as a preliminary step to creating the new program. Although
Queen's lacked the funding to achieve this second aim, for the next decade Goodwin kept the matter before his school's administration and
began corresponding regularly with Fernow about it. In 1901 the board
of governors at Queen's brought Fernow to Kingston to lecture on forestry and authorized the expansion of the School of Mining to include
a forestry program. Because this initiative required the approval of the
provincial government to be implemented, George Grant and William
Harty, respectively the principal of Queen's University and chairman of
its School of Mining, opened negotiations with the politicians.[3]

Initially at least, it appeared that Queen's would win its case. In 1901
the reigning Liberals enacted legislation that authorized the School of
Mining to add forestry to the courses it taught. When the Grits made
granting the funds needed to support the forestry course contingent on
Queen's financing the erection of a new building in which a number of
new courses (including forestry) could be taught, the university responded promptly. At the end of April 1902 Sir Sandford Fleming was among
those involved in laying the structure's cornerstone. In a speech to mark
the event, Richard Harcourt, Ontario's minister of education, heaped

laurels on Queen's for opening 'new fields of usefulness'; Harcourt was already corresponding with Fernow about the cost of hiring a forestry lecturer. To ensure the government did not forget its promise to ante up once the building was completed, the *Queen's Quarterly* published a series of articles throughout 1902 that described in detail the steps the university was taking to establish its forestry school. As a capstone to this public relations campaign, Queen's arranged for Fernow to deliver a series of lectures in early 1903 on various aspects of forestry. His talks captured national media attention, and the university saw them as the final step on its way to opening the doors of its forestry school in the fall of that year.[4]

By this time, however, another suitor for the forestry school was also knocking at the provincial government's door. James Loudon had been head of the University of Toronto since 1892, and by all accounts, he had been a bull-headed character for long before that. Loudon's biographer describes him as having been a 'failure in his public and personal relations,' and a 'stern, diffident, and unbending' man who 'was often loath to compromise.' Not surprisingly, news that Queen's was about to be awarded the forestry school set Loudon off. His primary motivation was not a fundamental belief in either the efficacy of forestry or the University of Toronto's suitability to host a school to train foresters. On the contrary, his animus was the desire to deprive a rival university of a prize for which it had long been vying.[5]

Loudon thus set out to convince the Ontario government to locate the forestry school at the University of Toronto, and presented his case in the fall of 1902. He insisted that this made perfect sense because the university already boasted most of the support that forestry would need, such as a Faculty of Arts, botany professors, and associated subjects like surveying. All that was required was the appointment of a professor of forestry and the purchase of the necessary equipment. Moreover, Loudon emphasized that there was another 'incidental advantage of no small importance connected with the establishment of such a school here,' namely, its proximity to the – albeit fledgling – provincial forestry service. As he put it, locating the forestry school in Toronto would mean 'the feasibility of co-operation between the proposed institution and the Provincial Department of Crown Lands in the work of Forestry.'[6]

In the meantime, Loudon took steps to knock out of the running another contender for the forestry school, one that seemed far better suited to hosting it than the University of Toronto. James Mills was president of the Ontario Agricultural College in Guelph, founded in 1874 and affiliated with the University of Toronto thirteen years later. Over the course of 1901–2 Mills began lobbying the provincial government. He argued, inter alia, that the OAC had been delivering 'forestry-type' courses since the early 1880s and was now offering one at the senior level that consisted of lectures and practical work on the plantations that students had set out over the previous few years. Despite the strength of its appeal, by December 1902 it appears that Loudon had co-opted Mills' campaign. Instead of fighting for the forestry school to be established in Guelph, Mills agreed to support Loudon's drive to locate it in Toronto in exchange for having the prospective forestry students attend the OAC for six weeks of instruction during their summer term.[7]

With Mills in his corner, Loudon recast his sight on the Ontario government over the winter of 1902–3, but to no avail. In one memorandum to the incumbent Liberals, he underscored the cost savings that would accrue from establishing the forestry school in Toronto. If it were based on the 'Yale model,' which was a graduate program, the new school would require only one professor initially (the other subjects would be covered by the university's existing faculty) and utilize the OAC's facilities and faculty for a 'summer school' session. Loudon argued that taking these steps would cost merely $5,000, but added that a 'necessary adjunct to the organization of the School would be the reservation of a suitable tract of wild forest land, to be placed at the disposal of the School, for practical work in Forestry to be done during the summer.'[8] Loudon then turned to the University of Toronto's senate for help. By January 1903 it had presented a statute to the Ontario government that called for the creation of a 'Professorship in Forestry' and a three-year curriculum, a move the Toronto press applauded. At the same time, Loudon publicized his campaign in the *Educational Monthly* and in the 'college section' of the *New York Post*. The problem for Loudon, however, was the Liberals' disinterest in facilitating his plans because of their standing commitment to Queen's.[9]

Fearing that victory was slipping from his grasp, Loudon stepped up his campaign in a most devious way. In succession, he called on his religious (Methodist) and political (Liberal) allies for assistance. As he cryptically put it in his memoirs, 'I decided to bring matters to an issue in the matter as between Queen's and Toronto.'[10]

In this regard Loudon benefited from the fact that post-secondary education, specifically its religious aspect, was a highly contentious issue in Ontario at this time: in 1901 Ontario's premier George W. Ross declared that 'the University question is the most dangerous one we have taken up.' Queen's was front and centre in this controversy. Since its inception in 1841 it had been a Presbyterian university, and the province did not fund non-secular, post-secondary institutions. In 1893, however, Queen's had ingeniously circumvented this obstacle by naming its science program a 'School of Mining and Agriculture,' which the government agreed to support. During the early 1900s Queen's pushed for the government to pick up the full tab for its operation, including the cost of establishing the new forestry school, but herein lay the problem. The Presbyterian Church was resisting a move by Queen's to become independent, which made it highly risky for any politician to back the move to provide Queen's with full public funding.[11]

And Loudon knew it. He thus chose to attack Queen's where it was most exposed by plotting a strategy that relied on the input of those occupying the upper echelons of the Methodist Church. His cabal included the Reverend Nathanael Burwash, president and chancellor of Victoria College in Toronto and president of the Methodists, and James Mills, fellow Methodist and the president of the OAC. Loudon and Burwash published a series of articles that questioned the state's inclination to fund a 'clerical' university. In addition, Loudon was aware that senior Methodists (including Burwash) were about to meet with provincial officials (including Premier Ross) to discuss the question of funding Queen's. Loudon saw it as an opportune moment to deliver a stern warning to the government on the matter and 'the only way to block the movement' to establish a forestry school at Queen's. Not only did the Methodists deliver the message, they did so on the eve of the premier's meeting with 'a large deputation' that lobbied for 'aid for [the] School of Forestry at Kingston.'[12]

With the battle between Toronto and Kingston intensifying in the press in the spring of 1903, Loudon tapped his political contacts to quash any remaining hopes Queen's entertained of landing the new forestry program. He and his wife had been instrumental in establishing the University of Toronto Alumni Association in April 1900 as a means of strengthening the University of Toronto's lobby for greater government support. Within a few years the organization boasted over twenty branches in Ontario, and Loudon knew full well the political weight they could bring to bear at Queen's Park. In mid-1903 Loudon directed this powerful ally to exert intense pressure on the Liberal government, which held only a razor-thin majority in the Legislature. Loudon sternly warned the Grits that, if they dared offer Queen's financial support for its science program, including a new forestry school, the 'liberal graduates of

Toronto and Victoria [College]' would turf them from office post-haste.[13]

Loudon soon learned that his stratagem had paid off, but the news came with a clear indication of how relatively unimportant the forestry school was to the government. In early June 1903 Loudon had written to Richard Harcourt, the minister of education, asking for the government to clarify its position on the matter. Loudon reminded Harcourt that only $5,000 would be needed to create a forestry program at the University of Toronto, and delivered a poorly veiled threat regarding the repercussions the Liberals would feel if they did not respond appropriately to his inquiry. Harcourt replied that his government would not fully fund Queen's under the present circumstances and explained that he was 'very anxious' to establish a department of forestry in the province because its importance 'having regard to the natural conditions of the Province cannot be questioned.' Harcourt had thus sought advice on the matter from Fernow, who had indicated that the project could begin by spending a mere $1,500 on a lecturer. While Harcourt admitted that such a start would be 'a very modest one,' he added that 'the way would have been prepared for gradual expansion, and we would soon have a fully equipped department. Not content with this,' Harcourt stated pointedly, 'your requisition called for $5,000 for the first year and contained an intimation that this sum would suffice only by way of commencement. To this requisition my Colleagues would not consent, especially since I had suggested that a Lecturer in the first instance, and as a mere beginning, would suffice.'[14]

While Loudon's political manoeuvring had undeniably played a role in achieving his goal (by mid-1903 the Liberals were publicly back-pedalling from their promise to grant Queen's the new forestry school), other factors contributed to precipitating the government's about-face. Representatives from the university in Kingston learned from their sources within the government's inner sanctum that the Liberals were not simply refusing to fund a forestry school at Queen's. The Grits were not keen to support *any* forestry school, independent of its location.[15]

This knowledge impelled Queen's to shift gears; it would continue its drive to obtain the forestry school because doing so would give it an immensely valuable bargaining chip in dealing with the Ontario government. Over the second half of 1903 Queen's continually reminded the Liberals of their previous promise to grant it the forestry school. It also demonstrated its continued commitment to the project by lining up a string of high-profile silvicultural experts to deliver another lecture series. Concomitantly, Queen's intensively lobbied the province for more money for the university as a whole. By this time it was desperate for financial support, especially for its science program, and its senior administrators realized that they could sacrifice the forestry school for the university's greater good. G.M. Macdonnell, Queen's University's solicitor and one of its most senior trustees, aptly captured the need to focus on the forest and not just the trees. As he advised W.L. Goodwin, director of the School of Mining and the spearhead in the movement to acquire the forestry school, 'we must let Forestry alone, just now having bigger game to follow.'[16]

Soon enough Queen's had bagged its 'bigger game' in a shrewd quid pro quo that was a real lifesaver to the cash-strapped institution. While the backroom negotiations are difficult to decipher, Edward J.B. Pense, the member of the provincial parliament for Kingston, announced in March 1904 that Queen's had withdrawn its claim for the forestry school in favour of Toronto's. Officials at the University of Toronto learned that Queen's had taken this step because 'the Government agreed to give Queen's, or the Mining School, $6,000 a year to be released from their [i.e., the Ontario government's] promise to establish the Forestry School there.' As D.D. Calvin notes in his 'original' history of Queen's, the upshot was a 'benevolent attitude of the Provincial Government to the School of Mining' that was 'of great help to the Trustees' over the next few years.[17]

While the University of Toronto now had clear title to the forestry school, there was still the problem of the provincial government's meagre interest in funding it. The Liberals again demonstrated their stand on the issue over the course of 1903–4, when officials from the University of Toronto were applying intense pressure on them in the hopes of landing public grants to fund new buildings for physics, chemistry, mining, and the new forestry program. The effort culminated in a powerful delegation (300 strong) of the university's supporters meeting the premier and presenting its demands (along with a petition signed by over 1,400 students). The lobby proved successful as far as the physics, chemistry, and mining initiatives were concerned, but not on the forestry front. As Premier Ross explained to the deputation in late March 1904, 'as to the School of Forestry you are not on the right line'; in his view, 'the country was not ready' for one. It would be far more important to develop a cadre of 'instructors' who could be sent to the existing forestry schools in the United States and Europe to learn about this new profession. This group could then educate farmers about woodlot management over the winter and 'look over' northern Ontario during the summer season. 'The Crown Lands Department,' Ross concluded, 'with its rangers, its forest reserves, [and] its relations with lumbermen could cover a larger area than any chair at a university.'[18]

While the government pleaded that its finances did not afford it an opportunity to fund the new forestry school, politically savvy observers recognized that this excuse was a red herring. The Liberals had, after all, just committed to spending $180,000 on the University of Toronto's new physics building, and only a fraction of this total would be needed to establish the forestry program. The BC Lumberman wondered aloud about the government's feeble rationalization. Noting that Premier Ross had insisted that, with regard to the forestry school, 'the state of the Provincial finances did not warrant an appropriation for that purpose. The fact is,' the magazine pointed out, 'there is nothing in it practically for the Government – no rake off for any of their supporters or future contributors to campaign funds – and a proposition which depends entirely on its merits without ulterior consideration of this sort, stands a poor chance with the politicians.'[19]

The Liberals thus went down to defeat in January 1905 without establishing the forestry school, but they had taken one auspicious step in this general field. Judson F. Clark was a native of Prince Edward Island who had earned his bachelor of science degree in 1896 from the Ontario Agricultural College in Guelph. He then obtained his master's and a doctorate in forestry under Fernow at Cornell University and began teaching there during his post-graduate studies. Clark had so impressed James Mills, the OAC's president, that Mills had begun lobbying in 1903 to have the Ontario government hire Clark to advise it on forestry matters. The effort bore fruit in May 1904, when the premier named Clark Ontario's first provincial forester with a mandate to act as the government's private consultant.[20]

While Clark's appointment and the electoral victory in early 1905 of the Conservatives under James Pliny Whitney, who promised to 'reform' the administration of the province's woodlands, seemed to bode well for forestry in Ontario, this was not to be the case. In fact, the Tories were more reluctant than their predecessors had been to take meaningful steps to manage Ontario's Crown forests.[21]

They demonstrated their penchant in this regard almost immediately on taking office. One of the Conservatives' first acts was to transfer the Bureau of Forestry, in which Clark worked, from the Department of Lands and Forests to the Department of Agriculture. This reorientation reflected the Tories' casting silviculture as 'farm forestry,' an activity that embodied almost exclusively reforestation in 'Old Ontario' and the management of farmers' woodlots. Edmund John Zavitz would play the leading role in achieving these ends over the next half-century, an endeavour that would earn him the moniker, 'the Father of Reforestation in Ontario.' In 1904 the Liberals had hired him to establish a small nursery at the OAC to raise seedlings that could be given away to farmers who agreed to plant windrows and the drier, 'blow sand' tracts of their properties. Zavitz left Guelph in the fall of 1904 to pursue his master of science in forestry at the University of Michigan, but the Conservatives rehired him upon his graduation the following spring.

The job the Tories gave Zavitz undermined hope that Ontario would ever establish a genuine forestry school. In May 1905 the Conservatives appointed Zavitz 'Lecturer of Forestry' at the OAC. This would minimize the expenses associated with teaching 'forestry' in the province (the OAC was already engaged in this endeavour), and the OAC's traditional focus had been on farm, not industrial, forestry. Moreover, Zavitz's main duties were to develop forest nurseries in southern Ontario and survey the region's larger waste areas. Fulfilling this latter task would result in his landmark *Report on the Reforestation of Waste Lands in Southern Ontario* in 1908.[22]

The government's support for Zavitz and reforestation in the south was a sharp contrast to its utter disinterest in Clark practising forestry in the north. Its decision to transfer Clark, the province's only forester, to the Department of Agriculture had conveniently left him in a bureaucratic straitjacket. Clark's administrative domain was now facilitating settlement in the hinterland, even though his expertise was in raising forest crops. When Clark proffered sage advice regarding forest management to the Tories over their first year or so in office, much of which recommended that foresters and not politicians set timber policy, they obstinately refused to entertain any of it. It will become clear just how deeply this experience grated on Clark.[23]

The Whitney government was a highly unlikely candidate, then, to establish the province's, let alone the country's, first forestry school, but a number of factors coalesced to cause the Conservatives to take this step. In the end, their actions spoke to their desire to further the University of Toronto's interests, and not forestry's.

J.P. Whitney, the premier, and his profound concern for the University of Toronto's vitality, were integral to this process; he has been described as having been possibly 'the greatest friend the institution has ever had.'

Shortly after he won power, Whitney took the unprecedented step of providing the University of Toronto with stable and sufficient funding. By the summer of 1905 he had established a royal commission to investigate the university's affairs through his – and not the minister of education's – office. He ensured that the commission's members collectively represented, as Whitney's biographer puts it, 'an exceptional group of public-spirited commissioners,' and that they undertook a comprehensive and far-reaching investigation. The product was a report that acted as a clear and effective road map for the university for the better part of the next seven decades. Not surprisingly, the Tories accepted nearly all of the report's recommendations, including one calling for the establishment of a forestry school in Toronto.[24]

The inclusion of this recommendation was a function of several forces. One of the report's key thrusts was the creation of a more diverse university that gave greater emphasis to practical training, specifically in fields germane to developing Canada's vast natural resources. Applied science and engineering were on this list, and B.E. Walker ensured that forestry was also there. Walker was one of the commission's five members, a leading Canadian businessman (the general manager of the Bank of Commerce), and a long-time proponent of establishing a forestry school in Toronto. He had warned Whitney shortly after the latter had ascended to the premier's office that the time for 'practical teaching' in forestry 'cannot well be further delayed.' Walker undoubtedly had a hand in ensuring that the commission into the University of Toronto's affairs consulted with both Fernow and Gifford Pinchot in the United States over the matter and adopted their view in favour of creating a school – instead of merely a chair – of forestry. He was also the most outspoken advocate of achieving this aim at the landmark Canadian Forestry Convention in January 1906; when his published paper prophesied that 'we shall soon see in the University of Toronto a faculty of forestry in itself,' he had the inside track to know of what he spoke.[25]

The wording of the royal commission's recommendation and the views it expressed regarding the new forestry school warrant special attention, because they are immensely significant in light of future events. The commissioners identified the pressing need to establish this type of educational institution in Ontario, and, more importantly, the key factors that would determine its fate.

The commissioners' report achieved these ends in the space of only a few paragraphs. They declared that the University of Toronto's status as the 'provincial' university behoved it to fulfil certain obligations regarding 'higher education,' and noted that this was 'eminently true of instruction in forestry.' The report also expressed surprise 'that Ontario with its rich areas of timber has hitherto failed to set up a school of forestry in its own university for the double purpose of providing technical training for young men in an important branch of science and of benefiting in the conservation of its forest wealth by their knowledge and skill. It would be difficult to mention a case in which the State's duty and interest go more completely hand in hand.' Although the OAC in Guelph already 'provided for instruction in agricultural forestry which meets the needs of farmers with woodlots to care for and develop,' the report underscored that the 'larger problem is that which touches the immense Crown domain urgently calling for the application there of the newest discoveries in forestry.'

The report then hit upon the crux of the matter, namely, that the forestry school would only succeed if the party in power backed it and its graduates' activities. The commissioners realized that Ontario was in an enviable position because, in contrast to the situation in the United States, the government controlled nearly all the commercial forests. This 'simplifies the situation,' the report asserted, in so far as both using the woodlands for the forestry school and applying the principles the school was teaching in the Crown forests. But, the report contended, this made political support crucial to the undertaking's success. As it poignantly stated, 'there is no doubt that a great work in forestry can be done in this Province by the University, provided it receives the co-operation and encouragement of the Government.'[26]

After the Conservatives accepted, in May 1906, the commission's recommendation to establish the forestry school in Toronto, the focus turned to realizing the project's practical aspects. From the outset there was no question that Dr Judson F. Clark, the P.E.I. native and Cornell forestry graduate who was now Ontario's provincial forester, was *the* man for the job. As early as 1903 a chorus of voices had begun calling for his appointment to head the province's forestry school, wherever it ended up. James Mills, the OAC's president, informed Whitney in 1906 that the matter had become urgent because Yale University was courting Clark. Mills added that his 'own judgement is that there ought to be a close connection between the outside work of the Government Department of Forestry and the more academic work of the University, and that Dr Clark would do the best work for the Government, the University and the country if he were appointed Dean of the School of Forestry in the University, with the salary of a full professorship.' Such ringing endorsements produced the desired effect. In late January 1906 Whitney informed the University of Toronto's president that his government now saw Clark as its man, too.[27]

While Clark was admittedly keen to take up the assignment, he was leery of doing so because his time under the employ of the Ontario government had made him acutely aware of its lack of interest in introducing effective forestry measures. Under the circumstances Clark felt it was useless to establish a forestry school, and he outlined his concerns in a candid letter to Whitney in mid-February 1906. He reminded the premier that he had been hired in 1904 to be the government's 'consulting forester,' and since that time he had been carefully studying the situation in Ontario's woodlands. Clark felt that the province was now at a crossroads. It must commit to practising sound silviculture and open the forestry school to provide the personnel to achieve this aim, and there was no point in doing one without the other. 'If there be not opportunity to apply the knowledge gained for the benefit of the Province,' Clark argued, 'the training of foresters at the University will be a sheer waste of money, just as opportunities for systematic forest management cannot be taken advantage of without trained men to lead the way.'[28]

Events soon confirmed in Clark's mind that his future lay not in Ontario. He addressed the Canadian Forestry Convention in Ottawa in January 1906, for example, and pointed out that Ontario had set aside thousands of square miles of forest reserves that had the potential to produce immense stumpage revenues far into the future. The problem, he stressed, was that the politicians had refused to reinvest any of this income in managing these areas.[29] Not long after delivering this speech, an editorial in the Toronto *Globe* lamented Clark's 'retirement' as provincial forester and labelled it a 'decided loss to the province.' While the details remain fuzzy, it appears that the Conservatives had run out of patience for Clark's penchant for offering unwanted forestry advice. At the very least, they had sternly reprimanded him, but it appears more likely that they had simply left his status with the government in abeyance. The problem was that the Tories needed Clark to achieve one of their broader policy objectives, namely, establishing the new forestry school. Whitney admitted as much when he declared in the Legislature in early March that his government was 'keen to retain Clark's services.'[30]

So, over roughly the next half-year, Clark kept his official title as provincial forester but devoted most of his time to acting as the de facto head of the soon-to-be-established forestry school. In April 1906, for example, Clark wrote the premier and the provincial secretary to impress on them the need to establish a 'university forest' for the University of

Toronto. The commission into the university's affairs had recommended that the government set aside a million acres in northern Ontario as an endowment for the *entire* university, and Clark resoundingly endorsed this proposal for two reasons. It would generate a perpetual source of revenue for the University of Toronto and produce numerous benefits for forestry in the province by providing both the requisite 'practice forest' for the forestry school's future students and an 'unsurpassed opportunity' to demonstrate to the public the utility of proper forest management. Into the late summer Clark continued drawing up plans for the new forestry school. His correspondence with Dr C.A. Schenck, head of the Biltmore Forest School in North Carolina, revealed Clark's determination to produce graduates who were neither strictly 'theoretical foresters' nor 'lumber Jacks,' but a hybrid of the two.[31]

In the meantime, Clark had brought another matter related to Ontario's forests to the provincial government's attention, and it was one that was certain to raise the politicians' hackles. Lumbermen paid stumpage for the sawlogs they cut from Crown lands based on the volume of wood extracted. To determine this figure, the government used a 'rule' or formula that estimated the 'board feet' (i.e., a piece of lumber measuring one-inch thick and one foot by one foot) contained in a log. Since 1879 the formula Ontario had used was the 'Doyle Rule.' While it was generally accurate when used to compute the volume of lumber in large-diameter logs, it grossly undermeasured the volume of wood in small-diameter logs. The Doyle Rule had thus been an appropriate 'scale' to use when it had been implemented because the lumbermen were still operating in stands of relatively large trees, but by the late 1800s this was no longer the case. As a result, the provincial treasury was mulcted out of hundreds of thousands of dollars in stumpage dues each year.

Clark had studied this issue intensively and devised a simple and effective solution to it. Over the summer of 1905, while he was acting as Ontario's provincial forester, he had studied white pine logs in the Ottawa River valley and generated reams of data pertaining to their diameters, tapers, and the board feet they produced. After an untold number of calculations, Clark developed a new formula for measuring saw logs: the 'International Rule.' After he published his findings in the *Forestry Quarterly*, his formula became widely accepted as 'the most accurate of the current rules.'[32]

Despite his devotion to this project, one for which the Ontario government had paid, Clark learned yet again that he was solving a problem that the Ontario government was not interested in addressing. In April

1906 he sent Whitney a memorandum that forthrightly called for the Conservatives to abolish the Doyle Rule, which Clark termed 'absolutely antiquated as a measure of log contents.' Instead, he suggested returning to the old, pre-1879 rule, or adopting one of the formulae in force in the other major lumber-producing provinces in eastern Canada. Clark estimated that this would generate at least 20 per cent more in annual stumpage dues at a time when timber revenues represented over one-quarter of Ontario's yearly income. While it was axiomatic that implementing this change would serve the public interest, Clark's message emitted an unbearable dissonance to the provincial politicians. In a nutshell, ending the Doyle Rule was anathema to the province's lumber industry, and in an era during which the lumbermen were still 'king' when it came to Canadian politics, the Conservatives knew that implementing Clark's recommendation would be political suicide; R.C. 'Bob' Hosie (2T4) later remarked that Clark's efforts 'got [him] into trouble with the lumbermen.' It is little wonder that the government kept the Doyle Rule on the books until 1952, despite successive foresters' attempts to abolish it.[33]

It appears that the Ontario government's refusal to amend the Doyle Rule was the last straw for Clark. By the fall of 1906 he had seen enough to know that it was time he moved on. Clark resigned as provincial forester and the unofficial leader of the still-to-be-created forestry school, and accepted an enticing offer from industry in British Columbia. In a heartfelt letter to C.A. Schenck, head of the Biltmore Forestry School, Clark provided insight into his decision to 'retire' from Ontario's civil service. 'I did not think I would be happy being head of a department for training foresters,' Clark stated forthrightly, 'if the provincial forest administration refused to cooperate in the work and employ our graduates.' To shed further light on this situation, Clark included excerpts from his resignation letter to the Ontario government that also revealed the degree to which he had chafed at the bit during the previous year. 'I know a provincial forest school would be immensely beneficial for Ontario,' Clark stressed, and 'hope that I could be of real service in this connection led to the refusal of a very tempting offer from Yale University last winter.' But now, Clark declared in exasperation, 'I have ... come to the conclusion that it would be a sheer waste of money to train foresters at public expense unless there was an intention on the part of the forest administration to use the men so trained in the public service. I have failed to note any such disposition on the part of the present forest administration.' Ultimately, Clark had resigned his position with the

Ontario government because, as Fernow later put it, 'he saw that there was nothing expected of him except to talk.'[34]

The media attached great symbolism to Clark's departure, but for the Ontario government, his exodus simply created a pressing practical problem. The *American Lumberman* described Clark's decision to leave the province as marking the 'abandonment of anything like a practical forestry policy in dealing with public lands' in Ontario, and the *Canadian Forestry Journal* echoed these sentiments. For the government, which was still committed to creating a forestry school at the University of Toronto, it was in dire need of finding someone to spearhead the project.[35]

All eyes turned to Fernow to lead the way, and he seemed on the face of it a star candidate. As North America's pioneer forester and a veteran of nearly three decades of ground-breaking silvicultural work in the United States, Fernow's name was practically synonymous with forestry at this time. Moreover, from the moment the idea of establishing a forestry school in Canada had been mentioned, back in the mid-1890s, Fernow had been counselling the various parties who had been attempting to realize this aim.

At the same time, however, several traits made him a most undesirable candidate for the job of creating the new forestry school. Although he knew full well that politics shaped forestry as much in Canada as in the United States, it will become clear that, over the dozen years he served as dean, Fernow acted as if he were completely oblivious to this reality. Perhaps this was a function of his overflowing self-confidence that came with his quasi-celebrity stature. Whatever the cause, his approach to carrying out his duties in Toronto redounded, more often than not, to the detriment of the forestry school he would found.

More importantly, the fact that Fernow and not Clark would head up the new forestry school at the University of Toronto dictated that it would be established on shaky ground, to say the least. The eighteen months Clark had spent toiling as Ontario's provincial forester had taught him an object lesson in the provincial government's lack of concern for achieving his goal of implementing meaningful forestry reforms, and as a result, he declined the job of starting Toronto's forestry school. Fernow, on the other hand, came to the table with no such first-hand understanding. He had certainly run into political problems in the United States in general and New York State in particular, but in his mind, it appears he conceived – at least initially – of Ontario as being a completely different milieu. Fernnow was convinced that, because the provincial government controlled nearly all of the commercial forests and had a fiduciary duty to manage them, it would be keen to adopt silivicultural practices. Thus, Fernow picked up the reins Clark had dropped, naively believing he could effect the same fundamental reforms as Clark had wanted by using the forestry school as his base of operations.[36]

The major illusion under which Fernow was operating was glaringly apparent from the moment he began negotiating in early 1907 to organize the forestry school in Toronto. By this time, the University of Toronto had determined that it would establish a 'faculty' and not simply a 'chair' of forestry, and Maurice Hutton, the university's acting president, had officially approached Fernow about taking on the task of realizing this plan. Fernow explained, in his mid-February reply, that the State College of Pennsylvania had hired him on contract to help organize a forestry school there, and that he had the option of converting this position into a permanent one. For this reason, Fernow declared that he would only take the job in Toronto if it entailed creating 'a forest school or forest department of the first order, worthy of my endeavour. If such a proposition is involved and you have in view the means for such an undertaking, I would be pleased to be considered as a candidate.' He also insisted that the 'proposition must be of a wider scope, or at least as favorable as is offered me by the State of Pennsylvania.' By mid-March the university's board of governors had determined that Fernow was its man, and he visited Toronto to discuss the matter.[37]

What came next drove home the point that Fernow's notions regarding the forestry school bore no relation to those held by either the provincial government or senior administrators at the University of Toronto. Fernow responded to an invitation from Hutton to elaborate on his 'conceptions as to what the needs of a forest department in your University would be' by composing a fourteen-page treatise on the subject. Fernow emphasized to Hutton that he felt his lengthy reply was warranted because 'a true appreciation of the needs, as I see them, can be formed only by considering the objects to be attained.'

Instead of confining his comments to the practical considerations involved in setting up the forestry school, as Hutton had asked, Fernow's memorandum painted with far broader strokes. Entitled 'An Educational Forest Policy for the Province of Ontario,' it began by listing the 'reasons for a broad-gauged forest policy on the part of the government of Ontario, including the subject of education.' These included the desirability of sustaining the large income that the government annually derived from the Crown woodlands and the fact that a large part of Ontario was only fit for raising crops of trees. Fernow then practically insisted that the government both implement fundamental forestry reforms and consider the new school the cornerstone of a strategy designed to achieve this aim. To demonstrate his point, he remarked that 'technically educated men are now employed and found desirable in all engineering works, in mining, and lately even in agriculture, [and] the only field which has remained behind in an educational direction is that of the exploiter of the woods.' The time was thus nigh, in Fernow's view, for professional foresters to oversee all aspects of the province's timber operations. But, he stressed, even more was needed to ensure the plan's success. All those involved in logging Crown forests must be educated about the principles of prudent timber management. '*Not until all the men of different grades that are actually dealing with the woods have more or less education regarding the practice of Forestry,*' Fernow underscored, 'and the technical forester is allied with the practical lumberman to secure the same end, will successful forest management become possible. Upon the basis of this practical attitude towards the problem of establishing a forest policy in the Province I would base my educational policy.'[38]

Fernow did not stop there. He described how the forestry school would become the centrepiece in a revolutionary approach to timber administration in Ontario. Because of the University of Toronto's status as 'the Government's own institution of learning,' Fernow was adamant that his forestry school should not merely teach students. Rather, it ought to educate all the province's residents about its work through public lectures and an intensive public relations campaign. As far as the new forestry school's faculty members were concerned, Fernow argued that they should form a 'Board of Advice' upon which the government would call for silvicultural counsel.

Halfway through his opus, Fernow finally turned to the subject about which he had been asked: setting up the forestry school. He recommended that the program include four years, with the first two devoted to 'fundamental' subjects and the latter two to 'professional' ones. He added that the University of Toronto already taught most of the non-core courses his students would require. At the same time he emphasized that 'forestry is a practical profession,' and thus an important element in educating students would be 'early experience in practical work.' For this reason he advocated including a limited amount of 'work in the woods' (in both the woodlands and milling ends of the business) in the program.

Fernow closed by discussing the price tag for the forestry school that he envisioned, and did his best to justify the expense. He stated that at least $12,000 would be required annually to support the enterprise, although 'for a beginning a smaller amount will suffice.' The school would need, at a minimum, one full professor and one associate or assistant professor and two instructors to carry out the teaching and extension work. He pointed out, however, that the instructors would not be essential initially. This would reduce the 'start-up' budget to roughly $7,000, the second year to $10,000, and $12,000 each year thereafter. Moreover, after the school was well under way he added that it would need more funding for experimental and investigative work, and these areas would 'finally crown the effort of the University in inaugurating a complete system of forestry education.' Recognizing that the university's board of governors and the provincial government would scoff at his budget estimates, Fernow closed by putting the cost of the undertaking into perspective. 'When it is considered that such a system of education in forestry would require less than one-half of one per cent of the income now derived from the timberlands of the Province,' he pointed out, 'the financial question would, indeed, appear paltry in comparison to the result which must necessarily follow from the existence of a corps of trained foresters.'

While Hutton responded promptly to Fernow's submission, the university's president was predictably cool. In a short note to Fernow, Hutton offered him the deanship of the new forestry faculty at an annual salary of $3,000. In agreeing to establish a separate faculty for forestry instead of adding it as a department to the Faculty of Applied Science and Engineering, the University of Toronto was following the paradigm that had been established in Europe, where forestry was typically a distinct academic entity.[39]

Fernow wrote to Hutton again on 25 March (only two days after he had first submitted to Hutton his grand scheme for forestry in Ontario), and because he was clearly concerned by Hutton's reaction to it, he desperately sought an explicit guarantee from the university's president that the university was fully supportive of it. Fernow advised Hutton that he needed 'some more definite assurance that something in the line of the propositions which I elaborated for you yesterday is actually contemplated.' Leaving nothing to chance, Fernow enclosed another copy of his 'An Educational Forest Policy for the Province of Ontario' in order to allow Hutton to 'more readily ... acquaint [his] Board of Governors with my attitude. Indeed, unless the means for a really efficient department, worthy of my effort, are actually in sight I could not be induced to undertake the building up of the department by uncertain steps.' At the same time, Fernow wanted to make certain that, whatever happened, his annual stipend would be equal to the university's best-paid faculty, which was $3,500. 'I have the same attitude towards the salary question,' he told Hutton. 'I am too old and my reputation too well established to accept anything but the current maximum rate, which is indeed less than I received here and very much less than I can make leisurely in private business.' He also assured Hutton that 'I am somewhat indifferent to salaries, provided opportunity for effective work is offered.'

Fernow then lobbed the ball into the university president's court. If Hutton were prepared to meet Fernow's conditions, Fernow declared, 'I would be pleased to have you place my name before the Board.' Fernow closed by pushing Hutton to act post-haste. If the State College of Pennsylvania agreed to fund its new forestry school to the extent that Fernow had recommended before the University of Toronto had offered him the post in Toronto, he would feel a 'moral obligation' to take the job in the United States.[40]

Hutton again reacted promptly to Fernow's letter, but he did not present an accurate picture of the situation to the board of governors. At the board's meeting on 28 March, Hutton submitted a report from its special committee on forestry that recommended Fernow be appointed dean of the forestry faculty at a salary of $3,500. Conspicuously absent from the discussion, however, was any mention of the string of other conditions that Fernow had attached to his acceptance of the position in Toronto. Hutton's oversight in this regard was understandable. The board of governors would have rejected nearly all of Fernow's 'propositions,' especially the financial ones.[41]

What made matters worse was that Hutton did not clarify the situation for Fernow, and Fernow did not press the president to do so. After Hutton informed Fernow of the board's decision, Fernow immediately accepted the offer without getting written confirmation that the university – and indeed the Ontario government – would meet all his conditions. 'Professors are notoriously bad business men,' Fernow's 30 March acceptance letter to Hutton began. While 'for business reasons I should insist that the conditions I had elaborated as a basis for my acceptance of the Deanship should form part of the contract and be "nominated in the bond," I take it for granted that your Board was cognizant of them and that my appointment was made in general accord with these conditions. With such an understanding,' he declared, 'I shall be pleased to accept the appointment and devote myself, heart and soul, to the task of building up a worthy forestry department.'[42]

If ever a professional faculty had been established on an unstable foundation, this was it. Fernow had taken on the task of creating an avant-garde forestry school in Toronto, and was determined to make it the beacon for a radical and sweeping reformation of Ontario's forestry policy. He had clearly and succinctly outlined his plan to Hutton, the University of Toronto's president, and asked that his ideas be transmitted to the university's board of governors to guarantee that its members understood his scheme. He also made his acceptance of the new post conditional on the board explicitly pledging its support for his groundbreaking undertaking. Apart from Hutton, however, no one on the board of governors ever learned of Fernow's 'conditions.' Instead, the board had simply hired Fernow to create what was, in the context of the myriad matters with which it was dealing, a new but relatively miniscule academic division that addressed one of the royal commission's many recommendations. The University of Toronto was motivated by expediency, not by a burning desire to overturn the province's approach to administering Crown timber and turn Ontario into North America's poster child for progressive forestry. When Fernow accepted the job under the assumption that the board fully backed the many tangents of his larger initiative, he had cast his fate. He would go forth wrongly believing that the University of Toronto, the Ontario government, and Ontarians in general supported his profound commitment to revolutionize forestry in the province. He could not have been more wrong. Fernow's error went a long way to explaining the plethora of problems that the Faculty of Forestry at the University of Toronto confronted not only during his dozen years as dean, but during the reigns of his successors as well.

Chapter 2

'The Child of My Creation,' 1907–1919

A few months before the Faculty of Forestry was set to open at the University of Toronto, Thomas Southworth, a long-time civil servant in Ontario's Department of Crown Lands and outspoken champion of forestry, delivered a speech to the Canadian Institute that summed up the daunting challenges that the new school faced. The title of his address asked rhetorically, 'Do We Need a Forestry College?' While he responded that many would be quick to answer 'we do,' there was good reason to believe otherwise. Having canvassed Ontario's leading lumbermen to learn if they would employ graduates of the new school, he revealed that all but one of them had stated categorically that they would *not*. Southworth then explained that the government was responsible for managing the Crown woodlands, and logic thus dictated that the 'first large employer of properly trained foresters and forestry students should be the state.' But the Ontario government had shown little inclination in this regard. 'Unless the men controlling the forested lands of Canada can be brought to this point of view,' Southworth concluded in reference to those who believed a forestry college *was* desirable, 'I cannot see that we have any particular need' of such an institution.[1]

Southworth's inauspicious assessment would be true for practically the entire time that Bernhard E. Fernow served as dean of the Faculty of Forestry, and unfortunately, it applied to more than just the lumbermen and the Ontario government. The pulp and paper industry enjoyed meteoric growth in Canada during the first few decades of the twentieth century, but it was still under the impression that its fibre resources were far greater than they would turn out to be. This gave it little incentive to hire foresters. Moreover, the Dominion Forest Service (DFS)[2] was still in its embryonic stage and able to count its foresters on one hand, and its growth during the early 1900s would be steady but slow. Finally, the lack of interest in forestry also extended to the University of Toronto's attitude towards the Faculty of Forestry. For the years during which Fernow was dean, the university's administration was disinclined to provide him with more than a subsistence level of support.

Fernow was cognizant of most of these difficulties, but he was up to the challenge of creating the country's first forestry school. With fewer than a dozen foresters working in Canada in 1907, he recognized that his graduates both specialized in a field for which society heretofore had had little use and faced bleak job prospects as a result. 'Only a radical change in attitude,' he postulated shortly after the Faculty of Forestry opened, 'a realization that forest conservation is a present necessity, and that existing methods are destructive of the future, will bring forward the needed reform.' The existence of his school gave cause for hope, he reasoned, as increasing the number of professional foresters in Canada would heighten awareness of the efficacy of their work. The upshot, he suggested, would be that 'Ministers of Crown lands will be forced by public opinion to divorce technical administration from political influence.'[3]

Fernow devised a number of strategies to achieve this end during his twelve years as dean (1907–19). His profound dedication to producing the best possible graduates – as much in terms of character as technical knowledge – became the Faculty of Forestry's trademark until well after the Second World War. His ability to realize this goal meant that his first crops of foresters were able to lay a firm foundation on which their successors could build. It also helped the Faculty of Forestry enjoy consistent growth until the First World War, with its enrolment rising from six in 1907 to nearly fifty twelve years later. By the same token, Fernow's penchant for speaking his mind – even if it meant locking horns with the Ontario government – undoubtedly backfired. 'Yes, talking in public can be dangerous and you will have to take the consequences sometimes from quarters least expected,' Fernow told a friend in 1909. The dean admitted that he was, 'of course, constantly "putting my foot in it," but I am now old enough not to care what the consequences may be as long as I am arousing the public in general to see what the duty of the Government is.' The problem was that Fernow's outbursts aroused the politicians far more than the public, and his forestry school only suffered as a result.[4]

Almost immediately upon being appointed dean of the University of Toronto's Faculty of Forestry, in March 1907, Fernow set out to create it. The most pressing matter on his docket was curriculum. While a few years earlier he had called for Canada's first forestry school to be set up as a graduate program, an approach for which there had been widespread support, he recognized that circumstances made this unfeasible. With only limited resources available to him and forestry still in its infancy in Canada, Fernow opted to draw up an undergraduate program that would provide his students with the broadest possible forestry education. This entailed including several liberal arts subjects in the course offerings, and focusing the first two years of study on fundamental sciences and the final two on professional forestry. Incidentally, in 1909 Fernow created a six-year option for students wishing to graduate with a 'double degree' (a bachelor of arts and a bachelor of science in forestry), but its failure to attract a large number of students led to its cancellation twelve years later.[5]

The curriculum that Fernow set may not have appeared extraordinary, but he was determined to ensure that his faculty attracted only exceptional students. Recognizing the immense challenges that his graduates would face in their working lives and how their employment opportunities would be few and far between, he was determined to produce professional foresters of the highest quality.[6]

His approach to achieving this aim was multi-pronged. He asked the university's board of governors – and it agreed – to set the Faculty of Forestry's entrance requirements above those required by the rest of the university. Initially, this included senior matriculation (the future Grade 13), two modern languages (German and French), and honours standing in English. As Fernow told Frank B. Robertson (1T4) in 1909, the forestry faculty was committed to keeping the classes small and 'the entrance requirements high in order to exclude all but the best.' Over the next half-dozen years, Fernow would intermittently raise the standard further, all in the hope, he argued in his *Annual Report* for 1916, 'of securing a better class of men.' Fernow also believed this approach would draw the most mature students to the Faculty of Forestry. He knew firsthand how in Europe, where forestry had a respected history and was being practised on a broad scale, freshly minted forestry graduates had the occasion to apprentice in their field. Because there was no such opportunity for his understudies in North America, he felt they had to be as worldly as possible when they left his charge.[7]

For those who made it into the Faculty of Forestry, including his fellow staff members, Fernow demanded that they always strive for *and* achieve excellence to protect the collective reputation of foresters and forestry.

He had precious little patience for those whom he felt were not living up to his ideals. When he discovered that a handful of students were skipping the new business law course he introduced shortly after the Faculty of Forestry opened, for example, he sent the guilty parties a curt letter. 'Allow me to say to you,' the dean's admonishment to one of the perpetrators read, 'that I consider it decidedly impolite on your part to cut lectures without excuse, and especially the seminar of Mr Falconbridge, who as an outsider must be naturally offended by such neglect and receive a false impression of the character of our institution.' Likewise, when Fernow found out that George Sydney Smith (1T4) was cutting classes and labs without just cause during the 1910–11 session, the dean advised him that 'for the good of the Faculty, if not of yourself, you will have to mend or retire.' When Smith continued his wayward drift, Fernow fired off another letter in the summer of 1911, warning him of the dean's determination 'to clean out from the Faculty all those who did not attend faithfully, because we cannot afford in our profession any laggards.' Fortunately for Smith, he heeded Fernow's final warning.[8]

The lofty bar Fernow set for those within the Faculty of Forestry produced a predictably high attrition rate among his students. It was not uncommon for 40 per cent of those who entered the school each year to fail. The class of 1913, for instance, jocularly reported in the student yearbook how it had begun with sixteen first-year students but 'the hand of the Examiner was heavy upon us, and but eleven survived, more or less battered, to enter the second year.' Two more would fall before the class graduated.[9]

Fernow also applied this eminent standard to his colleagues. Alexander Herbert Douglas Ross had earned his master's degree in forestry from Yale University in 1906. He had been working for the Dominion Forest Service at its nursery in Indian Head, Saskatchewan, when Fernow offered him the chance to teach forest mensuration and forest utilization in the Faculty of Forestry in April 1907. Ross became the school's first lecturer, and after one semester, Fernow recommended to the university's president that the university offer him a long-term appointment and a major raise. Fernow was motivated not by the awesome nature of Ross's work but rather by a recognition that the lecturer had filled the position 'acceptably' and might be lost to the DFS if the University of Toronto did not provide him with more job security and a higher wage. For roughly the next six years, Fernow recommended that Ross be granted the standard raises due lecturers of his rank, even though the dean was growing increasingly dissatisfied with his performance.[10]

In the summer of 1913, however, Fernow decided the time had come to terminate Ross's employment with the Faculty of Forestry. In mid-August, Ross had written Fernow regarding his salary because he had not been granted his customary increase. Assuming this had been merely an oversight on Fernow's part, Ross asked the dean to correct it.

Fernow responded by making it clear that he had not erred and excoriating Ross for what Fernow believed were the latter's many significant shortcomings. Informing Ross that 'the failure to raise your salary was deliberate,' Fernow advised his soon-to-be-former colleague that this decision was based on the dean's view that Ross's 'work for the students is not satisfactory. As you may remember I pointed this out to you a few years ago, and hoped that by an encouraging rise in salary your efforts might be stimulated to improve your teaching. I regret that the anticipation has not been realized.' Fernow took steps to assure Ross that the dean had not rendered this judgment hastily. 'I have had complaints from employers and employed that the men are short of knowledge or the ability of using it in your lines,' Fernow pointed out, and noted that 'I take such complaints by students always with a discount, but when such good men as we graduated last year and this year find themselves handicapped in the practical field, as they inform me, I cannot but believe that there is true cause for complaint.' Fernow closed by insisting that he was compelled to confront Ross about the issue 'for the good of the Faculty.'[11]

Fernow knew full well that he was essentially showing Ross the door. The dean forwarded Robert Falconer, the university's president, a copy of his letter to Ross and a note expressing his hope that the president would 'approve of this manner of making a resignation on his [i.e., Ross's] part easier.' Falconer replied to Fernow by stating the obvious, namely, that 'in view of such a letter, this should ... be ... the last session for him [i.e., Ross] as a member of the Staff in Forestry.' It was. Ross left the university in the early spring of 1914, although he would always contend that he had become 'so dissatisfied with conditions and poor pay at Toronto, that I resigned.' Ironically, Fernow's successor as dean would dismiss Ross's replacement roughly two decades later.[12]

Fernow was stereotypically 'old school' in his heavy-handedness, yet he balanced this trait with a soft touch that belied his gruff countenance. If you respected his rules and played by them, he bent over backwards to make you feel part of the forestry family, at times quite literally. Students frequented the Fernow home on Avenue Road for a variety of reasons. His wife, American-born Olivia, taught them German, fluency in which was essential because most of the forestry literature was in Dr Fernow's mother tongue. Olivia also knew the students by name, and hosted an 'open house' for them each Sunday that made certain they would eat well at least once a week. Similarly, Fernow allowed 'needy' students to board at his home in order to save money while at school. Such generosity had been the only reason that Reginald D.L. Snow (2T8), for example, who had entered the Faculty of Forestry in 1913, was able to afford his first two years in Toronto.[13]

Fernow would also show tremendous patience and understanding to his understudies who committed to achieving his standard, no matter how badly they had heretofore missed the mark. 'I do not know how often "the vilest sinner may return,"' he preached to one floundering student, 'but I am always willing to give anyone another chance if he recognizes the depth of his guilt.' On the occasions when Fernow realized his mentoring was insufficient to right a sinking ship, he would despatch an 'update' to the concerned parents to offer them an opportunity both to learn more about the situation and facilitate a recovery before it was too late. C.R. 'Charlie' Mills (4T3), who took six years to complete his four-year degree, was one of the many who benefited from this mentoring. Having entered the Faculty of Forestry in 1909, Mills failed his first two years, during which time he had repeatedly ignored Fernow's warnings to shape up or risk being kicked out of the faculty. The dean had informed Mills' father of the situation, and the latter had begged for the dean to take his son back. Fernow acquiesced to the distraught father's appeal, and future events would vindicate this decision. Mills went on to enjoy a lengthy career with the Ontario Department of Lands and Forests before serving for two decades as the inaugural secretary manager of the Ontario Forest Industries Association.[14]

While some undoubtedly bridled under Fernow's tough love, it earned the undying respect of most of his charges. Letters of appreciation from his graduates express their profound gratitude for the life lessons he taught them. Ernest H. Finlayson (1T2), who went on to enjoy a prolific career with the Dominion Forest Service, left no doubt about his feelings when he wrote to Fernow at the time of his convocation. 'I go out into the world, Doctor,' Finlayson's heartfelt thank-you letter read, 'with respect and gratitude to you for all you have done for me and I trust that my work will prove me a worthy disciple of your teachings. I bless the day I first met you at Joe Lake four years ago for it was that canoe trip that sowed the first seeds of my desire to enter the forestry work. With kindest regards to Mrs Fernow, I remain faithfully your pupil.'[15] Likewise, James

Kay (1T9) reflected on his life-changing experiences with Fernow. 'I have often thought and felt,' Kay remarked on the occasion of Fernow's passing, 'that the most precious and permanent thing that remains with us long after we leave the University, have [sic] not come from deep study of books or lectures, but an indefinable something that is passed on, or radiates from the men who try to guide our aspirations.'[16]

Undoubtedly, a major part of that 'indefinable something' was the remarkable camaraderie that Fernow was able to foster among the students and staff; this became the Faculty of Forestry's hallmark. In some ways, it was bound to happen. This was a relatively tiny professional school in which the students took practically all the same courses at the same time for each of their four years. They also lived and worked side by side on their annual field trips, during which they gained practical experience. The inaugural class spent time in Rondeau Provincial Park in the summer of 1908, for example, and subsequent ones ventured north to board with various lumber companies at different times of the school year (either during Christmas holidays or in the spring). The upshot was a profound bond among all in the Faculty of Forestry, especially those in the same year. As forestry students wrote in 1913 in the University of Toronto's yearbook, 'we have studied together, scraped through Exams together, tramped, worked and slept together, have feasted and gone hungry together.' These ties were further cemented through the creation of the Foresters' Club in 1909, to which all staff and students belonged. It met fortnightly during the school year and facilitated lectures and addresses by guest speakers.[17]

B.E. Fernow did whatever he could internally to develop foresters of the highest distinction, but the Faculty of Forestry's success ultimately rested on two external forces: the University of Toronto and the provincial government. Unfortunately for Fernow, neither supported his work in a meaningful way. To be sure, they rarely missed an opportunity at a public ceremony to wax poetic about their wholehearted belief in the efficacy of and urgent need for forestry education and reforms in Ontario, but seldom did they back up their words with actions.

In dealing with the university, Fernow quickly learned that it would be a struggle to gain every inch of ground for his faculty. He had accepted the deanship on the understanding that the university would support his work adequately, specifically in terms of staff (a minimum of one full and one associate or assistant professor and two assistants) and funding ($7,000 the first year, $10,000 the second, and a minimum of $12,000 each year thereafter). Almost immediately after Fernow assumed the job, however, the administration demonstrated that it would not uphold its end of the bargain. Receiving the teaching complement for which he had asked was a crucial issue for Fernow, as he envisioned himself as instructing only part-time. He felt he could best serve the school by publicizing the forestry movement to foster broad public support for the cause. But when the Faculty of Forestry opened its doors in the fall of 1907, he received only one lecturer (Ross) and one class assistant (Edmund John Zavitz), and the latter was available to teach only one day per week in Toronto because of his work at the Ontario Agricultural College in Guelph. In early 1908 Fernow vehemently protested being short-staffed to Robert Falconer, the university's new president, particularly in light of the 'the plan that I proposed and was accepted by the Board of Governors.' In response, the board expressed its desire to help but 'to take no further steps than may be necessary by the immediate pressure of students from Session to Session, as the expenses of the University are mounting very rapidly.' It agreed to provide only one more lecturer.[18]

This decision prompted Fernow to change tack slightly in order to make his case more palatable to the board. In mid-February 1908 he reminded Falconer that he had accepted the deanship on the condition that the funding for his second year in Toronto would be $10,000. 'Urgent need of economy, I understand now, makes it desirable to cut down this budget,' Fernow explained, but he proposed that 'subjects that border between forestry and botany (like dendrological subjects) be turned over in part to that department.' He pointed out that this created a challenge because he could only achieve this end by hiring someone whose speciality was both 'dendrology and forest geography.' Fernow had fortunately found such a man, Clifton Durant Howe, and interested him in the job. Howe was then employed as the assistant director of the Biltmore Forest School in North Carolina and had superb credentials. More importantly, from Fernow's perspective, he argued that it 'would be perfectly fair' to charge Howe's entire salary 'to the Botanical department as all the courses expected of Dr Howe might with propriety be given by that department.' Fernow added that Zavitz's increased commitments at the OAC in Guelph would end his availability to the University of Toronto, and thus the Faculty of Forestry needed to replace him as well. Again, the dean had found the ideal candidate. Harold Reginald MacMillan, a Canadian who was a graduate of OAC and about to earn his forestry degree from Yale. Fernow closed by indicating that hiring these new staff members, with one as a cross-appointment, would mean that the Faculty of Forestry's budget would come in at $9,000 for 1908–9, $1,000 less than the original amount for which he had asked.[19]

Fernow's strategy proved successful. In late May 1908 the board of governors agreed to fund 'for forestry and botany' the two appointments for which Fernow had asked. While Howe came, MacMillan did not, as he chose instead to accept an offer to work for the Dominion Forest Service in western Canada. Instead, James Herbert White (0T9), who was the Faculty of Forestry's – and Canada's – first forestry graduate in 1909, was offered the post. His acceptance marked the beginning of a teaching career at the faculty that would span the next four decades.[20]

Despite this breakthrough, over the rest of his tenure as dean, Fernow repeatedly reminded the university that it was breaking the terms of the deal he had brokered with it and that this refusal was hindering the Faculty of Forestry's progress. While the lack of a permanent practice forest in which the students could be taught was a pressing matter, far more serious were the faculty's grossly inadequate facilities. When it had opened its doors in the fall of 1907, its six students had attended class in the west wing of University College. Late the next year the school had moved to a brick house at 11 Queen's Park, a site it shared with the Department of Botany for roughly the next two decades. From the outset the space was insufficient in terms of laboratories, offices, and lecture halls, and the situation worsened as the faculty expanded (its total enrolment in its second year was 23 students). Fernow repeatedly stressed these shortcomings to the university's administration. His annual report for 1908–9, for example, described how the cramped conditions left him no choice but to use his office to conduct lectures and seminars. When the university ignored his pleas for new quarters and his calls for funding to hire guest lecturers to present in their fields of expertise, Fernow declared that 'the original avowed intention of the Board of Governors to make this a first class course in forestry cannot be realized without these additions.' By 1912 his annual report lamented this list of problems and despondently concluded that, 'altogether, after the first quinquennium of its existence, it cannot be said that the Faculty has reached a permanent form.'[21]

While some might suggest that Fernow's years as dean (1907–19) were not ones during which the University of Toronto was in a position to fund his program adequately, this analysis is faulty. It is true that the university's euphoria over the royal commission's report in 1907 quickly dissipated when the funding formula it recommended did not solve the university's financial woes. By 1910 the university was running a major deficit because of higher-than-expected inflation and lower-than-expected income from the Legislature. Martin Friedland writes that, thereafter, the University of Toronto 'once more would have to go cap in hand to ask for yearly grants from the government.' These constraints did not, however, prevent the university from providing several departments with relatively extravagant support. The engineering and medical faculties enjoyed dramatic growth, for example, during the first decade and a half of the twentieth century in terms of both new staff and facilities. In contrast, the university's administration showed no such benevolence towards forestry. Even though the Faculty of Forestry grew dramatically during the first twelve years of its existence, in the eyes of the school's decision-makers it simply did not rank as a high priority.[22]

The same could be said for the Ontario government, although Fernow was at least partially responsible for the forestry faculty's malaise. Judson Clark had learned first-hand in the early 1900s that the province had little interest in implementing forestry reforms. This knowledge had caused him to decline the opportunity to become the Faculty of Forestry's first dean; he saw little use in administering a forestry school if the politicians would neither hire its graduates nor allow them to practise what they had been trained to do. Fernow had definitely not learned these lessons as well as Clark had, but still he seemingly understood the importance of currying favour with the politicians in a milieu like Ontario in which they controlled the commercial forests and also held the strings on the public purse that funded the Faculty of Forestry.

Ample evidence attests to Bernhard Fernow's cognizance of this dynamic. Shortly after his appointment as dean in Toronto, for example, W.L. Goodwin, the director of Queen's School of Mining, wrote to Fernow to congratulate him on his new job. Goodwin also suggested that Fernow would be able to advance the forestry movement in Canada without facing the political interference that he had endured at Cornell University. Fernow's response indicated that he knew that crossing the border would not make his work any easier. 'There are some opportunities here which I hope I shall be able to utilize,' Fernow informed Goodwin, 'but I am afraid that great expectations placed on my acceptance of the position may not fully be realized, for I find that political considerations are as much in the foreground in shaping policies as in the United States.'[23]

As much as Fernow understood the political aspect of his job, he never came to grips with several basic facts surrounding it. The first was the Ontario government's refusal to hear the forestry message that he was preaching. During the Tories' nearly fifteen-year reign (1905–19), they did little to improve the manner in which they administered Crown timber. Occasionally, they would proclaim their intention to implement progressive policies, but rare were the occasions when their actions supported their words.

This incongruity was poignantly illustrated during the first year of the Faculty of Forestry's existence. Over the course of late 1907 and the first half of 1908, Frank Cochrane, Ontario's minister of lands, forests, and mines, announced that his department was going to 'inaugurate a more rational and conservative policy regarding the treatment of timber resources.' The plan included hiring Fernow's graduates to manage the provincial forest reserves according to 'forestry principles,' monitor cutting, improve the forest-fire ranging system, and reforest the wastelands in southern Ontario. In an article in *University Magazine*, Fernow exclaimed that this news meant that 'the Faculty of Forestry has received a testimony of justification which will rejoice every forester's and every patriotic citizen's heart.' This moved him to take two steps. When the dominion government inquired as to whether any of the Faculty of Forestry's handful of students would be available for summer work, he replied that he felt compelled to 'reserve our men for home service.' Fernow then submitted the names of his best students to the Ontario government for the work it was apparently planning to undertake. In the end, however, nothing came of Cochrane's bombast. In fact, Fernow was actually left in the embarrassing position of having to beg the Ontario government to hire his students as summer interns after the faculty's inaugural session![24]

In addition, Fernow never accepted that the government perceived of his role in the province in very different terms than he did. The politicians saw him strictly as the faculty's dean whose presence they would have to grin and bear. In sharp contrast, he felt that, as a civil servant, he had a duty to strive to achieve 'the greater good,' even if it went against the wishes of the elected officials. 'Fernow believed passionately,' his biographer writes, 'that the professional forester acted as an agent of the state and developed policies for forest administration for the benefit of the public based on scientific principles. This was the German model of the civil servant, professionally trained and serving the greater interests of the state.' This was the reason Fernow 'took the position that Ontario had first claim on the school's and students' services,' even after the government repeatedly demonstrated that it did not want them. A few months after accepting the offer to come to Toronto, for example,

Fernow replied to an inquiry that asked for advice about a pulpwood exporting policy for Newfoundland. Although eager to offer his views, he declined because he felt that 'no doubt, it will soon be necessary for me to do so for the Ontario Government, who is now looking for me to advise them in regard to their forest policy.' How he drew this conclusion is unknown, for there is no evidence that the province's politicians ever knocked on his door to solicit his advice.[25]

Fernow's conception of the role he ought to play as the Faculty of Forestry's dean created a third major problem for him vis-à-vis the Ontario government. He lacked even a modicum of savoir-faire in dealing with the elected officials, and this greatly diminished his chances of success in pushing both the faculty's and forestry's agenda. His behaviour towards the politicians during this period stands as an object lesson in what *not* to do when endeavouring to lobby effectively. The unfathomable part is that he recognized that 'political considerations' would determine the fate of his work, yet he repeatedly went out of his way to antagonize and bait politicians and senior civil servants. As one of his obituaries euphemistically put it, 'he enjoyed a controversy as a mental exercise in clarifying ideas.' Furthermore, at times it appeared that he delighted at being the fly in the ointment. When William Saunders, a colleague at the Central Experimental Farm in Ottawa, asked Fernow about the Canadian Forestry Association's 1909 annual gathering, Fernow replied that 'the Meeting went off not quite without a storm, in which I am afraid I played not an unimportant part, going for Aubrey White [Ontario's deputy minister of lands and forests] and for Senator Edwards in a pleasant enough manner, but I am afraid not to their liking.'[26]

Fernow's demeanour thus resulted in a string of nightmarish run-ins with members of Ontario's Conservative government that made him persona non grata in its inner circle. Ironically, nearly all the incidents involved Fernow's ground-breaking forestry work in Ontario that should have been heralded as major achievements. One of the first involved the Commission of Conservation, which the dominion government had established to provide it with advice regarding the development of the country's prodigious natural resources. The impressive credentials of the commission's 'scientific' members gave it significant stature. Fernow had been appointed as its forestry representative, and he saw this as a golden opportunity to further his cause. Over the course of the 1910s he oversaw various pioneering forestry projects, including the first large-scale reconnaissance of a province's woodlands (in Nova Scotia in 1912) and regeneration surveys in Ontario's pulpwood forests (in the Lake Abitibi region in 1919 and the Goulais River valley north of Sault Ste Marie in 1920).[27]

While much of this work was potentially invaluable in improving forestry in Ontario, Fernow could not help but deliver the conservation commission's findings in a manner that raised the provincial government's ire. In 1912, for example, Clifford Sifton, the commission's chairman, directed Fernow to carry out a cursory survey of the conditions prevailing on northern Ontario's Clay Belt. Fernow spent roughly a fortnight travelling by railcar along the National Transcontinental Railway's line and recording his observations, and he presented them formally in a very brief report in 1913. Its main message was that the Clay Belt's agricultural potential was good, but that it was imperative to identify the promising farmland before allowing settlement to proceed. Fernow warned that 'unless the early colonization is properly directed, disappointment will be experienced through the irresponsible settlement of good, bad, and indifferent locations.' His findings were hardly revolutionary, as numerous officials had been offering the Ontario government the same advice for years. It was unprecedented, however, for someone occupying such a high-profile position to deliver this news, which was anathema to provincial officials who had long boasted that the Clay Belt represented Ontario's new Arcadia. More importantly, Fernow presented his observations at the commission's annual meeting in January 1913, and they made headlines in the country's major dailies. The day after the *Globe* provided its readers with details about his address, its editorial insisted that the provincial government act on Fernow's advice.[28]

This touched off a maelstrom, one that amply demonstrated that Fernow's superiors at the University of Toronto fully understood the importance of fostering good relations with Queen's Park even if the dean did not. William H. Hearst, Ontario's minister of lands, forests, and mines, was incredulous at Fernow's apparent temerity. Hearst immediately responded with a lengthy official statement that dismissed Fernow's trip as a 'rapid inspection' and declared that the dean's 'conclusions arrived at from premises so slender are valueless.' Fully supported by the premier, Hearst emphatically defended the government's northern colonization record and insisted that Fernow was simply wrong about the hinterland. At this point, Robert Falconer, the university's president, intervened to undertake damage control. He wrote Hearst to inform the minister that he (Falconer) had already spoken to Fernow about the issue and learned that 'the extracts in the papers do not do justice to either his [i.e., Fernow's] views or his report. I told Dean Fernow,' Falconer assured

Hearst, 'that it seemed to me that the Conservation Report should have been first sent privately to the Ontario Government, though of course that is not any concern of mine. I also asked Dr Fernow to write you sending you a full report, and stating that he made the report at the request of the Chairman of the Dominion Conservation Commission.' Fernow dutifully followed this directive, apologized for the troubles he had caused, and affirmed that he was not 'condemning the Clay Belt.' While Hearst responded by thanking both Falconer and Fernow for their explanations, he also warned Falconer about the seriousness of the dean's offence. 'I certainly feel very keenly in the matter,' Hearst reported, 'from the knowledge obtained from the newspaper reports of Dr Fernow's address and his alleged interviews with the papers, for if these are at all correct, a very grave wrong has been done to an important part of the Province and a rather serious charge has been made against the administration of this Department which has no foundation whatever in fact.'[29]

Unfortunately for Fernow, the matter did not die there. Robert R. Gamey, the controversial Conservative MPP for Manitoulin, asked his own government whether it would investigate Fernow's comments and if it believed them. This drew another pointed editorial from the *Globe* that further steamed the reigning Tories. 'If we are to judge from certain questions placed on the order paper of the Legislature regarding Dr Fernow's report on New Ontario resources,' the piece surmised, 'it will shortly be regarded as treason to express an opinion of any sort of variance with that of a Minister of the Crown.' It closed by defending the dean for having told the truth, and insisted that 'Fernow has done nothing worthy of censure.'[30]

This brouhaha was just dying down when Fernow precipitated another flare-up by single-handedly transforming what should have been a major milestone for Ontario's forestry movement into a public relations disaster. The Commission of Conservation was concerned inter alia about water management, and its attention had been drawn to the Trent River valley in eastern Ontario. This important watershed was suffering badly from the effects of wanton deforestation, and so the commission turned to Fernow to investigate the conditions affecting its health. Over the summer of 1912 Faculty of Forestry members J.H. White and C.D. Howe, accompanied by three assistants (two of whom, Herbert Read Christie (1T3) and Frederick McVickar (1T3), were their students) surveyed roughly 4,200 square miles in the watershed north of the Kawartha lakes. Their report, which called for a broad forest management program, including reforestation, to rehabilitate the area, was heralded by many as a crucial landmark in chronicling the state of Ontario's forests and suggesting measures to administer them prudently.[31]

While the quality of the report spoke for itself, Fernow was associated with publicly presenting its contents in a manner that touched off a wave of anger. Speaking at a dinner hosted by the Canadian Club of Timiskaming in late November, Fernow reiterated his recommendation that the government classify the Clay Belt's soils before opening them to settlement and then used the experience of those in the Trent River valley to demonstrate what would occur up north if the politicians did not heed his advice. 'As an example of what happened when settlers were allowed to go on poor lands,' the Globe's front-page story explained, 'Dr Fernow said that on 600,000 acres in the Trent Valley they had surveyed, there were 190 farms for sale for taxes at six cents an acre, and there were 150 families of degenerates who did not have money enough to move away. These people,' the article continued, 'were as poor as Job and lacking in enterprise, and were physically and mentally defective.' The next day the Globe elevated Fernow's speech to its lead story whose caption read, 'Poverty and Wretchedness in Trent Valley District.' It provided details as to how, in the Peterborough district, 'a rural slum condition exists which for sordidness and degeneracy, judging from reliable information received by the Globe yesterday, has not a parallel in any city in Canada.' While the Toronto daily made it clear that its source was an informant and not Fernow, the implication was that the dean's speech had conveyed the same message.[32]

For the next few weeks the story attracted attention to Fernow and forestry but for all the wrong reasons. One of the county councils in the Trent River valley admitted that its bailiwick was suffering from environmental damage, but zealously defended its residents against the allegation of degeneracy. In another story that made the front page, one especially irate representative derided Fernow's disparaging depiction of their region as 'utter rot.' When C.D. Howe, Fernow's colleague at the Faculty of Forestry, attempted to set the record straight, he seemed to throw fuel on the fire. Revealing that he had been the Globe's informant, Howe argued that the story's author had taken poetic licence in reporting the views that Howe had expressed. Howe categorically denied having used the word 'degenerate,' and instead insisted that he and Fernow had employed the terms 'defectives' and 'subnormals,' noting that 'no stigma is attached to [these] terms.'[33]

And on it went. Fernow was seemingly doomed to turning opportunities to further the forestry movement into occasions on which he would retard its progress. Even though he was seemingly acutely aware of the deleterious impact his behaviour was having, it was as if he could not help but commit gaffes that continually incited the wrath of the Ontario government. As he remarked to a fellow professor at Purdue University in Indiana, his historic reports kept bringing him 'more trouble than glory!'[34]

Fernow's approach had a profoundly detrimental impact on Ontario's forestry policy in general and the Faculty of Forestry in particular. It undoubtedly made the provincial government less inclined to initiate bona fide silvicultural reforms, and it created a major chasm between Fernow's forestry school and the bureaucrats responsible for administering Crown timber. As J.H. White, Fernow's colleague at the faculty, explained to H.R. MacMillan in early 1918, 'up to the present we have had no relation to the Provincial [Forest] Service, despite the fact that we are a state supported school due largely to the relation established between our Dean and the Ontario Crown Lands Department.' By the end of First World War Ontario had no forestry policy to speak of, and the provincial government had hired only one of the Faculty of Forestry's graduates, Frank S. Newman (1T3). Judson Clark's worst nightmare had been realized. As it has been written elsewhere, the Toronto forestry school 'was producing silviculturists whom the politicians refused to employ.'[35]

Near the end of Fernow's reign there were signs of progress on this front, but they had nothing to do with the dean. White had been ascending the Faculty of Forestry's academic ladder, graduating from 'class assistant' to 'instructor' in forestry and botany immediately after he had earned his bachelor of science in forestry in 1909. During the mid-1910s he cemented his tight relationship with E.J. Zavitz, whom the government had appointed as Ontario's second provincial forester in 1912, by working in Zavitz's office. In 1917 Zavitz arranged for White to be seconded from the faculty in order to reorganize the government's forestry bureaucracy. For roughly the next thirty months White diligently toiled away at this project, the product of which would be the 'district' system introduced during the early 1920s.[36]

The efficacy of his work apparently impressed the elected officials, especially G. Howard Ferguson, the minister of lands, forests, and mines (1914–19) and future premier of Ontario (1923–30). Although Ferguson was arguably the most corrupt minister who ever held this portfolio, he was paradoxically open to improving the government's administration of Crown timber if it did not interfere with his political machinations. White quickly came to realize the opportunity before him. He noted roughly one year into his tenure with the Ontario government that he now appreciated 'what Ontario's problem is + how much constructive work can be done. And the outlook to me looks good. A province that uses 1,000 rangers + spends $450,000 on fire protection without the slightest hesitation is fruitful ground. The present Minister is very sympathetic + while we have some political troubles this interference is much less than I had expected ... So long as Zavitz is at the head + I am connected with either him or the school,' he predicted, 'we are assured of friendly cooperation for we both have the same interests at heart.' These were clearly auspicious omens for the forestry faculty.[37]

But before the faculty could capitalize on these promising signs it would first have to endure the trying years of the First World War. The conflict had a profound impact on Toronto's Faculty of Forestry, and it felt the war's effects in numerous ways.

The war tightened the ties that bound the faculty's students to each other and their professors, and it demonstrated just how closely knit this group really was. Fernow, and his wife Olivia, kept a constant stream of correspondence flowing from the dean's office to 'their boys' who had enlisted. Likewise, those in uniform faithfully responded to the Fernows' requests for updates on how they were faring overseas.

One of Olivia Fernow's despatches to the front in March 1917 poignantly captures the deep-seated emotion. 'Dear Boys,' her letter opens, 'you certainly are the finest fellows in the world. Would you believe it? We have heard from 21 of you since Christmas. Every letter counts, whether it comes to Mr White, Mr Fernow, or me. Whoever gets one rushes to tell the rest. I want to write each one of you a long, long letter, but I mustn't. I must just content myself with sending you a few items of news about our forestry boys at the front.' After light-heartedly reporting that Albert Bentley (2T1) was interviewing Germans 'in their own tongue' and that 'it looks as if he speaks a better German than they,' she turned to the tragic news of Douglas Aiken (1T6), who had died on the battlefield in early November 1916. 'He was so brave,' she explained, 'that he went by the name of the "fire-eater." I almost smiled, though the tears were in my eyes. Our lovely, gentle student!' With this in mind she described her keen concern for those who were at the front. 'Oh! Dear boys, you can't imagine how lonely the Forestry Building is, and how we are longing to have our undergraduates back for work and our graduates for a greeting anyway,' her conclusion states, 'Good-bye! Write again! We all love your letters!'[38]

As much as the Fernows agonized over the safety of their students and graduates overseas, Dr Fernow had to worry about his own security during the war. The problem was his Prussian heritage. Although he had become a naturalized citizen of the United States in 1883, the fervent 'anti-Kaiser' sentiment that pervaded Canada from the outbreak of the war cast aspersions upon anyone and anything remotely associated with Germany. This generated significant pressure on Falconer, the university's president, to eliminate from his university professors with German roots. This was particularly true when members of the provincial government, upon whose largesse the university depended, endorsed this cleansing. Only a few months after the outbreak of hostilities, Falconer had already succumbed to this campaign by removing three staff members.[39]

Attention soon focused on Fernow, but he successfully countered the campaign. Dr Forbes Godfrey, the Tory MPP for West York, trained his sights on the dean in January 1915. Godfrey denied that he was motivated by 'personal vindictiveness' in charging that it was 'dangerous when men of this type [i.e., German] are permitted to impart knowledge to the youth of this country.' Not only did he declare that he would use the approaching legislative session to 'protest against the Government rendering financial assistance to the University of Toronto until it has guaranteed that it will discharge every German on staff,' he pointed out that Fernow had not been asked to resign even though the dean was not a Canadian citizen. For this reason Godfrey demanded that the University of Toronto release Fernow, especially because the dean was 'very familiar with the resources of the Dominion and he has a wealth of information at his command which would be regarded as somewhat valuable to the Kaiser.' Fernow immediately responded to the attack by categorically defending his loyalty to the Allies. In a statement to the *Globe*, Fernow declared that 'all Canadians should fight. All the young men of the country should join the army to go and fight for Britain.' He emphasized that he had 'urged every member of his classes to take up military training, and has helped to make it convenient to do so.' Fernow also denied any lingering connection to the German army, in which he had served as an officer during the Franco-Prussian War, and explained that he had not taken out Canadian citizenship because he expected to be returning to the United States shortly (his term terminated automatically in 1916 when he turned 65). Fernow closed by arguing that he was a pacificst, and if he had his way, 'this war should stop at once.' The pressure for the University of Toronto to fire Fernow mounted with each passing month of the conflict, and he responded to each new attack with an ever more zealous defence of his loyalty to the Allies.[40]

Fernow's performance during the war was truly remarkable for several reasons, one of which has remained hidden until now. His tireless endeavour to encourage 'his boys' to do their duty to their country worked, as he retained his job despite all the efforts to have him fired. He lost his campaign to retain German as one of the Faculty of Forestry's compulsory subjects, however, as it fell by the wayside in mid-1918.[41] But Fernow's behaviour during the war was even more extraordinary because he had managed to conceal on which side of the line his loyalties really lay. In the mid-1950s Ralph S. Hosmer, one of Fernow's long-time friends and former head of Cornell University's forestry department, recounted how in 1917 Fernow had travelled to Washington, DC, to attend a meeting of the Society of American Foresters. During a lunch break Fernow had sat down with a group of his senior colleagues. Naturally, their conversation turned to the war and Fernow's feelings about it. Hosmer described how the group rose to leave the restaurant when 'Dr Fernow said: "Gentlemen, I have said more in the past hour than I have in the last two years. My heart is with the Fatherland. But my head, with the Allies."' As Hosmer put it, Fernow's remarks stood as a stark 'indication to some of us of the tact which he had exercised in those difficult years in Toronto.'[42]

Fernow's 'tact' in aggressively supporting recruiting efforts during the war most likely contributed to him retaining his job, but it also probably redounded to the detriment of his faculty in a highly predictable way. Even without his campaign, his forestry students were predisposed to enlist, a function it seems of Fernow's selection of 'mature' students who were drawn to his field by their desire to perform a public service. 'The forester is a patriot by profession, for his business is to provide for the future of the nation in peace time,' the dean's message to the class of 1918 reads, 'but he is to be found as patriotic in war time, when the sterner demands on his citizenship call for sacrifices.' In addition, the forester's skills, such as surveying and timber production, were highly valued by the war effort. But Fernow's continual exhortations to his students to enlist undoubtedly contributed to the disproportionately high numbers of them who did so. By war's end nearly all the Faculty of Forestry's students and graduates who were eligible had enlisted, and fifty-two of its fifty-five graduates had served on the front. Fernow repeatedly boasted about this exemplary level of commitment to the cause, but as he noted, 'it is thus that the Faculty of Forestry has suffered perhaps more than any other by the call to arms.'[43]

And suffer it did. Fifteen students lost their lives, a fatality rate far above the 10 per cent estimated for the university as a whole. In addition, dozens more were physically injured in battle. The postwar years would also reveal that many of those who returned without visible scars had nonetheless suffered severe wounds that would never completely heal.[44]

The tragic toll that war inflicted on the Faculty of Forestry compelled Dean Fernow to take a more lenient attitude – at least temporarily – towards his students, but true to form, he did so only reluctantly. As a general policy, the University of Toronto granted students who served overseas credit for one complete year of university. Fernow recognized the valuable service the students were providing, but he was definitely uneasy with this means of rewarding them. To his mind it had the potential to undermine his goal of creating the best possible graduates. It meant, for example, that students like Wallace A. Delahey (1T5), whom he had asked to 'defer' continuing in the forestry faculty because of weak academic performance, metamorphosed into professional foresters on completing their military service. In Delahey's case, the university's compassion was well placed. After working for the Ontario government and industry during the 1920s and 1930s, Delahey served as the assistant timber controller in charge of pulpwood operations during the Second World War, the Ontario Forest Industries Association's first manager, and executive vice-president of The Great Lakes Paper Company thereafter. His culminating achievement was the University of Toronto's decision to grant him a doctor of laws, honoris causa, in 1957 on the occasion of the Faculty of Forestry's golden jubilee.[45]

With little choice but to accept the university's policy, Fernow took steps to safeguard the forestry faculty's reputation against this unique class of students about whom the dean entertained grave doubts. He insisted they not look a gift horse in the mouth, and take steps retroactively to bring their knowledge up to par. As he gruffly explained to John L. Simmons (1T5) in the spring of 1915, the faculty had 'reluctantly come to the conclusion that in view of your enlistment, although your record is poor, we may give you the degree *honoris causa* for you have hardly earned it by your work. I must advise you, however, that upon your return you ought deliberately to take another year to fit yourself for life work, for with your present equipment you will hardly be able to secure or hold down any position in the forest service, and we would hardly be able to recommend for such.'[46]

It was during Fernow's tenure as dean that certain threads became anchored in the Faculty of Forestry and started weaving their way through its history. These included strong international connections (in terms of a culturally diverse student body and staff and professional work and research), familial ties among the faculty's students, and their remarkable ability to distinguish themselves in and out of the forestry field.

From the time of the Faculty of Forestry's establishment, it enjoyed strong international connections on several levels. As far as its professors and students were concerned, they reflected the breadth of the lands in which forestry was practised at the time. Fernow was of Prussian heritage, after all, and had come to Toronto via the United States. In the context of early 1900s Canada he was truly a foreigner, something to which the difficulties he endured during the First World War attested.[47] Likewise, two professors who would become fixtures in the faculty during its first few decades, C.D. Howe and W.N. Millar, were Americans. Fernow's students had far more diverse backgrounds. Americans Leonard R. Andrews (1T3) and Samuel 'Sammy' S. Sadler (1T3) came from California and Pennsylvania respectively, George W. Bayly was born in southern India and schooled in Switzerland (he entered the faculty in 1910 but only graduated in 1927 for reasons that will be explained), and James D. Aiken (1T6) and William Kynoch (1T4) came from Cape Town, South Africa, and London, England, respectively.

Otto Nieuwejaar (1T7) was a particularly interesting example of the Faculty of Forestry's culturally rich student body. Born in 1895 in Bergen, Norway, his father was a shipping magnate, who was clearly prepared to finance his son's lengthy academic career. After spending one year studying forestry in his native country, Nieuwejaar came to Toronto in 1915, completed his bachelor of science in forestry two years later, and earned a master's degree in economics from the University of Toronto in 1919. Nieuwejaar then returned to Europe to round out his forestry training in Germany and Norway. By the mid-1920s he was a professor in the University of Latvia's forestry school, a position he left in 1928 when he ran into difficulties with the republic's native-language policies. He then recrossed the Atlantic, earned a doctorate from Yale in 1930, and taught forest economics in the United States for the next four decades.[48]

When it came to the Faculty of Forestry's research interests, they were undeniably Canadian-centric under Fernow, but the school also had a strong and indelible impact on international forestry. Fernow was convinced, and rightly so considering the incipient nature of forestry in the Dominion, that his school ought to focus its energy on addressing domestic issues. At the same time, however, the faculty's staff was communicating with foresters from around the world. J.H. White, for example, was discussing silvicultural issues with representatives from India, New Zealand, and California during the mid-1910s. Moreover, Fernow demonstrated remarkable prescience regarding forestry's future direction in a letter he wrote to Robert S. Woodward, the president of the Carnegie Institute, in March 1911. Acknowledging that one of his former students had been the inspiration for the idea, Fernow explained that he wished to bring to Woodward's attention the idea of studying 'tropical forest conditions and forest ecology, with a view of their eventual use to civilizations as regards wood supplies ... You are, of course, aware,' Fernow told Woodward, 'that forest resources of the temperate zone are everywhere rapidly dwindling, and so far the belief is that the hard tropical woods are not fit to supply our needs. But we do know that in the tropics there are soft-wooded species and medium-hard species in relatively small quantities that would answer. The tropics, on account of the rapid growth of vegetation there, may eventually become the producers of the forest products of the world, if biological knowledge is developed by which change of composition can be secured.'[49]

Fernow's graduates would become increasingly involved in international forestry as this field developed in lockstep with their careers. Donald R. 'Sandy' Cameron (1T1) personified this trend. As a student he spent his summers working in the Canadian hinterland on various projects, including supporting the contingent of Indian Affairs officials who paid treaty money to First Nations down the Albany River valley and around James Bay in 1907. Immediately after graduating from the Faculty of Forestry, the Dominion Forest Service hired him, and he worked his way up to assistant director by 1925 and director eleven years later. While assisting with the government's efforts in the Second World War, he was chosen as Canada's inaugural representative on the Forestry Advisory Committee of the (interim) Commission of Food and Agricultural Organization (FAO) of the United Nations in 1943 and at the FAO's first conference in Quebec City in 1945. Two years later he was appointed chief of the FAO's European Forestry Office, and thereafter advanced the interests of international forestry.[50]

While it took Cameron more than thirty years to take his forestry work cosmopolitan, one of Fernow's graduates would have a far greater impact on the world's woodlands within much less time. Leon M. Ellis (1T1) spent roughly a half-dozen years after graduation working as the assistant superintendent of the Canadian Pacific Railway's Forestry Department. This was his last assignment on native soil, as he enlisted in 1916 and rose to the rank of captain and assistant chief forest officer of the Forestry Division of the Canadian Forestry Corps in France. It was there that he met Colonel J. Sutherland, who was a senior forestry official in Scotland. Sutherland convinced Ellis to go with him back to the British Isles to act in an advisory capacity, but Ellis soon capitalized on another opportunity within the Empire. In 1920 he was appointed New Zealand's first director of forestry.[51]

In his own, inimitable way – he later described his professional achievements as demonstrating 'creative ability, initiative, leadership and clean aggressiveness' – Ellis literally changed the face of that country's forests and the industries dependent on them. Within short order he had drawn up comprehensive legislation for managing New Zealand's forests – one of his obituaries describes the statutes as 'amongst the best of their kind in the English speaking world' – and a forest service to administer them. He also initiated a forest inventory and a broad investigation into the silvics of the country's native species. The results of the former convinced him that New Zealand would be best able to meet its long-term timber demands through the government undertaking a massive reforestation program using 'exotic' species. Michael Roche, Ellis' biographer, notes that this 'would also gain time for the intricacies of indigenous forest regeneration to be understood.' This explains how *Pinus radiata*, a species originally native to a nook in California and probably brought over in minor quantities during the gold-mining rushes of the 1860s, became such a conspicuous part of New Zealand's commercial woodlands. Ellis 'transformed forestry in New Zealand,' Roche writes, not only by converting it from being an importer to a major exporter of pulp and paper products but also because 'his legacy [i]s apparent in the extensive plantation forests of the central North Island and the *esprit de corps* of the State Forest Service.'

Ellis would also leave his mark in another part of the southern hemisphere. In 1928 he abruptly quit the New Zealand Forest Service over problems with his compensation, moved to Sydney, Australia, and set himself up as a forestry consultant to take advantage of what Roche describes as 'the afforestation boom in New South Wales.' At the same time, Ellis pushed for both Australia and New Zealand to make much 'fuller use of their God-given resources,' specifically the eucalyptus in Australia. In 1936 he joined the Australian Paper Manufacturers Limited, and helped effect a fundamental revolution in that country's industry. He ended its dependence on imported fibre by overseeing the production of pulp from indigenous hardwoods. When he died at the age of fifty-four in 1941, his life's work stood as a truly astounding example of how one of the Faculty of Forestry's earliest graduates had permanently affected woodlands thousands of miles from Toronto.[52]

Several other long-term trends established themselves at the Faculty of Forestry under Dr Fernow. One was the school's adamant opposition to admitting female students. Although women faced many hurdles to gaining a university education at this time (attempts in the 1880s and early 1900s to establish a women's college at the University of Toronto had failed), Fernow was convinced that they should *never* become foresters. To one young woman who inquired in 1918 about employment prospects in his field, Fernow replied that 'there are some occupations for which women are not specially fitted and forestry is one of them, at least in Canada, on account of the rough life in the woods which it entails.' The Faculty of Forestry would continue to adhere to this view for more than another four decades, making it the last department on campus to admit women.[53]

Another pattern that established itself was the inclination on the part of families to send more than one generation through the forestry faculty's doors. George W. Bayly (2T7) first entered the Faculty of Forestry in 1910, and his son, George H. 'Terk' Bayly (3T9), followed exactly one-quarter century later. Likewise, thirty years separated the dates on which John F. Turnbull (1T9) and his son, Norman J. Turnbull (4T9), graduated from the faculty.

Finally, the trend that would define every dean's term in office was the ability to graduate foresters who enjoyed distinguished careers. Several of Fernow's students gave exemplary service in promoting 'government' forestry. At the national level, Ernest H. Finlayson (1T2) blazed an extraordinary trail. He joined the Dominion Forest Service (DFS) immediately after graduation, and began climbing up its ranks. After he worked himself to near exhaustion acting as the secretary to the Dominion government's Royal Commission on Pulpwood in the early 1920s, Ottawa's reward was naming him the DFS's third director of forestry in 1924. Over the next dozen years, Finlayson zealously promoted forestry awareness across the country and pleaded for governments to manage this resource wisely. The stress he experienced in performing this work undoubtedly contributed to the serious mental health issues from which he suffered during the late 1920s and early 1930s; his life came to a tragic end when he disappeared without a trace in a snowstorm in Ottawa in February 1936. On the research side, William M. Robertson (1T9) had a stellar career. His first job after graduation was working with the Dominion government's Commission of Conservation as its forest research specialist (he helped carry out one of Canada's earliest studies into the impact of harvesting on the regeneration of pulpwood forests). In 1921 Robertson was retained by the DFS when Ottawa disbanded the commission, and there he remained for the next several decades. Over this period he was the first head of the Petawawa Forest Experiment Station and rose to become the DFS's chief of research. One of his most enduring accomplishments was designing form-class volume tables that are still relevant to cruising work.[54]

At the provincial level, Fernow's students made equally significant contributions. Ernest C. Manning (1T2) was hired by the Canadian Pacific Railway and the DFS to perform forestry work after he graduated. In 1918 he joined the British Columbia Forest Service, and by 1936 he was its chief forester. Manning's work in this capacity was cut short when he died in early 1941 in Trans-Canada Air Lines' first fatal accident, which represented the worst crash to date in Canadian aviation history.[55]

Three of Fernow's graduates had particularly noteworthy impacts in Ontario. Peter McEwen (1T6) devoted his life to improving the work done by the Department of Lands and Forests (DLF). He was involved in some of its earliest forest surveys, and he had a hand in creating Ontario's inaugural three forest districts (he was given charge of the one for Parry Sound). While serving at this post McEwen built what was apparently the province's first steel fire tower near MacTier. Then in January 1943 he presented a paper to the Canadian Society of Forest Engineers in which he argued for a forest ranger school to be established in Ontario, a proposal on which the government immediately acted. Frank S. Newman (1T3) followed a very similar path. He spent his entire career working for the DLF as the superintendent of the St Williams Forest Station on the north shore of Lake Erie. Last but not least among this group is Reginald 'Reg' A.N. Johnston (1T7). Having served as a bomber with the Royal Flying Corps during the First World War, he returned to Ontario and literally launched a lifelong career with the DLF; just after the war Johnston piloted the first plane that reported spotting a forest fire and airlifting a crew to fight it. When the DLF established a new research division during the Second World War, Johnston was appointed its head and chose Maple (just north of Toronto) for its base. He oversaw this work until his retirement in 1964.[56]

Fernow's understudies also made their mark while working for Canada's forest industry *a mari usque ad mare.* John D. Gilmour (1T1) led the way for foresters in the pulp and paper industry. One year after graduation he joined the Anglo-Newfoundland Pulp and Paper Company at Grand Falls, at a time when Newfoundland was not yet part of Canada. For roughly the next two decades Gilmour revolutionized the firm's approach to managing its woodlands. He hired professional foresters to oversee the company's harvesting operations and carried out a slew of investigations into methods of cutting and disposing of slash that would foster natural regeneration. He turned to consulting in 1940, work that took him from South America to Scandinavia, and after the Second World War he established himself as one of Canada's leading forest economists.[57] Robert 'Bob' W. Lyons (1T6) worked under the legendary Ellwood Wilson at the Laurentide Paper Company in Grand'Mère, Quebec, for just over a decade after graduation. During this time, Lyons set up Canada's first commercial-scale reforestation program that was annually planting roughly three million seedlings by the late 1920s. In 1928 Kimberly-Clark hired Lyons to be the woodlands manager for its subsidiary, Spruce Falls Power and Paper, in Kapuskasing, Ontario. There, he introduced numerous innovations, including company camps and mechanized harvesting operations, but he will be best remembered for his silvicultural initiatives. After overseeing several studies during the mid-1930s that indicated spruce was not reproducing in certain parts of cut-overs, Lyons was involved in establishing the first sustained, commercial-scale reforestation program in northern Ontario. The University of Toronto recognized his outstanding career when it awarded him a doctor of laws, honoris causa, in 1957.[58] Whitford J. Vandusen (1T2) made his mark in British Columbia, where he enjoyed a six-decade professional association with the iconic lumberman, H.R. MacMillan, with whom he had first worked when the latter had acted as that province's chief forester.[59]

Finally, William Kynoch (1T4) merits special attention because he was both an outstanding graduate and someone whose career foreshadowed the Faculty of Forestry's future involvement in the field of 'wood science'; as a native of London, England, the fact that he captained the Varsity Blues ice hockey team also dispels the notion that such talents are innately Canadian! After leaving the forestry faculty, Kynoch began work on establishing the Dominion Forest Products Laboratory, which was originally associated with McGill University in Montreal. He built an experimental wood-preserving plant there, in which he developed a process to impregnate railway ties with creosote to prevent rot. Kynoch's project earned him in 1918 the faculty's first forest engineer degree, which was awarded to a student who held a bachelor of science

in forestry 'upon furnishing evidence of three years' practical employment in forestry work and the presentation of an acceptable thesis.'[60]
Thereafter, Kynoch became the director of the Forest Products Laboratory. After ten years of service he was appointed a wood science professor at the University of Michigan's School of Forestry and Conservation,
where he remained for the next quarter-century. During these years his
list of contributions to the 'utilization' end of forestry included the lami-
nated tennis racquet. It is noteworthy that Kynoch advised the University
of Toronto's president and the Faculty of Forestry's dean in 1929 that
it would be prudent to develop a 'wood science' program in Toronto,
whereupon he was told that the demand 'for that kind of work' was not
sufficient to justify funding a forest products laboratory.[61]

By 1916, if not earlier, B.E. Fernow had lost patience with what he saw as
the University of Toronto's relative indifference to the plight of his faculty. To illustrate his case, he pointed out that American forestry schools
were staffed with four to six professors, whereas he made do with three.
Moreover, the former taught their students for forty weeks of the year because they had permanent practice forests in which they could run field
sessions. His faculty did not enjoy this luxury, however, so his students
attended classes for only twenty-five weeks and were expected to gain
their practical experience by working as 'interns' during their summers.
Rectifying these two deficiencies was essential to the Faculty of Forestry's
prosperity, Fernow insisted to the university's board of governors. 'It
would appear that the first or pioneering stage of the Faculty is coming
to an end,' he commented in the late 1910s, 'and that reorganization
on line suggested in former reports, contemplating increased staff and a
permanent practice camp, must soon be inaugurated.'[62]

Fernow's increasing frustration after 1916 was not simply a reflection
of the University of Toronto's attitude towards his faculty, however. By
this time he had reached the superannuation age of sixty-five, and his
more than four decades of dedicated service first to planting and then
nurturing the forestry movement in North America had taken their toll
on him. In no uncertain terms he was ready to name his successor.

But this was not a simple matter. The war had decimated the foresters' ranks in Canada, making it a most inopportune time for Fernow to
vacate the dean's office. He thus agreed in early 1916 to remain 'for a
time at a reduced salary.' When he learned a short while later that he was
ineligible for a retirement allowance from the Carnegie Foundation because he had not accumulated sufficient university teaching experience,

Fernow had little choice but to remain at his post until a replacement could be found.[63]

Just as there had been no question in the early 1900s that Judson Clark was the best person to act as the Faculty of Forestry's inaugural leader, so, too, was there general agreement about who ought to be its second dean: H.R. MacMillan, better known as 'HR.' A native of Newmarket, Ontario, he graduated from the Ontario Agricultural College in Guelph with a degree in biology in 1906 and from Yale with a master's in forestry two years later. MacMillan's performance was so impressive that the school's dean described him as 'one of the most competent students to have passed through' and he was voted his class's valedictorian. By this time he had already begun working as a summer student with the Dominion Forest Service, and he accepted a full-time job with it after leaving Yale. In 1912 he was appointed British Columbia's first chief forester, and he set about modernizing that province's forest service. During the First World War the Dominion government sent MacMillan overseas to help market Canadian timber. By 1916 MacMillan had had his fill of public service, which probably accounts for his rejection of an offer to become the founding dean of the University of British Columbia's forestry school. Instead, he went to work for one of B.C.'s largest lumber producers. One year later, he was back on the public payroll as assistant director of aeronautical supplies for the Imperial Munitions Board. For the remainder of the war MacMillan procured Sitka spruce – British Columbia was the only place in the British Empire where it grew – for the construction of aircraft.[64]

Over the course of mid- to late 1917 Fernow made it clear that MacMillan was the *only* person to whom he trusted handing over control of the Faculty of Forestry. In a letter to Robert Falconer, president of the University of Toronto, Fernow sketched HR's impressive credentials, stressed that MacMillan had 'travelled the world,' and noted how he was 'young and energetic.' The dean concluded that he was the 'most eligible Canadian for the place.' At the same time, Fernow warned Falconer that MacMillan 'cannot be had at a low salary, and he would probably be unwilling to work with as niggardly appropriations as I have been willing to work. Incidentally, I may add that the best forest schools in the States dispose of $20 to $30 thousand budgets.' Fernow closed by advising the president that, if the university wished to procure MacMillan, it should act 'very promptly.'[65]

J.H. White, Fernow's long-time understudy and colleague at the Faculty of Forestry, shared the dean's perspective on MacMillan and felt it was crucial for forestry in Canada that HR accept the post. Over the course of late 1917 and early 1918 White repeatedly presented his case to MacMillan. He predicted, for example, that the University of Toronto would likely soon begin providing the forestry faculty with far more funding than 'the mere pittance we have been getting along with,' forestry would 'make more progress in the five years after the war than in the previous ten,' and Ontario and British Columbia would lead the charge. He also stressed how Fernow's failing health (the dean was suffering stroke-like symptoms) made it imperative to replace him immediately. The nub of the matter, White contended, was that anyone could 'direct the technical training of our students,' but the situation demanded a leader like MacMillan who was intimately familiar with the 'Eastern Canadian temperament, education, economic problems, methods of government, ways of doing things + ... all the hundred and one things with which our forestry movement is bound up.' In wrapping up his plea, White threatened to remain with the Ontario government and not return to the forestry faculty if 'an outsider takes charge of our school,' and begged MacMillan to come to Toronto 'even if you should afterwards decide to go back to B.C.'[66]

Despite the urgency of the situation, MacMillan refused to confirm his interest in the job in Toronto. Throughout 1918 Fernow corresponded with MacMillan about the position and repeatedly informed him that it was 'still open to you at any time you choose to take it.' Fernow also updated MacMillan on his 'undiagnosed muscular trouble,' and that he was simply 'holding on' to the deanship until his replacement could be crowned. Concomitantly, George Creelman, the OAC's president and member of the University of Toronto's senate, lobbied MacMillan to accept the deanship.[67]

Suddenly, it appeared that the logjam was breaking. MacMillan agreed to 'come east' to discuss the job with Falconer, in December 1918. When that meeting was cancelled, it was rescheduled for April 1919.[68]

But then fate intervened. MacMillan's hectic April schedule prevented him from leaving British Columbia. The next month, while on board a ship near Powell River, he spent the evening speaking with Montague Meyer, a leading timber salesman in the United Kingdom. As HR's biographer describes it, four years earlier in London, England, MacMillan had begun discussing with Meyer 'the need for a lumber export company based in Vancouver.' They soon formulated a simple business plan, whereby 'MacMillan would contract for British Columbia timber which Meyer would sell in the United Kingdom.' Within a month the H.R. Mac-

Millan Export Company was born. No doubt HR's decision to reject the offer from Toronto was a function of his enthusiasm for the new venture he was initiating at this time. It also probably reflected his cognizance of the daunting political challenges forestry still faced in Ontario, the details of which Fernow had constantly relayed to MacMillan over the previous few years. On one occasion in 1916, when the dean had informed MacMillan that Ontario continued to appoint forestry officials on a patronage basis, HR had remarked that he expected nothing less and 'there must be something lacking in the average Ontarian's head.' Nevertheless, MacMillan had planted a seed in British Columbia that he would grow into an international reputation that made his name synonymous with Canada's forest industry.[69]

With that, Toronto's Faculty of Forestry had suffered an inauspicious déjà vu. Judson F. Clark – *the* best candidate – had declined the opportunity to become its first dean in 1906, and thirteen years later, HR – *the* best candidate – had passed on the chance to become its second dean. Although the impact of these 'missed opportunities' is difficult to gauge, they undoubtedly hindered the faculty during its formative years.

With no available heir apparent and Fernow incapable of carrying on, the Faculty of Forestry had to scramble to find a dean. Of the faculty's three remaining professors, White had already expressed his intention to take up full-time work with the Ontario government if MacMillan did not come to Toronto. William N. Millar was the staff's junior member, having joined it only in 1914 when he had replaced Ross. That left Clifton Durant Howe, whom Fernow had cleverly added to the Faculty of Forestry in 1908 by having him cross-appointed to the Department of Botany, and he had been promoted to the rank of associate professor in mid-1918. Recognizing that the faculty was in a lurch, Howe agreed in mid-1919 to be its 'acting' dean.

Fernow's parting correspondence to Falconer revealed much about the challenges that had fettered his progress and defined his decade and a half as the dean of the Faculty of Forestry. In May 1919 he wrote to Falconer to ask for a retroactive salary raise and the honour of being named professor emeritus. He supported his case by referring Falconer to the letters he had written back in March 1907 that had set out the conditions under which he had taken up the position in Toronto. The major one had stipulated that the forestry school was to receive at least $10,000 of funding in its second year, and at least $12,000 annually thereafter. 'Because of the poverty of the university,' Fernow declared to its president, he had exercised 'the utmost economy by burdening myself with double work and managing with less than $10,000, saving the University at least $20,000 for the last ten years.' In addition, he had donated soon after his arrival his private library and collection of 'lantern slides' that were worth at least $2,500, and 'Mrs Fernow taught German to our students for nine years at the rate of eight to nine hours per week without any compensation whatsoever.' He closed by promising to earn the honorary title by making himself 'useful to the Faculty, the child of my creation' (the University of Toronto granted Fernow professor emeritus but apparently not the raise). B.E. Fernow had thus overseen the establishment and growth of the Faculty of Forestry, even though it had received far less support than he believed both he and it had been promised. Although his successor would take up the post without the same list of demands and expectations, he, too, would grapple with the issues accruing from running a forestry school that neither the University of Toronto nor the Ontario government seemed particularly inclined to nourish.[70]

Chapter 3

'We Cannot Progress in Forestry Very Much Ahead of Public Opinion,' 1919–1929

Canada and its forest industry had an uneven experience in the decade after the First World War. The country enjoyed general economic growth during the second half of the decade but only after suffering through a recession that lingered for most of the first half. Moreover, all did not share equally in the good times. Some industries sputtered, while others roared. Pulp and paper, specifically newsprint, fell into the latter category. Its productive capacity tripled between 1920 and 1929 (this represented a six-fold increase from 1914), and Ontario participated fully in this phenomenon. At the same time, however, the industry began suffering from overcapacity by the mid-1920s, and the situation worsened through the rest of the decade. This disparate pattern also marked the country's lumber industry. H.R. MacMillan's export business boomed in British Columbia, whereas eastern Canada's traditional producers grappled with dwindling pine supplies and competition from 'HR' and his western rivals.[1]

The forestry profession suffered through an equally checkered decade after the war. While the pulp and paper makers were far more inclined than the old lumber barons to hire foresters, only a few newsprint makers committed to practising forestry. Dean Howe's graduates thus increasingly found work in the private sector, but seldom did they practise the profession for which they had been trained. It was much the same story in the public sector in Ontario. Foresters found jobs, but the government generally went out of its way to prevent them from applying the knowledge they had gained at the University of Toronto's Faculty of Forestry.

Clifton Durant Howe, the Faculty of Forestry's new dean, identified the problem as being two-fold. First was the provincial government's aversion to investing in forestry. In 1924 Howe described how he believed the more progressive pulp and paper firms in eastern Canada were 'making a sincere effort to get on a sustained yield basis.' What was lacking, he contended, was 'gentle pressure and guidance from above, that is, from the provincial government. It is the government that lacks the vision, or if it has any vision, lacks the courage to try it out.' Howe believed this trepidation reflected a paucity of public support for forestry, the problem's second dimension. The electorate simply did not consider it a priority to invest in silvicultural measures that would not pay dividends for decades. Howe repeatedly lamented this state of affairs throughout his twenty-two years as dean. In the early 1920s he despaired that, although the public occasionally lifted its head for long enough to think years in advance, right now 'the "to hell with the future" crowd is in the ascendancy,' a situation that would not change for decades.[2]

Howe pursued a well-defined strategy in an effort to address these two challenges, and although he made some headway, his plan failed to achieve its ultimate goal. On the one hand, he carried on his predecessor's tradition of moulding model graduates whose level of professionalism was intended to convince voters and politicians alike that forestry work was necessary. 'Our late and lamented Dean, Dr Fernow,' Howe told an audience in 1926, 'held that the first and most important thing in education was to make good citizens. That policy is being and will be carried on by the present staff.' On the other hand, unlike Fernow, who chose to make war with the Ontario politicians, Howe felt that making peace was a far more effective route to winning over the government. 'As soon as you begin to criticize an individual or a government severely, his back comes up and you may drive him to a position of extreme conservatism which he would not otherwise have held,' Howe postulated in the early 1930s. 'Through long experience in dealing with men and governments, I find that one can get better results by praising them for what they have done so far but at the same time suggesting that there is one more step that they should take.' To be sure, Howe's amicable approach to dealing with the provincial politicians placed him front and centre in Ontario's forestry developments during the 1920s. He had the ear of the government in the early part of the decade, and was consulted repeatedly by elected and unelected officials alike during its latter stages. Moreover, the provincial bureaucracy welcomed his graduates with open arms. But in the end, Ontario was no closer to implementing a forestry policy on the eve of the Great Depression than it had been ten years earlier.[3]

The Faculty of Forestry fared only slightly better in dealing with the University of Toronto because the university still saw forestry as a low priority. The only occasions on which the university responded favourably to Howe's requests for assistance were those on which he played 'hard ball' with it, and his success in these instances helped put the faculty on a firmer foundation during the 1920s. The profession's brighter job prospects led to slow but steady growth in the school's student body, with total enrolment ranging between the low forties and mid-sixties during these years. Although it was still a minor player at the university, the Faculty of Forestry was far stronger than it had ever been.

C.D. Howe came to the dean's office with very different lineage and fundamentally different qualifications than his predecessor. Bernhard E. Fernow was from noble, European stock, and he had spent more than a quarter-century in the public eye as the beacon of the forestry movement in the New World. Howe, on the other hand, descended from a middle-class New England family. Before taking up his post in Toronto in 1908, as a lecturer in botany and forestry, he had devoted most of his professional career to carrying out field research alone and far from the spotlight.

Howe's parents had both been avid naturalists and inspired him to study this subject, one in which he became an expert. After graduating with a science degree from the University of Vermont, he earned his master of science from the same institution and a doctorate in botany from the University of Chicago. During these years Howe's specialty became 'plant succession,' a principle that his supervisor at Chicago, H.C. Cowles, was developing. Howe recognized that exploring this concept required meticulous, costly, and time-consuming data-gathering, but he was utterly convinced that the investment was worthwhile. Howe began applying his approach in 1904 when he was appointed a lecturer at the Biltmore Forest School in North Carolina, where he was responsible for classroom and field instruction (he conducted the latter in the school's practice forest). Four years later he took the job at the Faculty of Forestry in Toronto. For roughly the next decade, Howe spent his summers working for Canada's Commission of Conservation, carrying out field research across the country into different aspects of forest succession; the commission's landmark reports on the condition of these woodlands owe much to Howe's exacting efforts. In addition, he developed the standard procedure for what his colleagues christened 'regeneration surveys' and 'permanent sample plots,' means by which he could monitor developments in the bush over long periods. By the mid-1910s Howe's pioneering contributions in this field placed him among Canada's foremost silvicultural researchers.[4]

Howe's reverence for assiduous, painstaking scientific methodology meant that, although he brought many of the same attitudes as Fernow to the dean's office, he also differed from his predecessor in several major ways. Undeniably, they both shared the view that Canada represented fertile domain in which forestry could take root because the government controlled nearly all the productive forest. Under this arrangement, industry could not be expected to invest in managing its woodlands because it did not enjoy secure tenure to them. As Howe explained to a leading official from the American pulp and paper industry, in 1923, 'as Dr Fernow said so often, the longer I live the more fundamental the statement appears to me: "Forestry is a function of the state."' And just like Fernow, Howe recognized that government control over timberlands was a mixed blessing. It certainly created a golden opportunity for prudent forest management, but it also meant that the elected officials would have to be convinced that it was in their best interests to invest in this work.[5]

It was here that Howe fundamentally diverged from his predecessor's understanding of forestry's political milieu in Ontario in particular and Canada in general. Fernow was convinced that he could single-handedly badger the politicians into changing their ways. In sharp contrast, Howe believed that the government would commit to practising sound forestry when gaining and retaining power depended, at least partially, on doing so. Only then would the politicians stop perceiving Crown woodlands as invaluable patronage resources and recognize them as assets that required prudent management. 'We cannot progress in forestry very much ahead of public opinion,' Howe told Roy MacGregor Watt, a forester with the Dominion Forest Service, in 1924, 'or at least can not maintain a policy for any length of time without public support. I fear we are still some distance from the point when the public will demand forest management and enforce their demands on governments.'[6]

This is where Howe's staying power came into play. Having served with Fernow at the Faculty of Forestry since 1908, Howe had seen first-hand both the politicians' relative lack of interest in effecting forestry reforms and how they had bucked when Fernow attempted to force feed them this agenda. Howe thus accepted that it would take time, years and perhaps decades, to achieve his end, and he was prepared to wait.[7]

Howe believed that Toronto's Faculty of Forestry had a very specific role to play in expediting this process. He shared Fernow's belief that their forestry school must produce the best possible graduates. But whereas Fernow aspired to this goal largely as an end unto itself, Howe posited that his school had to 'jealously guard the quality of the men graduated' because it was integral to the long-term strategy of fostering widespread support for forestry. Howe was convinced that his graduates must be, above all else, ideal citizens. Aware that the political milieu in which they were operating was hostile, sometimes openly, to the message they were preaching, he felt it was essential to cultivate foresters who would accept setbacks and have the tenacity to regroup and fight another day. This is how they would win over the politicians and public. To accomplish this goal, Howe hoped to capitalize on his school's favourable student-professor ratio (roughly 20 students entered the program each year during the 1920s) to inculcate in them traits such as perseverance and fortitude. 'We believe that this aspect of our work furnishes a greater return to the state for the money we expend than the knowledge of technical forest management,' Howe informed Robert Falconer, president of the University of Toronto, in 1922. 'We lay this stress upon mind training and the development of character because of insight into the details of the problems the men meet in their work after graduation … We are trying to establish a code of high ideals, inspire morale, and develop traditions in the school, since we consider these the biggest factors in turning out the type of man demanded by the forestry profession.' This was the impetus behind Howe's decision to raise his Faculty of Forestry's entrance requirements soon after he became dean, and the reason he had a clear conscience even though the incoming classes during the 1920s had an attrition rate of nearly 50 per cent.[8]

At the same time, Howe was as inclined as his predecessor to show remarkable kindness to those who deserved it. Howe was particularly soft-hearted immediately after the First World War, a period about which more will be said shortly, but compassion marked all his days as dean. He proudly referred to each of his students as 'one of my boys,' and acted as a father-figure and mentor to them, especially when times were tough.[9]

Howe's drive to build public support for forestry also shaped the role he performed at the faculty. Like Fernow, he believed that his most important function was to transcend his teaching and administrative duties in Toronto and work on converting the country's citizens into forestry partisans. To achieve this aim, Howe travelled throughout Canada – and the world – to preach his message. He also inaugurated an annual banquet that his faculty hosted, and its guest list included the continent's leading foresters and the province's most influential politicians and businessmen. During the early 1920s he brought a joint meeting of the American and Canadian forestry societies to the University of Toronto and the Imperial Forestry Conference to Canada for the first time, an accomplishment that resulted in unprecedented media attention for the forestry cause. Furthermore, he played a large role in initiating 'Save the Forest Week' in the mid-1920s. Forestry's profile in Canada rose during the interwar period largely because of C.D. Howe.[10]

Finally, Howe had a definite notion of the function his faculty ought to play in furthering its and forestry's agenda in Ontario. His school's principal function, he argued time and time again, should be to prepare men for employment with the provincial government's forestry service. To achieve this aim, and to foster a fruitful relationship with the politicians in the Legislature (who, after all, determined how much funding they would give to the University of Toronto each year and if they would hire Howe's students), he demanded that his staff and students refrain from criticizing publicly the government's forestry policy, or lack thereof. He determined that the Faculty of Forestry's interests would be best served by quietly accepting that the campaign to realize sound forest management in Ontario would be a protracted one.[11]

It soon became clear that Howe and the provincial government had tacitly agreed to a quid pro quo in this regard. His staff members, students, and graduates would refrain from publicly attacking Ontario's forest policy, and the province would hire his graduates. By the end of the 1920s the deal was paying off handsomely for Howe. The Ontario government permanently employed over thirty of 'his boys,' and the number was still rising yearly.[12]

Howe's adherence to this arrangement did not sit well with either him or his colleagues, but he zealously insisted that it be respected. He purged himself of his personal frustration over the lack of progress on the forestry front by venting to his confidants. On occasion, he would even publish an article or deliver a speech on the subject, but it would discuss the issue in such an oblique manner that no party could take offence. To the Royal Canadian Institute in 1925, for example, he bemoaned the fact that the 'forester is at a disadvantage in this country; he is not regarded as a producer but as a physician to attend a sick patient, and worse than that, he is not called in until the patient is very ill indeed; in fact, the patient is usually in the first stages of rigor mortis when the forester is summoned.' As frustrating as it was to operate under these conditions, Howe was committed to this modus vivendi.

But Howe's first priority upon assuming the deanship of the Faculty of Forestry in mid-1919 was coping with the fallout from the First World War. Like many other nations, Canada saw the return from the front of thousands of young men who needed jobs and help in readjusting to the world they had left behind. This was an especially daunting task for those who had suffered injuries. With the dominion government unwilling to play a leading role in rehabilitating the country's veterans, the Faculty of Forestry was left to fend for its own.

Howe's challenges were many. It has already been described how the University of Toronto had been granting credit for one year of university to those who had served in the armed forces, and how Fernow had been concerned that this might dilute the quality of the Faculty of Forestry's graduates. While Howe was obliged to continue this policy despite sharing his predecessor's reservations about it, he felt compelled to do much more for 'his boys.' Howe saw to it, for example, that the faculty's fifteen students who had died in the war were immortalized in the bronze plaque that bears their names and still hangs in the school's hallway.

While the loss of these men represented a major blow, a far more pressing matter was dealing with the students who had enlisted as vibrant and healthy young men but returned as mere shadows of their former selves. Howe bent the forestry school's rigorous and uncompromising pedagogy in reaching out to these veterans. Their stories stand as a testament both to Howe's humanity and their extraordinary struggle to regain their vitality.

The most poignant account involves George W. Bayly (2T7). Born in Mysore, India, in 1886, Bayly attended Kingston's Royal Military College, McGill University, and the University of Munich prior to arriving at the University of Toronto's Faculty of Forestry in 1909, where he hoped to apply his previous credits towards earning his bachelor of science in forestry in two years instead of the usual four. Bayly was in for a rude awakening, however, as he fared miserably at the Faculty of Forestry. After three years, his grades ranged between 30 per cent and 57 per cent. His abysmal performance left the faculty with no choice but to 'defer' granting him his degree. With his academic career in abeyance, Bayly had enlisted in the 3rd Battalion when war was declared in August 1914.[13]

Bayly's experiences thereafter – both during and after the war – can only be described as horrific. Years later circumstances dictated that Howe had to summarize the adversity Bayly had endured. 'His skull was fractured in the second battle of Ypres,' Howe wrote of Bayly in 1927, 'after which he spent six weeks in hospitals, three weeks on leave and worked in a records office one month, when he was sent back to France. After three weeks there he went to pieces and it was discovered that his skull had not yet healed. He spent three months more in hospital, then was sent to the Officers' Training School at Hyle. He had just completed the course, when he went to pieces again. He was finally discharged to the Reserve Officers' list and sent back to Canada. For the next four years he was subject to vomiting and dizzy spells. He is still subject to these, but at more infrequent intervals.'

The challenge for Howe after the war was Bayly's request for the Faculty of Forestry to grant him his degree despite his atrocious academic ·record. Howe had discussed the matter with James Brebner, the university's registrar, but the latter had not given Howe 'any encouragement' to believe that the rules could be bent far enough to accommodate Bayly. After Bayly found work in 1924 with the Ontario Forestry Branch's reforestation arm, he had written Howe again about obtaining his degree because it would have meant a far higher ranking in the provincial civil service. It was then that Howe stepped up to the plate. He arranged for his staff to tutor Bayly privately over the next three years – without extra remuneration – in order for Bayly to earn credit for the courses he was lacking. In early June 1927 Bayly was granted his bachelor of science in forestry, a degree on which he had begun working nearly two decades – and several lifetimes – earlier.[14]

If the post–First World War period brought heart-wrenching stories like Bayly's to the Faculty of Forestry, it paradoxically gave rise to tremendous optimism about the future. Many hoped that bringing 'the war to end all wars' to a merciful conclusion signalled the dawning of a new age in which the promise of a better world would be realized. In terms of forestry in Canada, auspicious signs abounded. Hundreds of thousands of returned soldiers needed work, and the war had demonstrated the strategic importance of natural resources and managing them wisely. Conditions thus seemed ripe for putting at least some of these veterans to work cropping the country's forests.

This was especially true in Ontario, where the upstart United Farmers of Ontario (UFO) under their leader, E.C. Drury, shocked the province when they won the 1919 election. One of the central planks in the UFO's campaign platform had been a commitment to 'clean up' and apply 'modern' forestry principles to the government's administration of Crown timber. To accomplish this end, Drury took Howe into his confidence and leaned on the dean for advice. Drury also struck 'the Timber Commission' to assist in this effort. When Howe excitedly informed John

D. Gilmour (1T1), in 1920, that 'things never looked brighter for forestry in Ontario,' he genuinely believed that the breakthrough for which the foresters in Canada had been anxiously waiting was about to occur, and Ontario would host the blessed event.[15]

But in short order Howe was to learn that the Ontario government would be unable to effect meaningful forestry reforms. Over the course of 1920–1 Drury had committed to achieving this end by removing the management of Crown woodlands from political control and placing it in the hands of professional foresters. To direct the revamped Department of Lands and Forests (DLF), Drury was determined to appoint one of the country's leading foresters. Howe closely advised the premier on this file, and Drury interviewed Ellwood Wilson, Laurentide Paper Company's veteran forester and engine behind that firm's pioneering reforestation program. Drury also acted on Howe's advice when he considered hiring two of the Faculty of Forestry's most eminent graduates, classmates Leon M. Ellis (1T1) and Gilmour (1T1), to assist with reforming forestry. Drury's plan foundered, however, largely because he ran into stiff political opposition, especially from the elected officials who despised Howe's influence in the process. 'I understand,' Howe confided to Gilmour in the early fall of 1921, 'that certain men in the parliament buildings resent what they call my "butting in," and if they are near enough to the Premier to influence him, it may be that I shall not be consulted again.' Howe was right, as Drury was forced to appoint a compromise candidate who was not a forester. Drury had also hired, in 1920, two top-notch foresters to recommend forestry reforms: Ernest H. Finlayson (1T2), who had risen to a senior position with the Dominion Forest Service, and Dr Judson F. Clark, the Yale graduate who had rejected the offer to head the Toronto forestry school in 1906 after learning that the provincial state had no interest in adopting sound forestry methods. The government shelved both of their reports, however. Sadly, when Finlayson commented in his submission that Ontario 'has less to show in actual forest development and organization than any province of the Dominion with the exception of Nova Scotia where the bulk of timber lands have passed into private ownership,' his views would accurately describe the situation for the foreseeable future. Howe could not hide his despondency over this turn of events. 'I fear that re-organization has been put off so long already,' Howe despaired to Gilmour, 'that the effective psychological moment is past. The old secret understandings between operators and office are apparently being restored.'[16]

Several Faculty of Forestry graduates did, however, convince the

United Farmers to take a few minor steps in the right direction during its four-year reign (1919–23). Drury's government began organizing the province into forest districts, for example, as a result of pressure exerted by Peter McEwen (1T6) and 'Reg' Johnston (1T7), two of the DLF's employees. A trio of Toronto graduates was given charge of the first units that were established in 1922–3: George M. Dallyn (1T6) in Tweed, Wallace A. Delahey (1T5) in Sudbury, and McEwen (1T6) in Parry Sound. The system would be expanded over the next decade until it covered the entire province. In addition, the United Farmers invigorated the government's lacklustre reforestation program in southern Ontario. The UFO established two new nurseries to complement the flagship one at St Williams, a seed collection centre at Angus, and enacted legislation to facilitate tree planting on wasteland and marginal farmland in southern Ontario.[17]

But this progress was more apparent than real. District foresters slowly gained jurisdiction over thousands of square miles of Crown forest, but they were unable to implement substantive forestry reforms. In addition, the focus on tree planting in southern Ontario reinforced the public's tendency to identify this activity as representing 'forestry' and the panacea for what ailed the woodlands. Although Howe supported the effort per se, he feared its drawbacks. As he lamented to Allison B. Connell (1T4), in early 1922, 'the lumbermen and lumbermen's associations are whooping it up strong for Drury's planting program. They are going to offer prizes to schoolchildren for the best essays on planting and the care of farm woodlots. This is all deliberate, to throw dust in the eyes of the public, to distract attention from what they are doing to the people's woodlands, from which the bulk of our future supplies, if we ever have any, must come.'[18]

This same pattern marked the rest of the decade during which the Conservatives controlled the Legislature (1923–34); they did not back up their promises to adopt 'modern' forestry practices with actions. During the early years of the Tories' reign, they dramatically expanded the province's forest-fire protection service. In the process, however, the politicians berated the foresters for their alleged inability to deal with the issue in order to justify spending tens of thousands of dollars on aircraft. The upshot was not a better but a more sensational protection program, about which more will be said below. Likewise, the reforestation effort in southern Ontario expanded under the Conservatives, and Toronto's Faculty of Forestry's graduates led the charge. At the same time, however, the province's foresters lamented how this program warped the public's understanding of their work and shielded the government from criticism about its inertia in the north.[19]

And it was in the north where the greatest discrepancy existed between the promise of the Tories' policy pronouncements and the lack of action to back them up. During the early 1920s Ontario's leading pulp and paper makers had undertaken significant silvicultural initiatives. Spanish River Pulp and Paper Mills, the province's and Canada's largest newsprint maker, had committed just after the First World War to managing on a sustained yield basis the pulpwood concessions it leased from the government in Ontario. The province's second largest newsprint maker, Abitibi Power and Paper Company in Iroquois Falls, initiated a major experimental reforestation project during the 1920s. Both of these firms were eager to carry their share of the burden attendant upon prudently managing the Crown forest, but they looked to the Ontario government to be an equal partner in this endeavour.[20]

Howe recognized that this was a reasonable request, but was sceptical that the politicians would take it up, for several reasons. He was cognizant that the pulp and paper makers saw that practising sound forestry was cost-effective, but he doubted the politicians shared this view. As he told Benjamin F. Avery, Spanish River's chief forester, 'the Government will have to be led gently and gradually on to the conception of managing the forests on a basis of continuous production. I fear they would shy at taking the whole dose at once.' In addition, Howe knew that the lumbermen, who had proven intransigent when it came to implementing forestry measures, exercised disproportionately large influence over the government.[21]

This great divide separated the pulp and paper industry from the Tories on the forestry front until 1926, when a propitious development seemed to bridge the gap. That year the Conservatives appointed William H. Finlayson as minister of lands and forests. If ever the elected official in charge of this portfolio was keen to reform the province's administration of Crown forests, it was Finlayson. His step-brother, Ernest, was a Faculty of Forestry graduate and the dominion government's director of forestry. Finlayson the forester boasted to Howe at this time that he had privately tutored Finlayson the minister before and after the latter had been sworn in. 'This resulted,' Ernest vaunted, in his step-brother 'stepping into the job with, perhaps, a greater knowledge of the principles of forestry than has probably been the case with any other non-technical man who has occupied a Cabinet post.' Howe was ebullient at Finlayson's appointment and his early discussions with him. 'He told me,' Howe informed H.R. MacMillan in reference to William Finlayson, 'that the province in the past has not made sufficient use of the graduates from our forestry school, but he intended to use them to a larger extent and to advertise to the public that he was using trained foresters in his work. This is the first time that any Minister of the Crown has publicly expressed any desire for cooperation with the school.'[22]

Over the next few years, the impression was that Minister Finlayson was committing the Conservatives to following a new and enlightened path to administering Crown timber, but behind the facade, practically nothing had changed. In 1927 the Tories established the Forestry Board to advise them in managing Ontario's pulpwood forests. Finlayson appointed Howe its chairman, and Howe's colleague, J.H. White, was retained as its research specialist. The board met frequently from mid-1927 to the fall of 1928, and drew up a short list of fundamental reforms that it felt were imperative for Ontario to implement. The Tories, however, ignored nearly all of them, and William Finlayson soon began finding flimsy excuses for avoiding any contact with the board. Howe was exasperated over the situation (this would not be the last time that one of the faculty's deans would sit on a forest advisory committee whose function was merely symbolic). Likewise, the Tories introduced a host of new forestry statutes during the late 1920s that enunciated prudent principles on which the future management of the Crown forests could be based. The problem was that the Conservatives did nothing to realize the potential of this legislation, and instead used it to further their own selfish, short-term political interests.[23]

The same paradigm applied to the area of forest research. During the first few years of the Great Depression, John A. Brodie (2T3) acted as the Ontario government's 'Forester in Charge of Research' and orchestrated a truly remarkable program of data-collection. He instructed the DLF's district foresters – practically all of whom were products of his alma mater – to prepare 'position papers' that recommended the 'best practices' for dealing with everything from land classification to slash disposal. In addition, Brodie's five-man staff carried out ground-breaking investigations into the silvics of the province's most valuable commercial species. The resulting reports both demonstrated that the present harvesting practices were depleting the supply of pine and spruce in the Crown woodlands and presented clear ideas about how to rectify this situation. Not only did the provincial government ignore these critical conclusions, it ensured that the information never saw the light of day.[24]

Howe's front-row seat to all these false starts reinforced his view that forestry lacked a constituency that was sufficiently powerful to make it at worst desirable and at best essential for the politicians to commit to bona fide reforms. Perhaps better than anyone else in the province, Howe knew that all the royal commissions, legislative initiatives, advisory boards, and research programs were meaningless because they were simply 'window dressing,' as one of his former students later put it. When one of Howe's colleagues in California in the late 1920s congratulated Howe on having been one of the driving forces behind the succession of new forestry statutes in Ontario, Howe thanked him for his kind words and expressed hope that they heralded a new age in silviculture. At the same time, Howe wittily noted, 'I am touching wood when I say this, judging by twenty years with politicians, whose goodly intentions often cannot be carried out for political reasons.'[25]

Although Howe quickly learned that the Faculty of Forestry faced many of the same challenges that it had in the past, he figured out a clever means of addressing them. His first few years as dean demonstrated that his cause lacked the requisite support from both the senior administrators at the University of Toronto and the politicians at Queen's Park. As a result, the only time he succeeded in achieving his ends was when he engaged in old-fashioned arm-twisting that made it expedient for the university – and often the government – to grant his wishes.

Howe kicked off his efforts to strengthen the Faculty of Forestry soon after he moved into the dean's office, in mid-1919, and conditions seemed particularly ripe. The Drury government had swept to power promising to revolutionize the administration of Crown timber, and it had established a commission in 1920 to determine a means of adequately funding the province's three universities. This inquiry had prompted the University of Toronto to ask all department and faculty heads for an assessment of their budgetary needs.

Howe had watched the university reject Fernow's repeated requests for additional staff and a new building since Howe had joined the Faculty of Forestry in 1908, and so he presented his demands for the faculty in a strategic manner in May 1921. To Robert Falconer, the university's president, Howe argued that he was at wit's end regarding these matters. He also just happened to describe, at length, the enticing offer from the United States Forest Service that had just landed on his desk. The USFS had asked him to be the director of its new forest experiment station in Asheville, North Carolina, and because research was Howe's forte, he found the opportunity particularly appealing. Howe pointed out that he was also strongly drawn to the job because he was so discouraged by the situation at the faculty. 'It can never perform its proper function in the Province and in the Dominion,' he declared defiantly to the president, 'without more adequate financial support.' He then listed the specific items for which B.E. Fernow had begged in his annual reports year after year – more staff, a practice forest, a new building, and increased laboratory space – and what the university's response had been so far: a 'meagre increase' in terms of lab space. The upshot, Howe pointed out, was that the other forestry faculties in Canada and the United States offered their students a far superior education than was available in Toronto, and Howe's graduates were thus at a serious disadvantage when they went to compete for jobs. Howe closed by making it clear that the university's attitude towards providing for his faculty would determine how he responded to the American offer.

Falconer could not afford to let Howe depart at this time. The university was just finding its legs again after the war, and the faculty had been hard-pressed to replace Fernow when he had retired in 1919; this situation had not markedly improved by 1921. Falconer's sensitivity to the bind in which he found himself was undoubtedly the impetus behind the University of Toronto's prompt move to meet Howe's demands. Two weeks after Howe had written to Falconer, the president informed Howe that the board of governors had agreed to grant his 'suggestions,' namely, the new building and $3,500 for new professors. 'I hope you will now see your way to remain with us and help us to develop this department,' Falconer's final line to Howe read; the dean did.[26]

Howe soon learned that promises from the university, just like the provincial government, rarely translated into tangible gains. As far as hiring the professors was concerned, he ran into serious problems, but these were not of the university's making. Howe's first choice was Herbert R. Christie (1T3), whom Howe described 'as one of the very best of our students.' But Christie had just begun organizing the forestry department at the University of British Columbia, which the administration had threatened to close if Christie left. Christie was desperate to join Howe's staff, but his dedication to the forestry cause compelled him to remain on the coast. Howe's second choice, whom he believed to be an equally strong candidate, was John Gilmour (1T1), whose storied career has already been chronicled. Although Gilmour, too, was keen to come to Toronto, to ensure he remained with the Anglo-Newfoundland Development Company, Lord Rothermere raised Gilmour's salary to $7,500 – an income that was over twice as large as the academic one Howe could offer him.[27]

The difficulties Howe encountered in hiring the new professors delayed the process for so long that it gave the university the occasion to close halfway Howe's window of opportunity; when he finally found a candidate to accept the post, the university agreed to hire only one – not two – new staff. In the fall of 1922 (just before the session opened), Theodore W. Dwight (1T0) accepted the university's offer to join the Faculty of Forestry as an associate professor. Dwight had graduated with the highest marks ever at the faculty, and had been acting director of the Dominion Forest Service. When he agreed to teach in Toronto, he embarked on a thirty-five-year career at the forestry school.[28]

Having been short-changed on the professorial front, Howe turned his attention to winning all the funding for the Faculty of Forestry's new building. Falconer had promised Howe a new building, in June 1921, but qualified his pledge by noting it would be realized only 'as soon as possible.' Nine months later Howe conveyed his disquiet on hearing rumours that 'again our new building and equipment will be postponed.' He reminded Falconer that the Faculty of Forestry had shared 'an adapted private residence' with the Department of Botany for over thirteen years, and this unsatisfactory state of affairs could not continue.[29]

With no progress on this front for over two years, Howe endeavoured to gain the University of Toronto's attention by again threatening to leave. In late 1923 he orchestrated events to see that word 'leaked' to the press about his negotiations with the University of Michigan at Ann Arbor to become director of its forestry school; one forestry observer described Howe's possible departure as 'a national calamity.' Howe followed up with letters to the president of the University of Toronto and to the provincial premier. They described how, sadly, the dean had 'come to the conclusion that a few more years of the present conditions will mean the passing of the [forestry] School. To accept an offer from another university would be against my inclination and I would leave this University with a sorrowing heart.' Howe then spelled out the specific conditions that would keep him in Toronto: another staff member and the 'immediate prospects' of funding for the Faculty of Forestry's new, $100,000 building.[30]

Howe's use of the media to publicize his cause clearly concerned Falconer. Speaking at the faculty's annual alumni dinner at Hart House, in January 1924, Falconer mentioned the 'rumors [sic] in the public press concerning the departure of Dean C.D. Howe of the Faculty of Forestry.' At the same time, the university's president pledged that 'every effort would be made to retain the Dean in Toronto.' As the *Canada Lumber-*

man recorded Falconer's speech, 'a new forestry building, the President explained, was the surest way of accomplishing this end.'[31]

But Howe had learned not to be reassured by words alone, and continued pushing the president on the matter. In a short note to Falconer in mid-March, on the occasion of the death of E.C. Whitney, the University of Toronto's munificent benefactor and veteran timber baron, Howe asked rhetorically whether it would 'be quite fitting that money left to the University by a lumberman go into a new Forestry Building?' When this proved to be a non-starter, a few days later Howe informed Falconer that the dean was again being head-hunted, this time by a 'pulpwood organization at an initial salary of $8,000 a year' (roughly double Howe's university wage). Howe insisted that he had 'decided that even such a salary should not entice me away from what I believe to be my mission in this world,' but that 'there will be a more joyous and sustained effort, as well as greater efficiency, in our endeavour to lead the public to an understanding of what forestry means to Ontario if the hope of a forestry building is fulfilled.' Falconer then thanked Howe for his continued allegiance to the University of Toronto, and expressed his hope 'that before the spring is over we may show our appreciation of your loyalty to the University.'[32]

Howe then took steps that he knew would practically guarantee him success. First, he called on the country's leading business groups, with which he had been in close correspondence since becoming dean five years earlier, to lend their support to his cause. He wrote the Associated Boards of Trade and Chamber of Commerce of Ontario, for example, to explain the situation and the need for immediate action. The association's executive promptly passed a resolution – and circulated it to the province's major media outlets – that demanded that the government provide 'adequate and generous' support to the Faculty of Forestry because proper forest management was essential to Ontario's future prosperity. Howe also strategically lobbied H.J. Cody, Premier Ferguson's long-time crony and the person whom the premier had hand-picked to be chairman of the University of Toronto's board of governors. Within four months, the Ontario government announced that it would allocate $125,000 for the new forestry building, a sum that exceeded that for which Howe had asked![33]

Howe basked in this news, but he was acutely aware of the reason for his triumph. When R.S. Hosmer, a forestry professor at Cornell University, wrote to congratulate Howe on culminating his faculty's fifteen-year lobby, Howe explained that it was in response to 'public opinion.' In other words, Howe pointed out that he 'had good support from the Associated Boards of Trade and from the Canadian Manufacturing Association.' Likewise, when Howe wrote to the former organization to express his gratitude for its support, he confided that, 'as you know better than I, governments respect the opinions of business men and act upon them very much more than they do those of university men. I think your help in the matter of the new building was the principal factor in bringing it about.'[34]

The threads that had begun weaving their way through the Faculty of Forestry's history during B.E. Fernow's term as dean continued to do so under C.D. Howe, and a few new ones also appeared. One that became more pronounced during the 1920s was the faculty's connection to the international forestry community in terms of the school's students and staff, and their career paths. The Faculty of Forestry attracted young men from ever farther afield during the 1920s. In addition to the usual representatives from the British Isles and the United States, an increasing number from more exotic destinations made up the faculty's roll call. At one point in the mid-1920s natives of India, Gibraltar, Trinidad, and Russia were attending the school at the same time, and they were followed a few years later by young men from Holland, Norway, and Finland.

Howe's students also looked increasingly to the international scene for careers, but they continued to encounter a serious hurdle along this path. One of the greatest opportunities for working abroad was in the British Empire, specifically in India, which boasted one of the world's largest forestry services. But a technicality rendered the Toronto's Faculty of Forestry's students ineligible for working anywhere in the Empire except Canada. A group of the school's students had learned about this glitch in the early 1920s when it had asked Howe about applying to 'the Colonial and Indian Service.' After looking into the matter, they learned that the rules regarding such appointments stipulated that candidates' degrees had to be conferred by Oxford, Cambridge, or Edinburgh universities. Howe tried but failed to have the University of Toronto added to the list.[35]

Howe and at least one of his staff members were also tightly connected to international forestry. He and J.H. White travelled to countries where forestry was being practised, specifically Finland, Germany, and Sweden, and were in constant correspondence with foresters throughout the world. White was especially active. He communicated regularly with forestry schools and their graduates in places like India, Japan, Poland, the

United Kingdom, and Yugoslavia. Moreover, the period's relatively lax import and export restrictions permitted him to exchange seeds – and not merely ideas – with his foreign contemporaries.[36]

At least one of Howe's students from the 1920s went on to make his mark in forests far from Canadian soil. William E.H. Munro (5T2) was from New Jersey and entered the Faculty of Forestry in the fall of 1927. As his student record euphemistically puts it, however, he withdrew in March 1928 'at the request of the Faculty ... for persistent neglect of attendance.' Undaunted, Munro headed south to Colombia, where he spent most of the Great Depression working for the Tropical Oil Company on its plantations. He returned to the Faculty of Forestry after the Second World War to complete his undergraduate degree.[37]

Connections of another kind – familial – also continued to wind their way through the faculty under Howe. There were a number of father-son teams whose senior member graduated during the 1920s. These patriarchs included Peter Addison (2T9), George W. Bayly (2T7), John A. Brodie (2T3), Ralph S. Carman (2T1), Horace H. 'Holly' Parsons (2T5), and Walter E. Willson (2T5). Another of Howe's students during the 1920s – Richard Boultbee (2T9) – cleared a path which his sibling would soon follow. The Irwins bridged both groups. Brothers Cecil H. and John C.W. Irwin (2T2) were classmates, and one of the former's offspring would follow in his father's footsteps in the years to come.[38]

The familial connections were expressed in another way as well. A handful of Howe's graduates from the 1920s was apparently drawn to forestry because its kin had been involved in bush work. George R. Lane (2T6) was the scion of the first superintendent of the government's nursery in St Williams, Ontario (he worked in this same capacity at another government nursery for a few years after graduating), and John M. Robinson (2T9) spent part of his childhood growing up in Algonquin Park, where his father was superintendent. Likewise, Gregory J. Thomson (2T5) came from Creemore, Ontario, which was near his father's sawmill that operated under the banner of Peter Thomson and Sons.[39]

The graduates of the 1920s also continued another trend that had begun under B.E. Fernow: achieving excellence in forestry, and beyond. Some of them were true pioneers in their fields. 'Holly' Parsons (2T5) and Frank T. Jenkins (2T3) were two such ground-breakers. They attended the Faculty of Forestry just as Ontario's Department of Lands and Forests was beginning to use aircraft to cruise timber. They were involved as students in its inaugural projects, including the first large-scale forest inventory that used aircraft (and ground personnel) in the James Bay watershed north of Cochrane, Ontario. From there Parsons and Jenkins were on their way. They became leaders in the field of aerial sketching, the former in the public sector (with the DLF) and the latter in the private sector (first with the esteemed James D. Lacey and Company and then as a consultant). Along the way, they developed the remarkable skill of being able to map and estimate timber stands from the air, often sitting in the nose of the plane with their heads jutting out and little but a pair of goggles and leather cap to protect them. Not only did they perform this dangerous work with aplomb, but they did so with a degree of accuracy that rivals 'modern' forest inventories.[40]

A host of Howe's 'boys' from the 1920s also rose to senior positions in the Canadian forest industry. William G. Wright (2T0) and John B. Matthews (2T9) took very different paths to becoming chief foresters of their respective firms. A native of England, Wright came to Toronto with a bachelor of science in agriculture, from Edinburgh University, a diploma from the Saxony Forest School, and the Military Cross he had won during the First World War. His previous credits allowed him to earn his bachelor of science in forestry from the faculty in only one year, and five years later it awarded him its second forest engineer degree for his work on taper as a factor in measuring standing timber. By this time he was already working for Price Brothers in Quebec, a firm that made him its chief forester in 1938. Matthews spent the first twelve years of his professional career working with Ontario's Department of Lands and Forests. In 1942 he jumped at the chance to become the chief forester for one of Abitibi's mills at the Lakehead. Four years later he was promoted to the same rank for the entire company, a position he held for the next two decades. During this time he was the spearhead behind the company's decision to establish its 'Woodlands Laboratory' at Raith, about 150 kilometres northwest of present-day Thunder Bay. Two others, Albert H. Burk (2T4) and Donald W. Gray (3T0), rose to be the woodlands manager for the Kalamazoo Vegetable Parchment Company, which reopened the mill in Espanola, Ontario, during the Second World War.[41]

Graduates from these years also distinguished themselves in businesses other than the forest industry. John C.W. Irwin (2T2) spent only a few years in forestry before entering the insurance business. In the late 1920s he joined the educational and medical book department of the Macmillan Company, where he worked with W.H. Clarke. In 1930 the two of them left to start Clarke, Irwin and Company, which concentrated primarily on publishing educational books. Although this was a most inopportune time to enter this industry (many publishing houses were going under), their firm survived the Great Depression and went on to become a dynamic force in Canadian literature (Clarke, Irwin and Company published Hugh Maclennan's first works).[42]

Academia and research also attracted a fair number of Howe's students from the 1920s. The Faculty of Forestry hired two of them, Gordon G. Cosens (2T3) and Robert C. 'Bob' Hosie (2T4), and more will be said about Cosens in the pages that follow. Hosie was a native of Glasgow, Scotland. Immediately after graduation he began working as a teaching assistant in the faculty, and remained on staff (he reached the rank of full professor in 1945) for the next forty-one years. Hosie devoted his summers to studying various aspects of regeneration in Ontario for both government and industry. Moreover, he was an integral force in uniting his fellow foresters for long enough in 1949 so that they were able to form the Ontario Professional Foresters' Association. Not only was he a charter member, the OPFA recognized his monumental contribution to forestry by making him its first honorary life member in 1968. Like Hosie, Gordon C. Grant (2T4) came to the Faculty of Forestry from Scotland, but unlike Hosie, returned to teach there after graduating. Louis R. Seheult (2T8) was a native of Port of Spain, Trinidad, who led his class during each of his four years in Toronto. He earned his master's degree from the faculty in 1936, and his thesis addressed the avant-garde subject of the use of trucks in logging operations. After spending roughly fifteen years working in industry, Seheult joined the forestry department at the University of New Brunswick. James L. 'Alex' Alexander (2T1) carried out extensive growth and yield studies for the British Columbia Forest Branch (BCFB) after leaving Toronto. He accepted a teaching position in 1927 at the University of Washington in Seattle but returned to British Columbia during the Second World War to work as a researcher. Alexander was working in that capacity when he died prematurely in 1951 from salmonella poisoning that he and many other BCFB personnel contracted at their annual dance. Finally, William R. Haddow (2T3) worked with the Ontario government until 1929, when he departed to pursue graduate work in forest pathology at Harvard under the legendary and former University of Toronto professor J.H. Faull. Thereafter, he worked in Ontario as a forest pathologist with the federal government.[43]

Carl C. Heimburger (2T8) was one of this era's more cosmopolitan and interesting graduates. He was born in 1899 under Tsar Nicholas II in St Petersburg, Russia, to a Finnish father and Danish mother. His family moved to what is now Helsinki, and he left to study in Denmark in the mid-1910s. By 1933 he had earned undergraduate and master's degrees in forestry from the Royal Veterinarian and Agricultural College in Copenhagen, his bachelor of science in forestry from the Faculty of Forestry in Toronto (in only two years), and his doctorate from Cornell University. Thereafter, he became the Ontario government's research specialist in tree breeding. Heimburger was best known for developing hybrid poplars that were fast growing, of high quality, and suited to the southern part of the province. In the 1950s he oversaw the establishment of the first commercial plantations of his 'engineered' trees by the Ontario Paper Company. Heimburger's ground-breaking work won him, in 1953, the distinction of becoming a Fellow of the Royal Society of Canada, the first professional forester to be so honoured, and he continued his research work until 1972. In that year he was replaced as the Ontario government's forest geneticist by Dr Louis Zsuffa, who was appointed an adjunct professor to the Faculty of Forestry during the 1970s and 1980s.[44]

A number of Heimburger's contemporaries from the Faculty of Forestry also entered the Ontario government's forestry service. John F. 'Frank' Sharpe (2T2), for example, began working for the department immediately after he left the faculty. He rose to become the chief of the division of timber management in 1941, and remained in that position until his retirement seventeen years later. Frank A. MacDougall (2T3) was one of the unlikeliest civil servants Ontario ever had, yet he established a record for longevity in this field. He began working for the department immediately after graduation. By 1928 he was given charge of the Sault Ste Marie District, where he was responsible for establishing the first plantations in what is now the Kirkwood Forest near Thessalon, Ontario. In 1931 MacDougall was appointed superintendent of Algonquin Provincial Park, a position he held for the next decade. In 1941 he was promoted to the department's highest unelected position, deputy minister, and became the first of several Faculty of Forestry graduates to ascend to this level of government. Immediately after his appointment, MacDougall began reorganizing the department into a modern bureaucracy. He remained at this post for the next quarter-century, making him the longest-serving deputy minister at Queen's Park. The irony was that MacDougall had literally begged Howe over the course of 1924 – during MacDougall's first full year of employment with the provincial government – to find him a job in the private sector because he was so frustrated by dealing with government red tape![45]

A few of Howe's protégés from the 1920s would become senior administrators in the civil service *outside* their fields of specialization. Their success stands as a testament to the skill set – most notably an exemplary standard of professionalism – that led governments, both provincial and federal, to view them as highly valuable assets. The few who established the trend at this time blazed a trail in which a disproportionately large number of their cohorts would soon follow.

William E.D. Halliday (2T6) was the pioneer. A native of England and graduate of Cambridge University, he worked for the Dominion Forest Service for much of his career and climbed the rungs of its bureaucratic ladder. In 1958 the federal government appointed him registrar of the cabinet in the Privy Council Office.[46] Allen W. Goodfellow (2T6) performed 'public service' of a different sort. Born in Fort Frances, Ontario, after graduation he started working for industry in New Brunswick. In 1939 he joined Fraser Company in the same province, and was soon working in Plaster Rock, where he was involved in managing the firm's freehold lands. Over the next four decades (he died in 1989), Goodfellow devoted himself to strengthening the town's economic lifeline: Fraser's sawmill. He sat on the local governing committee for more than twenty years, and was instrumental in modernizing Plaster Rock and transforming it into a model for other communities. In 1984 Goodfellow was recognized for his contribution when he was awarded the Order of Canada.[47]

The graduates from the 1920s also believed that their work entailed much more than merely cropping the forest. Howe's conversation with one anxious father provides keen insight into this realm. J.H. Porter wrote Howe in the late 1920s to ask about his son's suitability for the Faculty of Forestry. He stressed his son's keen interest in trees and how they grew, the birds and animals that lived in the woodlands, and his 'love of the outdoor life.' In response, Howe endeavoured to assuage the dad's concerns by pointing out that the young Alfred W. Porter (3T6) represented 'just the type of man we want for forestry.' Howe described how his twenty years at the faculty had taught him that 'the men who understand and appreciate all forms of nature are happiest in their work,' and 'love of nature' was crucial to their success.[48]

A few of Howe's students from the 1920s were given the chance to devote their lives to demonstrating their 'love of nature.' George M. Linton (1T9) spent his entire career, spanning more than four decades, as head of the Ontario government's nursery in Orono (just south of Peterborough). He grew millions of seedlings and performed countless hours of public relations work trying to convince locals to reforest their degraded

properties. Linton also recommended in 1921 that government manage the land that encompassed watersheds because of its importance,
an idea that presaged the establishment of conservation authorities in
Ontario after the Second World War. William B. Greenwood (2T5) took
care of the environment in a different way; he made it relatively accessible to the province's urban dwellers. He spent most of his professional
career with the Department of Lands and Forests, and served as the inaugural chief of its Parks Branch (1954–60). Under his rule, the number
of provincial parks grew from eight to over fifty. Peter Addison (2T9)
succeeded Greenwood and continued expanding this network of public
recreational retreats.[49]

Propitious circumstances gave one of Howe's graduates from these
years a chance to make a unique contribution to environmental stewardship. Howard H. Krug (2T6) had deep roots in the forest. His father and
uncles had established a furniture factory in 1886 in Chesley, Ontario,
not far from Owen Sound. After he graduated from the Faculty of Forestry, Krug worked for the family business as a timber buyer and began
managing the firm's timber lands on a sustained yield basis. This included initiating a reforestation program that lasted sixty years and planted
between 10,000 and 20,000 seedlings annually. During this time he and
his brother also amassed over four hundred acres of practically 'undis-
turbed' maple-beech forest at 'Kinghurst' in Grey County, and Krug was
committed to protecting it because it represented a rare example of
pre-settlement woodlands in southern Ontario. At the same time he carried out bird-banding work and led the successful effort to re-establish
bluebirds on the Bruce Peninsula. Krug's capstone achievement was his
decision to bequeath his forest to the Federation of Ontario Naturalists,
which recognized his decades-long commitment to conservation by making him an honorary life member. Today, the Kinghurst Forest Nature
Reserve is designated a provincially significant Life Science Area of Natural and Scientific Interest.[50]

But Krug and his classmates had an enormous hurdle to overcome before reaching these heights. By the late 1920s the newsprint industry's
fortunes had turned south, and several mills in Ontario had shut down
intermittently. Even when the signs of impending crisis began staring
Howe in the face at this time, he tenaciously clung to his belief that fostering broad support for forestry was *the* only way to further the cause.
Moreover, when George R. Lane (2T6) wrote to him just as the foresters'
world began crumbling around them on the eve of the Great Depres-

sion, Howe refused to accept that the future was as gloomy as others believed and insisted that staying the course would achieve their ends. 'Be calm,' Howe coolly advised Lane (2T6), who was worried about losing his job. 'Reach for a Lucky.' Unfortunately for Lane, this counsel could not save him and his colleagues from enduring the profound difficulties that awaited.[51]

Chapter 4

'Forestry's Darkest Hour,' 1930–1941

For the Faculty of Forestry in Toronto, ironically, one of its greatest problems when the Great Depression hit was a function of the success it had enjoyed during the 1920s. The steady growth in demand for foresters in Canada, especially from Ontario's Department of Lands and Forests (DLF) with which the faculty now enjoyed a cozy relationship, had attracted a healthy number of students to the school each year. So, too, had the Faculty of Forestry's reputation; most considered it the country's best. So when the economy slumped dramatically over the course of 1929–30, the timing could not have been worse. The faculty's enrolment was hitting unprecedented heights at the very moment when foresters began losing their jobs in droves. As Dean C.D. Howe lamented to Gordon C. Grant (2T4), in mid-1931, 'the bottom has dropped clean out of the market for our men. We have the largest graduating class, eighteen, in our history and the very poorest time to find positions for them.'[1]

Howe was acutely aware of why forestry, both in the private and public sectors, was particularly vulnerable at this time. Many of the major pulp and paper companies in Ontario and Quebec that had hired a fair number of his graduates in the decade after the First World War landed in the hands of cost-cutting receivers who gladly showed the foresters the door. Likewise, governments at all levels were equally committed to paring expenditures, and they trimmed the most in areas the public cared about the least. As Howe confided to Leon Ellis (1T1), in the winter of 1932, the elected officials were 'using the hatchet right and left, hitting the objects close at hand especially those that they think will lose them the least number of votes. So forestry is coming in for more than its share.' The result was devastating. The Dominion Forest Service (DFS), for example, cut two-thirds of its forty-eight forester staff over the course of 1930–1. This situation was exacerbated by Ottawa's decision at this time to transfer natural resources under its control in western Canada to the provinces in which they were located. The upshot saw the DFS, which had been the country's leading forest service, lose its land base and mandate. As Ellwood Wilson, the legendary silvicultural forester with the Laurentide Paper Company in Grand'Mère, Quebec, lamented to Howe in mid-1931, 'this, to my mind, is Forestry's darkest hour.' While the Ontario government initially retained its foresters, soon enough most of them would also be out on the street.[2]

It need not have been thus, and Howe knew it. In the United States, the Great Depression was a boon to forestry. A major plank in Franklin Delano Roosevelt's political platform in the early 1930s was the mobilization of tens of thousands of unemployed young men for work on 'forestry and other land betterments'; within a few years, nearly 500,000 men were employed by the Civilian Conservation Corps. Roughly half of them worked on forestry projects, which were often directed by professional foresters. These efforts included fire protection, disease control, tree thinning, and tree planting (both reforestation and afforestation). Although Canada also established a string of make-work camps in its hinterland during the Depression, virtually none of them was devoted to forestry. This stark contrast provided irrefutable evidence of how low forestry ranked as a political priority in C.D. Howe's adopted country.[3]

This knowledge, and the Great Depression's brutal impact, exacted an enormous toll on Howe. The period's wholesale firing of foresters caused the Faculty of Forestry's enrolment to drop precipitously. After peaking at nearly seventy students in the early 1930s, it bottomed out at thirty-one in 1937–8. These distressing conditions even drove Howe to change his and the faculty's guiding principles. He accepted that his forestry school's raison d'être was no longer to produce employees for the Ontario government because industry was hiring most of his graduates by the late 1930s. This new reality compelled him to increase his students' practical 'logging' training as much as possible, for there was no reason to think that forestry would be practised any time soon.

These dramatic changes came too late for one of Howe's colleagues, however, whom the dean had already fired for repeatedly breaking his cardinal rule of never criticizing the provincial government's forestry policy. Although this incident occurred in the early 1930s, it clearly haunted Howe until the day he left the faculty.

Most of the Faculty of Forestry's graduates who were the victims of the cutbacks – Howe called them 'the unemployed depression group of foresters' – naturally suffered immensely. Not only had they lost their jobs at a time when there was little hope of finding *any* other employment, but this process often occurred in a callous manner. Although there are many stories, Harry P. Eisler's is poignant. A graduate of the class of 1921, Eisler had moved up the ranks of the Dominion Forest Service to become a district forest inspector in Saskatchewan. His work was so impressive that both Saskatchewan and Manitoba offered him senior posts in mid-1930, after the dominion government had transferred control over their natural resources to them. Instead, Eisler had elected to remain with the dominion government, a decision he no doubt regretted in late April 1931 when he received a telegram from Ottawa notifying him that his 'services were terminated on that date. It sure felt like a bad dream,' he confided to Howe, 'and I wandered around the house in a daze for a couple of days. Then I went to Regina to try out the new Department of Resources – later tried to cajole two estate owners into letting me landscape their places but all without success.' He ended up working as a guard and gardener at the jail in Prince Albert, Saskatchewan, a position that paid him one-third his previous salary. Howe was at an uncharacteristic loss for words in trying to console Eisler, terming the situation across the country 'an awful tragedy.' To another of his graduates who turned to him for counsel at this time, Howe could only offer practical advice that spoke to the desperate state of affairs at the time. Because the man lived on a farm, 'where you can sleep and eat,' Howe advised him 'to stay there until the storm blows over.'[4]

While many others faced similar misery, a few stories had far happier endings. Good fortune undeniably had a hand in deciding their fates, but so, too, had the Faculty of Forestry. If it had taught its students anything, it was how to survive in an unfavourable environment. Howe's emphasis on developing graduates of superior quality who could stomach repeated rejections at the hands of the politicians had prepared his charges well for overcoming the ordeal that the 1930s represented. More than anything, their training at the faculty provided them with a much-needed springboard to myriad careers outside their fields of expertise when none was available within them.

Some, such as the eminently named John A. Macdonald (3T3), chose to leave the forestry profession for health reasons. Born in Saskatchewan, he was a superb student who participated fully in the university's extracurricular activities. A back problem prevented him from working in the bush, however, and prompted a switch to a career in accounting. He earned his chartered accountant papers in 1945, and won the gold medal from the Canadian General Accountants' Association thirteen years later.[5]

Nearly all the other graduates who pursued alternate professional avenues during the Depression did so because their prospects in forestry had disappeared. Some like Clarence Cooper (3T2) went into law. Although he left forestry very quickly, graduating from Osgoode Hall in 1936, Cooper retained close ties to both forestry and the faculty (he served as the secretary-treasurer of the Canadian Society of Forest Engineers from 1944 to 1948). Ralph E. Sewell (3T3) graduated the year after Cooper and took a slightly different route around the obstacles the Depression presented. Unable to find a job after leaving Howe's charge, he completed a master of commerce degree at the University of Toronto and immediately began working as a salesman for the Coca-Cola Company of Canada. He crowned his career with that firm by becoming its president. Cecil H. Irwin (2T2) moved to a piece of land that he had purchased near Carnarvon (northeast of Toronto), in the spring of 1933, after he was axed from his forestry job. The following summer he opened the Sherwood Forest Camp, which served as a recreational retreat for boys (summer) and adults (fall). Because the Faculty of Forestry still lacked its own practice forest and permanent bush camp, Irwin's facility hosted its spring and fall field sessions for several years in the mid-1930s. Adrian C. Thrupp (2T2) was Irwin's classmate, and the Depression brought his fourteen-year career with the Dominion Forest Service to a screeching halt. Thrupp moved to the United States to earn graduate degrees from the universities of Idaho and Washington, and then worked for the Boeing Company in Seattle in its material control section.[6]

One student from the 'Howe era' dealt with the lack of job prospects in forestry in a manner that set a trend that would become increasingly pronounced after the Second World War. Fred G. Jackson (3T2) attended teachers' college immediately upon graduating, taught high school in his native Toronto during the Depression, and spent several summers carrying out forest surveys. He taught part-time at the Faculty of Forestry after the Second World War, and then joined the Ontario civil service. In 1975, forty-two years after he had graduated from the faculty, Jackson donated the funds to create an eponymous award.[7]

These sanguine stories were glaring exceptions, however, for the foresters' typical tale during the early part of the Great Depression was one of desperation. Despite the unprecedented gloom, Howe still clung tenaciously to the view that the best means of furthering the forestry movement was avoiding direct criticism of any level of government. When word leaked out in 1930 that the Dominion Forest Service was going to fire most of its foresters, for example, Howe organized both a delegation to meet with the government and a letter-writing campaign to MPs to protest the move. But his dignified, orderly campaign bore no fruit.

What was most startling about Howe's behaviour during the Depression's darkest days was that he was prepared to adhere to his dictum, even if it meant being utterly ruthless in dealing with one of his colleagues. Unfortunately for the person in question, Willis Norman Millar, the upshot of Howe's actions would be disastrous.

The battle between Millar, whom B.E. Fernow had hired in 1914, and Howe was rooted in the early 1920s. The subject over which they first locked horns was the practice or demonstration forest and field camp that everyone associated with Toronto's Faculty of Forestry had deemed was essential if it were to carry out its mandate effectively.

Only months after becoming the faculty's dean in 1919, Howe made it clear to Robert Falconer, president of the University of Toronto, that the lack of a practice forest was its Achilles' heel. In a letter to Falconer, Howe pointed out that Fernow had continually stressed the need for a practice camp because the Faculty of Forestry's interim measure for providing its students with field experience was sorely wanting. 'Thus far we have been dependent upon the courtesy and good will of various lumbermen,' he told Falconer, 'harbouring ourselves in their camps for four to six weeks, in order that our students might receive some practical training in woods work. This arrangement never has been at all satisfactory and in some cases it has proved very inefficient in results.' Most importantly, it meant that the faculty operated at an enormous disadvantage compared with other forestry schools. Howe underscored that 'no amount of classroom instruction can take the place of such experience and contact with the actual problems of the forest; in fact, the general criticism of our graduates by their employers has been that they lacked training in the woods work and the more practical aspects of the profession.' He added that officers from Ontario's Department of Lands and Forests had also come to see 'the necessity of a forest ranger school,' and that such an institution would naturally become part of the faculty if it already operated a 'permanent practice camp.' Finally, he felt that the Faculty of Forestry could use its field facility to host 'short courses' for industry officials who wished to upgrade their skills. For all these reasons, Howe proclaimed that his faculty's 'most urgent need … is a permanent practice camp.'[8]

Although Howe's prognostications turned out to be both prescient and precocious, at the time he realized he was pitching his message to the wrong audience.[9] As a result, in early 1920 he laid the matter before the government, namely, the United Farmers of Ontario, and their premier, E.C. Drury. The UFO proved receptive to his idea, and directed him to identify an appropriate site for both a 'permanent practice camp' and 'experimental area' for the Faculty of Forestry in consultation with E.J. Zavitz, the provincial forester.

It soon seemed like Howe would achieve his aim. He ventured north in the spring of 1920 with his colleague, Millar, and found an ideal spot in the Temagami Forest Reserve. The site's eight thousand acres boasted a wide range of forest conditions, was relatively accessible from Toronto, and included several abandoned mine buildings that were still usable. The United Farmers reacted so favourably to this proposed location that the faculty felt the deal was a fait accompli; Millar held the school's practice camp on the tract in the fall of 1920. At this time Howe informed Frank Sharpe (2T2) that 'we have a verbal promise from the Government that this practice camp will be made permanent. We expect to have between 10,000 and 20,000 acres on which we can make scientific investigations and experimental cuttings.'[10]

But then major problems arose. An article appeared in the *Varsity* on 4 February 1921, and its title announced that 'New Practice Camp Obtained Last Year – Ten Thousand Acres from Temagami Reserve Granted by Province.' Immediately after it appeared, Howe wrote to Falconer to express his fear of the story's potential impact. 'This premature announcement,' Howe explained, 'may embarrass us if it comes to the ears of the Government. I hope, however, that it will not be noticed.' It is unclear whether the United Farmers saw the article as a slight, but within a few months the matter had moved to their back-burner.[11]

This news sent Millar into a dither. Whereas Howe believed the practice forest was important to his faculty's success, Millar saw it as a life-and-death situation. He taught silviculture, after all, and he felt it was ludicrous to try to educate his students about how the forest ought to be cropped if there were no concrete examples to which he could point to illustrate his lessons. 'If we who teach the subject are unable, by our example, to demonstrate that silviculture may be practised,' Millar exclaimed to Howe in a letter in mid-May 1921, 'even under the very best existing circumstances, which is manifestly what has been going on at the school ever since its inception, is it reasonable to look for any practices whatever under other conditions, by those who may, to my mind, be presumed to be less interested parties? To me it seems that there is only one answer to this question,' Millar continued, 'and that the action of the Yale Forest School, for example, in putting into effect at once a successful silvicultural forestry operation in the immediate vicinity of New Haven, is the object that should be before every forest school.'[12]

When the practice forest did not materialize over the next few years, Millar's frustration boiled over, and the result was a raucous staff meeting in February 1923 that demonstrated the deep fissure that now separated him from his colleagues over the practice forest in particular and the direction in which Howe was taking the Faculty of Forestry in general. The government had apparently begun offering the faculty a tract in Algonquin Park instead of the one in Temagami. As an alternative, Millar had found 'the Atkins property,' which belonged to a landowner near Bronte (just west of Toronto), who was willing to permit the Faculty of Forestry to use his extensive woods as its permanent practice forest.[13]

The fight was not so much over the forest's location but who would control it, with Millar waging a solo battle to ensure that his faculty would have the freedom to demonstrate forestry's efficacy unfettered by political interference. At the staff meeting in the late winter of 1923, Millar put forth a motion that favoured the Atkins property because, above all else, it would leave the Faculty of Forestry in complete control. The other staff members (Howe, J.H. White, and T. Dwight) rejected Millar's idea outright. Instead, they favoured entering into an agreement with the provincial government to operate 'a cooperative forest camp,' probably in Algonquin Park. In order to realize this aim, however, the faculty would have to agree to a number of conditions. Most importantly, the government would determine all forestry operations on the tract, and it was made clear that 'economic factors' would decide which activities would be carried out.

Millar was outraged by his colleagues' stance. He insisted that the Faculty of Forestry must demonstrate that managing a forest could be a viable commercial enterprise. When they countered by arguing that the faculty's most important function was to prepare students for employment with the government, and that handling forest properties was beyond their mandate, Millar could not contain himself. He agreed that the faculty should assist its graduates in finding work but that 'its activities as an employment agency should not be permitted to interfere in the slightest with its teaching policy or its attitude on any question of forest policy.' He added that 'the acceptance of a forest camp under the conditions outlined by Dean Howe constitutes an acceptance of such interference.'[14]

The issue of the permanent practice forest soon became academic. Although the government repeatedly promised that it would grant the tract, it did not. In fact, only when Howe was about to retire as dean, in 1941, did the Faculty of Forestry receive its 'university forest,' a subject that will be discussed in greater detail in the next chapter.[15]

But the more important issue that had arisen during the 1923 debate over the matter was the degree to which Millar's views were out of step with those of his colleagues. The next few years saw him battle them, and specifically Howe, over a number of issues. Common to each one of them was Millar's belief that Howe's policy of bowing to the provincial government by never attacking its forestry policy was immensely damaging to both their faculty and the general forestry movement.

No sooner had the commotion over the practice forest settled down than Howe and Millar tangled again. In the spring of 1924 Millar took issue with the Ontario government's decision to reorient completely its fire protection service. After nearly a decade of building up a fire ranger system based on a blueprint that the faculty's own J.H. White had drawn up, the reigning Tories deemed this approach outdated and derided the foresters who had been involved in it for their inability to render effective service. At the same time, they announced a 40 per cent cut in the fire ranging staff and the introduction of a fleet of aircraft that they argued was certain to produce better results. Millar was beside himself when he learned of this dramatic turn of events, and he was determined to stand up and be heard on the issue.[16]

Before he did so, however, he ran his ideas past Howe. In mid-April the dean had reminded his fellow staff members and students that a 'forester has absolutely no right to criticize the actions of his superior officer. If he wishes to make open criticism, he ought to give up his position.' Millar could not accept this perspective and wished to publicize his views, and he provided Howe with a precis of his thoughts. Howe was unequivocal in responding to Millar's letter. There would be no censuring of politicians on his watch, Howe declared, because he was convinced 'that the best interests of the School would not be promoted by criticism in newspapers of a specific Government policy.' He insisted Millar abide by his wishes, which Millar did on this occasion.

But the issue of forest-fire protection flared up again the next summer.

James Lyons, Ontario's minister of lands and forests, authored a paper that appeared in the leading American aeronautical journal. It compared the data from 1923 with that from 1924 and argued that the use of the airplanes instead of foresters in forest-fire detection and suppression work had drastically decreased both the volume of burned timber and the cost of the service. He attributed the turn-around to the use of aircraft in 1924 and the deficient work that 'technically trained foresters' had carried out on the ground the year before. Lyons' article attracted broad attention, and editorials across North America cited Ontario's decision to create the world's first permanent aerial firefighting force as proof positive that aircraft were the way of the future.

This high-profile censure of foresters burned Millar, and for good reason. Millar dissected Lyons' data, and uncovered that the minister had taken great liberties with his evidence. Natural conditions accounted for the vast discrepancy in the data between the two years: 1923 had seen some of the worst fire conditions on record, and 1924 some of the best. Foresters and airplanes had played little, if any, role in affecting the figures.

Millar's discovery prodded him to launch a counterstrike. He wrote to Howe to reveal the truth behind Lyons' article. He also argued that foresters' tacit acceptance of such attacks, especially such groundless ones, was preventing them from 'attaining any position of influence.' In Millar's view, this approach meant that the forestry movement 'does not deserve and is not at all likely to secure the respect of the people of this country.' Despite his anger, Millar agreed to keep his thoughts to himself. He confronted Lyons about the minister's devious bending of the truth, but acquiesced to Howe's central tenet to keep the matter out of the public eye.

Millar would break this commandment over the next few years, however, but his transgressions were relatively minor. On one occasion in 1927, for example, he penned a letter to the *Globe*. It demonstrated how the amendments the Tories had made to the Assessment Act, which they argued would entice property owners into properly managing their woodlots, would actually have the reverse effect because the government had merely shifted and not reduced the tax burden. A few years later, he followed up with another letter that reinforced his earlier views by citing the testimony of those who were suffering from the impact of the statutory changes.

It was in the context of the Great Depression's first few years, when foresters were losing their jobs in droves, that Millar and Howe engaged in what would be their final and most deadly tussle. Millar fired the first shot when he published an explosive letter in the faculty's *Annual News Letter* that challenged Howe's modus operandi. As it has been written elsewhere, Millar argued that 'forestry professors must develop students with agile, critical minds who were sufficiently confident, competent and courageous to disagree openly with their superiors if they felt conditions warranted taking such a stance.' Millar was fed up with Howe's insistence that foresters submissively accept the directives they received from their government employers, a *mentalité* that Millar denounced as 'abysmal intellectual apathy.' To demonstrate the pressing need for a new attitude, Millar recounted how several provincial forest services were boasting about the progress they were making in protecting the woodlands from fire, even though the evidence did not support this conclusion. In addition, he lamented how seldom Canadian foresters had openly expressed their views about the pathetic state of silviculture in the country.

Millar then heard that Howe had launched a movement to get him fired, prompting him to take pre-emptive action to save his skin. In an eleven-page letter to Robert Falconer, Millar let the university's president know that Howe had mentioned on several occasions over the previous half-dozen years that the dean 'would welcome my withdrawal from the staff' largely because Millar's views on forest policy had rendered him 'unpopular with certain influential employers of foresters.' After Millar described the empirical evidence on which he had based his criticisms of the Ontario government, Millar pleaded for Falconer to force Howe to present *his* case before the president decided Millar's fate. Having served the University of Toronto for nearly two decades, Millar asserted that 'it was only common decency that I be convicted of something other than merely independence of thought and expression in the field of my profession. By whose authority am I threatened with all this and by just what procedure am I to be guillotined?' Millar begged to know.[17]

Millar's latest outburst drew no immediate reaction from Howe, but Millar's next two moves did. In a pair of articles published in the *Pulp and Paper Magazine of Canada*, Millar excoriated two prominent figures. His May 1932 piece tore a strip off F.J.B. Barnjum, the country's self-appointed forestry expert who publicized the cause in a maudlin manner that Millar and many in the profession felt was both misdirected and counterproductive because of its penchant for sensationalism. A month later, Millar set his sights on Robson Black, the manager of the Canadian Forestry Association. Black was mounting a coup d'état in the association that aimed to eliminate its focus on forestry publicity. In taking Black to task, Millar pointed out that Black was paid the highest salary of any official in Canada who worked on behalf of forest conservation, and yet Black had little interest in presenting the subject in an accurate way to the public.[18]

Millar's latest paroxysms were to be his last. Howe was as critical as Millar of both Barnjum and Black, but Millar had levelled his savage attacks in the public arena. They had ruffled feathers and significantly embarrassed the university (ironically, Millar's brother-in-law and iconic fellow professor at the University of Toronto, Frank Underhill, had just begun his string of run-ins with the university's administration for speaking out on sensitive political issues). For example, Millar's vilification of Barnjum had prompted Ernest Finlayson (1T2), Canada's veteran director of forestry, to complain vociferously to Falconer about the undignified behaviour of 'a University professor.' Falconer's reply noted his regret at Millar's attack and the fact that it had 'caused a great deal of annoyance to Dean Howe.' Falconer closed by assuring Finlayson that 'I am giving very serious attention as to what should be done in the situation.'[19]

By this time, Howe had given Falconer something that deserved 'very serious attention' indeed. It consisted of a lengthy missive that offered a scathing analysis of Millar's recent conduct and recommended that the university ask for Millar's resignation. Howe based his request for Millar's head on the latter's habit of discussing the policy of the Faculty of Forestry, and the provincial and dominion governments, 'in a super-critical manner before his classes.' Howe also argued that Millar's 'lack of tact and violence of his criticism antagonizes his employers and others,' and that his strictures contained so many falsities 'as to disclose a type of mind unfitted to carry on the work of a professorship in a university.'[20]

Falconer dealt decisively with the situation. He advised Millar to leave the University of Toronto because 'such a divergence of opinion has arisen between him [Millar] and Dean Howe and other members of the Faculty of Forestry, and such a clash of types of character and ideals as to the conduct of the Faculty ... has developed that no reconciliation is at all possible.' Millar submitted his resignation in early June 1932, and the board of governors accepted it shortly thereafter.[21]

Millar's life would end soon after he left the Faculty of Forestry. His resignation took effect at the end of the 1932–3 academic year. The only job he could land during one of the Depression's darkest moments was in a 'relief' reforestation project with the Civilian Conservation Corps in his native United States. He lasted barely a week on the job, as he died of a heart attack in late June 1933.[22]

While Howe was convinced that he had acted in the Faculty of Forestry's best interests, the episode with Millar weighed heavily on his conscience. Immediately after learning of Millar's passing, Howe did everything within his power to care for Millar's family, which had been left, as he told H.J. Cody, the university's new president, 'in a very precarious financial condition. He [i.e., Millar] had a wife and four children from ten to twenty-two years of age,' and had recently allowed his largest life insurance policy to lapse. Although Millar owned a home and little farm in Oakville, he held a large mortgage on them, and it would be practically impossible to sell these assets at this time. Howe pointed out that Millar's resignation became effective 1 July 1933, and he had died two days before that date. 'I infer,' Howe indicated, 'that legally he was still a member of the University staff.' He thus asked for advice in securing a pension for Millar from the Carnegie Foundation on 'compassionate grounds,' an application the foundation ultimately rejected. In the meantime Howe had asked Cody if the university's board of governors would approve a motion to purchase Millar's lecture notes from his widow for $250. 'I might say,' Howe noted in an effort to buttress his case, 'that the University is saving three thousand dollars this year by not filling Professor Millar's place.' This time Howe's lobby achieved its aim, and then some. After the board learned of the Carnegie Foundation's decision, it approved the payment of $1,000 'as a compassionate grant' to Millar's widow.[23]

A few things were clear once the dust had settled on this internecine feud. The evidence strongly suggests that the Ontario government brought such intense political pressure to bear on Howe that he had little choice but to dismiss Millar. While Howe had tried to soften the blow for Millar's family, the dean's hands were indelibly sullied by his drive to eliminate his outspoken colleague from his midst. Millar's firing would cast a pall over the Faculty of Forestry that remained long after the Great Depression had lifted. In addition, it had also fundamentally weakened the faculty. The school had lost an expert in his field (even Howe openly acknowledged that Millar was 'an outstanding authority on forest fire protection methods'), and it was left short-handed for the 1933–4 session. Howe complained in his annual report for that year that Millar's death had translated into 'extra work for the already over-burdened staff.' As a stopgap measure, he increased the number of industry experts, including alumni John D. Gilmour (1T1) and R.W. 'Bob' Lyons (1T6), who delivered guest lectures on various practical subjects.[24]

In time, however, Millar's passing produced one enormous benefit to the Faculty of Forestry. In mid-1934 the University of Toronto agreed to fund his permanent replacement, and Howe hired one of his own under-studies to fill the bill: Gordon Gunn Cosens (2T3). A native of Toronto, Cosens had spent his summer internships and five years after graduation under the employ of the Laurentide Paper Company in Grand'Mère, Quebec. There, he had worked with Bob Lyons (1T6) in carrying out one of Canada's most progressive silvicultural programs. During this time he had earned the faculty's first master's degree in forestry, although it was a master of arts and not a master of science in forestry, because the University of Toronto's senate only authorized the faculty to grant such degrees in 1930–1 (his thesis focused on the biological foundations of forestry). In 1928 Cosens went to work for the Spruce Falls Power and Paper in Kapuskasing, Ontario, first as its chief forester and then as its assistant woods manager. This lengthy service working with these two forward-thinking firms made Cosens the Faculty of Forestry's first professor who had both sound technical knowledge *and* practical experience.[25]

There was one major caveat to landing Cosens, which Howe felt was worth accepting for the good of his forestry school. Cosens' impressive credentials made him a highly coveted commodity, and the $4,800 annual salary Spruce Falls was paying him in 1934 was one with which the University of Toronto could not compete. Its best offer was $3,600. Even if the university could up its ante, industry had the flexibility to exceed practically any stipend that the Faculty of Forestry dangled in front of Cosens. There was, however, one way for the university to address the money issue. It could allow Spruce Falls to retain Cosens as a 'consultant,' and essentially have the company subsidize his academic income. Cosens agreed to this unusual arrangement, and Howe was amenable to winking at this obvious conflict of interest.[26]

The Faculty of Forestry was still reeling from the devastating events of the early 1930s, namely, the savage cuts to forestry and then Millar's firing as the coup de grâce, when it received yet another body blow that nearly knocked it out. One area in which Howe's political sensitivity had paid off in spades was in cementing his school's relationship with the Ontario Department of Lands and Forests. Since Howe had taken over as dean in 1919, the DLF had been hiring an ever-increasing number of the students who had graduated from his program, peaking at roughly 50 per cent in the late 1920s. Although this had not translated into fundamental forestry reforms, Howe's students were at least guaranteed jobs that appeared to be theirs for life.

This understanding ended unceremoniously in late 1934. In June the Liberals, under Mitchell F. Hepburn, won control over the Legislature for the first time in thirty years, on a platform that promised to cut government spending and balance the budget; their covert mission was to replace Tory appointees in the civil service with their own partisans. They began slashing almost immediately after taking the reins of office, and brazenly auctioned the Tories' fleet of vehicles at Varsity Stadium in August as a public testament to their determination to achieve their ends. When they began trimming the civil service, it appeared that the DLF, and particularly its Forestry Branch, would be spared. The forest industry was starting to recover by this time, and the Liberals had hired Frederick Noad, a veteran lumberman, to investigate the DLF and recommend means to improve it. And when Howe heard rumours that Noad's study would lead to greater control for the foresters, the dean was guardedly optimistic about the future.[27]

These hopes were crushed in late 1934, and not because of the government's ostensible aim of saving money. At the time, Howe informed George R. Lane (2T6) 'about the slaughter that has taken place in the Provincial Forestry Branch.' Over the course of two days, Ontario fired eighteen foresters. When the carnage was over, twenty-one had lost their jobs, including five district foresters (each was one of Howe's 'boys'), and their replacements were as unqualified as they were loyal to the new political regime. This had, in Howe's words, 'practically wrecked the Ontario Forestry Branch' and left 'an awful mess in the Department.' The reason for the firings, Howe found out, was not to cut costs, but in each instance 'some Liberal want[ed] the job.' This marked a new nadir for forestry in Canada. Howe informed Lane that 'this is the first time in the twenty-five years that I have lived in this country that any forester has been fired for political reasons. As you know,' he continued, 'there have been cases of transference but they never have lost their jobs entirely.' Howe was absolutely devastated by this unseemly turn of events. 'What I think about the situation of course,' he told Kelvin A. Stewart (2T2), 'isn't fit to dictate to a lady stenographer.'[28]

Howe was so aggrieved by the sackings because they had grave implications for those who had been fired, the forestry movement in Ontario, and his faculty. 'It also hurts me a great deal that the dismissals should have been so brutal,' Howe told Frank MacDougall (2T3), who retained his post as superintendent in Algonquin Park because of his loyalty to the Grits. 'Take a man like Kel Stewart,' Howe continued, 'who served through the war, who lost a brother while on duty in the provincial ser-

vice, who has given the ten best years of his life in building up an effective forest protection service – it hurts me as I say that all of this service
goes for nothing and he is told to quit his job within forty-eight hours.'
He then cited the government's treatment of another of his graduates
who was recently married 'with a baby three months old, his house engaged for six months, again dismissed at forty-eight hours notice with no
apparent investigation of what he was doing and the significance of his
work in the development of a forest policy in this province.' In Howe's
view, there was a far greater issue at stake in this housecleaning. 'A principle in regard to the profession is also involved,' he declared. 'The Government policy seems to assume that the work can be carried on just as
well or better by men without a technical training. This knocks the props
from all the standards that the profession has been trying to build up in
the past twenty-five years. If this idea is accepted by the Government and
by the public at large, then there is no need for a forestry school – all the
work and expense in the education of foresters has been a mistake.'[29]

The Liberals' brutal treatment of his foresters drove Howe to depart,
for one of the rare exceptions during his more than two decades as dean,
from his long-time policy of passively accepting such defeats. He did not
endorse openly alienating the government or abandoning the push to
cooperate with the DLF, but he did support 'the presentation of dignified protests to the Government as an exhibition of antagonism and it
cannot be considered as such by an organization sure of its position.'
First, he contacted 'some old time Liberals who were connected to the
University' to see if they could prevail upon the Hepburn regime. When
'they all said they had no influence with the present outfit,' he shifted
tack. He drafted a letter in early May 1935 to Peter Heenan and Hepburn, respectively Ontario's minister of lands and forests and premier, in
which he protested the cuts to the government's forestry staff. In it Howe
shrewdly pointed out that the Quebec government had not fired a single
forester during this period, and that Ontario could still make amends by
rehiring those whom it had dismissed. After H.J. Cody, the university's
president, vetted and approved the letter, Howe sent copies of it to both
Heenan and Hepburn.[30]

Howe did not stop there, and neither did those who wished to get
their jobs back, and the dean's decision to go on the offensive worked,
at least in a modest way. He laid the matter before the public by sending
versions of his protest letter to the press. No sooner had he done so than
Hepburn himself ordered that the DLF reinstate John A. 'Steve' Brodie
(2T3), who had been in charge of the government's forestry research

program. Brodie had helped his own cause by arranging for several 'influential men to speak to Hepburn on his behalf.' Thereafter, the Liberals began hiring back some foresters, but not in senior positions.[31]

This success was relatively minor, however. Not only did it fail to repair the damage that had been done, it did not make the foresters any less vulnerable to future attacks from the politicians. Howe complained to a fellow forestry professor at Cornell that 'as yet there is no indication of any change of heart or interest in forestry on the part of our provincial government. Our premier is quoted in the newspapers as saying that reforestation in Ontario is all nonsense; our problem is to get rid of mature timber as soon as possible and get money for it. Not much hope for forestry when an ignoramus like that has almost dictatorial powers.'[32]

This appalling situation prompted those engaged in forestry education in Canada to question the efficacy of their work and suggest possible new directions for it. Herbert Christie (1T3) was head of the forestry department at the University of British Columbia. His annual report for 1931–2 described how the 'quite limited' prospects of future employment in forestry and the province's difficulties in financing post-secondary institutions 'raise the question of whether there is now unnecessary duplication in forest schools as there undoubtedly is in other specialized courses. I think that one good forest school would serve all Canada, outside of Quebec,' he argued, 'and that it would be much better to have one strong than several weak ones. The logical place is Toronto, where the largest and best equipped school is now situated.' In a note to Howe, Christie added that there would also be a place for a number of 'short course ranger schools' to provide supplementary forestry education.

Howe wholeheartedly agreed with his colleague's diagnosis and remedy, and he had already begun creating a different kind of Faculty of Forestry in Toronto. He had been aggressively pushing for increased support for the faculty's graduate program since the mid-1920s, when it was just getting started. In responding to Christie, Howe noted – rather prophetically – that 'perhaps eventually, also, a school like ours might become entirely a graduate school.'[33]

Even before the pronounced downturn in the economy, Howe had begun re-evaluating the Faculty of Forestry's curriculum in light of the criticisms that were being directed at his graduates and forestry practitioners in Canada. The faculty's vulnerable underbelly was in the area of practical training, a weakness that Howe attributed to the lack of a permanent practice forest. The Great Depression, with its dramatic contraction in foresters' employment opportunities, gave him added incentive to reas-

sess his faculty's pedagogy. To achieve this end Howe drew up a question-
naire and sent it in late 1931 to nearly two hundred alumni who were
now engaged in both the private and public sectors. The respondents
overwhelmingly called for Howe to develop more practical skills in his
students, such as the ability to lay out harvesting operations and direct
road and dam construction. Malcolm Ardenne (2T4) succinctly summarized the impetus to change in December 1931. He informed Howe
that the students must learn the technical aspects of logging because 'it
appears that at the present there is not much prospect of forestry being
practised in Eastern Canada.'[34]

In light of this reality, in the early 1930s Howe adopted a more practical curriculum that was geared towards logging instead of managing the
forest. The faculty added new subjects, such as hydraulics and analytical
geometry, which helped fill its engineering lacunae. In addition, it dramatically increased the time its students spent in the field (in the late
1920s Howe had begun arranging to have his students spend their entire
third year working in the field, but the Depression derailed his plan).
Instead of a single, six-week stint at the practice camp at the beginning
of fourth year, they would now be given three weeks at the end of first,
second, and third years. Howe had also moved the site of fieldwork from
Algonquin Park to one of his graduate's summer camps in Haliburton
County because it was more accessible and offered 'a greater variety of
forest conditions.' Finally, Howe leaned heavily on the faculty's newest
member, Gordon Cosens, to develop the students' practical skills.[35]

This reorientation reflected Howe's realization that it was no longer
his school's primary purpose to produce foresters for the Ontario government. Previously, this understanding had led him to believe that they
could get by with little bush experience, because presumably they would
be more involved in policy formulation and technical work; Frank Mac-
Dougall (2T3) informed Howe, in early 1935, that the Faculty of Forestry
had a reputation for turning out 'white collar men and office men ...
[who] look down upon actual manual labour in the woods.' In the mid-
1930s, however, Howe had recognized that his dream of producing an
elite vanguard of foresters was foolhardy. 'The present state of affairs in
forestry employment,' he admitted in his annual report for 1934–5, 'indicates that we must give our students a broader and more thorough training in practical woods work better to fit them for private employment.'[36]

While Howe was forced to realign the Faculty of Forestry in the 1930s
in order to maintain its relevance, the school's well-established trends

continued unabated. The faculty still developed its ties to international forestry, for instance. During the Great Depression it attracted students from around the world, including natives of China and South Africa. Moreover, a few graduates went on to apply their craft overseas. Peter M. Morley (3T6), for example, spent nearly thirty years working in the pulp and paper industry in Ontario and Quebec before turning his attention to more distant lands. During the 1980s he advised a Canadian firm, Consolidated-Bathurst, about its plans to establish a newsprint mill in Brazil (it would be supplied with plantation-grown pine) and conducting forestry operations in Iran.[37]

Familial attachments also still wove their way through the forestry school between 1930 and 1941. Allan F. Buell (3T1), James W. McNutt (3T2), and Duncan R. Young (3T5) fathered sons who would graduate from the Faculty of Forestry, and Arthur S. 'Art' Michell (4T0) was unique in this regard. He joined the faculty's staff immediately after the Second World War and would later form the older half of the school's only father-daughter team. Finally, Oscar G. Larsson (3T7) would watch as his brother, Harold C. (4T2), followed in his footsteps a few years later.[38]

Despite the doom and gloom of the Great Depression, C.D. Howe still succeeded in graduating students whose accomplishments attested to the high-quality education he gave them. Their achievements spanned many realms, including forestry. Stanley 'Stan' T.B. Losee (3T1) struggled to find work during the Depression. In the late 1930s he became the understudy of the legendary Harold Seely, who was developing a method for estimating timber volumes for the federal government by interpreting oblique aerial photographs; it would become known as 'photogrammetry.' Losee left Seeley's side to join Abitibi after the Second World War and oversee one of the largest photogrammetry departments in Canada (it inventoried the thousands of square miles the firm controlled in Manitoba, Ontario, and Quebec). Losee contributed significantly to the development of his field, specifically improving the accuracy of this work, and his dedication eventually caused his eyesight to deteriorate prematurely. Together, his and Seely's pioneering efforts in this field set the stage for the first national forest inventory after the Second World War, a landmark achievement that allowed for the planning of sustained yield management even if its practice was still in the offing.[39]

Other graduates from this period made mammoth contributions to improving forest management in Canada. Desmond 'Des' I. Crossley (3T5) was at the head of this pack. He earned a master of science degree from the University of Minnesota five years after he left Toronto and went to work with the Dominion Forest Service. After serving in the Royal Canadian Air Force during the Second World War (he rose to be a squadron leader), he returned to Ottawa's employ and earned a reputation for being an innovative researcher with the DFS in Alberta. When North Western Pulp and Power Limited was building Alberta's first pulp mill in Hinton, in 1955, it hired 'Des' as its chief forester. From the moment the operation was established, he saw to it that its woodlands were managed on a sustained yield basis (the plan included a comprehensive reforestation effort). During and after his twenty-year career at Hinton he continued publishing technical papers in his field. The University of Toronto honoured Crossley in 1982 by awarding him the degree of doctor of laws, honoris causa, on the occasion of the Faculty of Forestry's seventy-fifth anniversary. Likewise, James B. Millar (3T1) and Edward Bonner (3T4) were on the cutting edge of progressive forestry. Both carried out landmark studies into the silvics of spruce on the northern Clay Belt for the Spruce Falls Power and Paper Company in Kapuskasing. Bonner subsequently set up and operated the first sustained, commercial-scale forest tree nursery in northern Ontario (in the town of Moonbeam just east of Kapuskasing).[40]

Two of the graduates from this period rose to senior managerial positions in the private sector, one in the forest industry and the other outside it. 'Al' Buell (3T1) began working for the Longlac Pulp and Paper Company (Kimberly-Clark) in 1937 and helped establish its pulp mill in Terrace Bay, Ontario, during the mid-1940s. In 1954 he left to become the woodlands manager for the E.B. Eddy Company, and seven years later he was appointed a director of the firm and its vice-president of woodlands. George D. Millson (3T5) completed a master of commerce degree after leaving the Faculty of Forestry, went to work for the Coca-Cola Company of Canada just before the Second World War, and returned to the soda maker after serving in the navy. By the mid-1960s Millson was the firm's vice-president, serving in this capacity at the same time as his fellow alumnus Ralph Sewell (3T3) was its president.[41]

Howe's Depression-era graduates did especially well in academia and public service. As far as the former is concerned, Anthony W.A. Brown (3T3) set the standard. Born in England, Brown attended Oxford and Cambridge before coming to the Faculty of Forestry at the University of Toronto in 1929. Not only was he at the top of his class every year despite his heavy involvement in extracurricular activities, by 1936 he had earned a master's degree in biology and docorate in biochemistry.

After studying at the University of London on a Royal Society of Canada Research Fellowship, he returned to Canada to supervise Ottawa's Forest Insect Survey in the late 1930s. During the Second World War, he worked for the Department of National Defence in the Directorate of Chemical Warfare and rose to the rank of major. In 1947 Brown joined the Department of Zoology at the University of Western Ontario as an associate professor, and two years later was appointed head and full professor; he served in that capacity for the next two decades. During this time he became an international authority in the field of insects that bite humans, the use of pesticides to prevent the bites, and the means by which insects become pesticide resistant. Brown's expertise in this area led the World Health Organization to hire him as a consultant during the 1950s and as a full-time employee after he left Western in 1968. He spent five years in Geneva, and then returned to work at Michigan State University as the John A. Hannah Distinguished Professor until 1976. Brown's outstanding professional career earned him a string of awards, including being elected a Fellow of the Royal Society of Canada in 1961 and recipient of the Entomological Society of Canada Gold Medal for Achievement in 1963.[42]

The career of Duncan A. MacLulich (3T1) followed a similar path. After earning a doctorate in biology from the University of Toronto, his career path wandered through the pathology department of the Mountain Sanitorium in Hamilton, Ontario, and Algonquin Provincial Park (as its bacteriologist). In 1962 he was asked by Waterloo Lutheran (now Wilfrid Laurier) University to organize its science departments and become the head of the biology department. Although he officially retired in 1974, MacLulich remained active in his field – population ecology – for the next two decades. His research provided the data for the 'Lynx-Hare Cycle' (the cycling of predator-prey population sizes), which became the standard textbook example of this principle. His colleagues recognized his profound commitment to Laurier by establishing a gold medal in his honour and naming its zoological museum after him.[43]

A number of other graduates of this era also entered academia. Both Fred Jackson (3T2) and 'Art' Michell (4T0) worked at the Faculty of Forestry, the former for six years and the latter for over thirty. John E. Bier (3T2) earned a master's degree and doctorate from the University of Toronto, and after working with the federal government as a forest pathology researcher in British Columbia for one decade, he joined the Department of Botany at the University of British Columbia.[44]

A handful of graduates from these years answered the call to enter

Ontario's civil service. Walter R. Grinnell (4T0) and Quimby Hess (4T0) personified this group; in the process the former earned his master's and the latter his 'FE' (forest engineer) degree from the faculty. Charles H.D. Clarke (3T1) was awarded his doctorate in biology from the University of Toronto four years after he graduated from the Faculty of Forestry, and then worked as a researcher for the DLF's Fish and Wildlife Branch. By 1970 he was its chief, and thereafter Clarke was loaned to the Canadian government to act as a wildlife consultant in Kenya.[45]

Alvah S. Bray (3T1) continued the trend of Faculty of Forestry graduates becoming senior managers in the civil service in fields other than forestry. Bray worked for the DLF after graduation until he was fired in 1935, and then returned in 1943 as district forester in Cochrane. Over the next fifteen years he was appointed regional forester for northwestern Ontario and chief of the DLF's Lands Branch. In September 1964 he was appointed deputy minister of industry and tourism, and he remained in that position until he retired eight years later.[46]

G.H. 'Terk' Bayly (3T9) rose to an even higher post within the provincial bureaucracy, yet his career spoke to more than just professional success; it attested to the forestry graduates' love of 'the woods' and their commitment to protect it. The political environment in which they worked was rarely receptive to this conception of their jobs, however, especially during the 1930s. Dean Howe's graduates were thus forced to subordinate their own views of nature to those of their employers (their other option was to look for work in another field), and this often turned them into little more than glorified planners of timber harvests. Bayly was one of the exceptions to this rule. After being awarded a Distinguished Flying Cross for his service in the Second World War, he spent the next four decades working for the DLF as an environmental steward. He succeeded MacDougall as deputy minister in 1966, and served five years in that position. He was then promoted to secretary of the management board and deputy provincial secretary, one of the most powerful unelected positions in the Ontario government. After his retirement, Bayly served with the Ontario Heritage Foundation and was chairman of its Heritage Trust Committee. In this capacity, he built up the foundation's 'natural' heritage program and was appointed as the Niagara Escarpment Committee's first chairman. For his tireless efforts on behalf of natural and cultural conservation, Bayly was awarded the country's highest honour in this field – the Canadian Parks Service Heritage Award – in 1989. An avid outdoorsman and an accomplished canoeist, his affection for nature's beauty was attested to by his creation of a public trail system through a property he had rehabilitated near Meaford, Ontario (this was also the site of his annual teddy bear picnic, when he hid the stuffed animals for local children to find). Bayly's obituary succinctly summarized the centre of his world when it remarked that 'trees were his life.'[47]

A few other forestry graduates from these years were also able to accomplish much in the field of conservation. Kenneth M. Mayall (3T5) was a native of Scotland and graduate of Cambridge University. Before taking up his studies in Toronto he cut timber in western Canada and acted, as he put it in a letter to Howe, 'as a private tutor in the family of an Ojibwa Indian' in Temagami, Ontario. A few years after graduation, he drew up a landmark inventory of the land in King Township (north of Toronto), and a plan for rehabilitating it and the wildlife it had formerly supported; it had suffered so severely from the effects of wanton deforestation that it was derisively referred to as 'the King desert.' Although the idea to restore it came from Aubrey Davis, president of the Ontario Hunters' Game Protective Association, Mayall provided the project's technical expertise. Thereafter, he devoted his life to conservation and ecology. Mansiel 'Manny' R. Wilson (3T4) spent most of his professional life with Canadian International Paper in Quebec. In this capacity he oversaw the establishment of CIP's Harrington Forest Farm, which was managed to demonstrate the efficacy of 'integrated forest resource management': forestry, wildlife conservation, public education, and recreation were carried on simultaneously. Later, Wilson ensured that Harrington was opened to the public as a 'Nature Centre.' A lifelong conservationist, he was a founding member and director of the Canadian Wildlife Federation and won a string of awards – including the Roland Michener Conservation Award – for his commitment to prudent management of the environment.[48]

Most of these achievements were still a long way off when the 1930s were drawing to a close, and fortunately for C.D. Howe and the Faculty of Forestry at the University of Toronto, by this time forestry's prospects were finally brightening. The Ontario government's policy of facilitating the export of pulpwood and the establishment of new pulp mills, and the rebound in the province's pulp and paper industry, were generating a strong demand for foresters. Moreover, the Ontario Department of Lands and Forests had reconsidered its need for foresters, and was beginning to reach out to the faculty in a manner that had not been seen since the 1920s. Howe noted in February 1939 that the provincial government was 'gradually coming to see the error of its ways.' It was effecting a minor reorganization of the DLF, rehiring some of the district foresters whom it had fired, carrying out experimental cutting operations on a minor scale, and reforesting a few tracts of old pine lands in northern Ontario. Together, these developments meant that Howe's graduates would fill jobs they once held with the DLF, and new ones to boot.[49]

At the same time, however, Howe knew that forestry still faced an uphill battle in Canada in general and Ontario in particular. The fundamental problems – political control and lack of public support – that had plagued it since the Faculty of Forestry had first opened its doors in 1907 continued to do so on the eve of the Second World War. This point had been driven home to Howe by his trips to see how the Europeans practised forestry; the contrast between the 'New World' and 'Old' left him green with envy. After his journey through Scandinavia in the summer of 1935, for example, Howe remarked that it was truly wonderful to see what foresters could do if they 'have a free hand in its [i.e., forestry's] development and when forestry is supported by public opinion.' Howe's frustration with the troubling situation at home was best expressed by one of his students, Andrew 'Andy' P. Leslie (2T9), who had been so disheartened by the cutbacks to forestry during the Great Depression that he had abandoned the profession. 'Sometimes I miss Forestry,' Leslie wrote in 1937 to 'Bob' Hosie, 'but soon console myself with the thought that I never did any real forestry work and it is unlikely that any of the sort I am interested in will be done while one is young enough to take part in it.'[50]

Howe was also sensitive to a novel challenge that was rising to confront forestry, one that would only grow in time and create ever greater difficulties for the profession. The shift in demographics (Canada's 1921 census showed that, for the first time, more residents were living in urban than rural areas, a trend that would accelerate thereafter), the generally unhealthy state of cities and towns, and the growing leisure time that many Canadians enjoyed as their country became more industrialized, were three factors that drove urban dwellers to seek 'natural' recreational opportunities within easy reach of their homes. This placed a new demand on the forests that were closest to or seen as most desirable by city folk. They often laid claim to these woodlands, and demanded that the provincial government take steps to prevent the forest industry from harvesting trees that the 'recreationalists' considered more valuable if left uncut.[51]

What alarmed Howe about this new constituency was the political clout it exercised. In discussing the matter with 'Des' Crossley (3T5) in the mid-1930s, Howe recognized how 'the development of tourist activities, especially the campers in the forest,' would militate against the practice of forestry. Even worse was the relative ease with which this lobby could derail sound forestry initiatives. 'Just before the Depression hit us Mr Finlayson [i.e., Ontario's minister of lands and forests] and the Forestry Board [of which Howe was chairman] had made arrangements for experimental cuttings in pine [in Temagami],' Howe recounted to Crossley. 'Although the area was out of sight of the lake over a ridge there was a great protest from the campers. The present government has sold the same area [to a lumber company] and it will be cut non-experimentally.'[52]

By the late 1930s, however, Howe was acutely aware that he did not have the energy to lead the troops to tackle this new demon or any of the old ones. His profound commitment to his school and forestry had left him enervated by the eve of the Second World War. Several times during the Depression his failing health had forced him to take extended leaves of absence from the Faculty. By the late 1930s his doctor had restricted him to working only two afternoons per week. Howe had typically retained his wry sense of humour through these challenges, commenting to Cody, the university's president, in late 1937 that 'I am not really ill, only receiving some preliminary warnings of approaching old age.'[53]

Dean Howe was thus compelled to turn his attention to anointing his successor. He discussed the subject at length with Cody in March 1938, and began by recommending his colleague, J.H. White. 'He deserves this from the standpoint of service, teaching capacity and standing in the profession throughout the country,' Howe declared. The problem was that White intended to retire at age sixty-five, which he would turn in but two years. In any event, Howe named another candidate he felt would be best suited to succeed him as dean: Gordon Cosens (2T3).

Howe was adamant that Cosens was *the* man for the job, and for one strategic reason. By the mid-1930s industry had replaced the provincial government as the main source of jobs for 'his boys.' He told Cody in 1938 that it was his 'well considered belief that the future employment of our graduates lies chiefly with the pulp and paper companies. In fact the governmental field of employment has become so restricted that the School could not live without the employment of graduates by private companies.' By this time Cosens had roughly two decades of experience working with two of Canada's most forward-looking newsprint companies. More importantly, even though Cosens had joined the Faculty

of Forestry in 1934, the University of Toronto had agreed to allow him
to remain on a retainer from the Spruce Falls Power and Paper Company in order to supplement his less-than-stellar professorial stipend.
Consequently, Cosens continued to enjoy an intimate relationship with
Kimberly-Clark, the American pulp and paper behemoth that owned
the newsprint plant in Kapuskasing for which Cosens worked. Having
Cosens as the faculty's dean, Howe thought, would facilitate finding jobs
for its graduates for the foreseeable future and give it a crucial conduit to
industry in general and this forestry-friendly firm in particular.[54]

As a result, Howe was fully supportive of Cosens remaining on a retainer with Spruce Falls as a condition of him becoming the third dean
of the University of Toronto's Faculty of Forestry. Howe, Cosens, and
Cody discussed the matter over the course of 1938–40, and by early 1941,
the deal was done. Just so there would be no misunderstanding, Cody
wrote Cosens in March of that year to inform Cosens that the board of
governors had formally appointed him as the faculty's head. 'It is dis-
tinctly understood,' Cody added, 'that during the summer vacation and
at such other times as may be deemed advisable, you are free to maintain
your connections with the lumber [sic] industry through any company
you wish to make arrangements. The Board feels that it will be a distinct
advantage to the Department [sic] of Forestry that you maintain these
outside contacts.' In reply, Cosens expressed his pleasure at his appointment. 'As an industrial connection is valuable to this Faculty from the
viewpoint of student and graduate employment as well as in actual teaching,' Cosens noted, 'I appreciate your help in arranging so that I may
remain with Kimberly-Clark Corporation and its Canadian subsidiaries.'
When Howe heard the news, he informed Cody of his regret at leaving
the faculty but also his 'greater satisfaction knowing that the headship of
the school will be turned over to the capable hands of my proposed successor.'[55]

As C.D. Howe left the university in mid-1941, he clearly did so with
conflicting emotions. It was a great comfort to him to learn that the Faculty of Forestry had, for the first time in its existence, secured *the* man to
be its dean, and that the University of Toronto was going to recognize
Howe's lifetime of 'outstanding services to Forestry in this Dominion'
by bestowing on him the honoris causa degree of doctor of laws. At the
same time, however, Howe was not blind to the stark truth about his
time as the Faculty of Forestry's head. Some of the problems he had
encountered were not of his own making, but he also recognized that
maybe his modus operandi had not been the best one for the faculty

during the interwar period. In reference to the honorary degree that the University of Toronto awarded him, Howe told Cody in June 1943 that it had 'eliminated the memory of certain frustrations caused by my own poor judgement as well as by the ignorance of politicians.' Howe did not indicate which of his actions he would have classified as reflecting 'poor judgement' (did firing Millar still weigh on his mind?), but this was no time to dwell on past mistakes. He moved to the southern United States 'for health reasons' after his retirement, and died in February 1946. Fittingly, four of Howe's six pallbearers were his former students, 'his boys,' to whom he had given so much and for whom he had tried so hard. While some of his decisions are vulnerable to criticism, no one could argue that C.D. Howe, who had not been *the* candidate for the Faculty of Forestry's deanship, had devoted his heart and soul to furthering the forestry cause and the faculty's interests in the process.[56]

Chapter 5

'The Present Pressure for Registration in Forestry Is Temporary,' 1941–1947

The period during which Gordon G. Cosens (2T3) served as dean of the Faculty of Forestry at the University of Toronto, from 1941 to 1947, was propitious for forestry in Canada. The Second World War drove home the importance of natural resources, and governments at both the provincial and federal levels exerted unprecedented control over managing the country's affairs, including woodlands. Industry also seemed ready to embrace this new spirit. The war and the boom that came in its wake heralded a highly profitable period for nearly all the players in this field; they could now easily afford to invest in practising silviculture. Moreover, the lumber industry was prepared to drop its obstinate opposition to adopting forestry measures, something to which the actions taken in early 1946 by W.J. LeClair, the secretary-manager of the Canadian Lumbermen's Association, attest. He wrote Sydney E. Smith, the University of Toronto's president (1944–57), and admitted that 'only recently ... the lumbering section of the forest industries have [sic] shown any interest in the employment of professional foresters, yet the lumber interests constitute practically fifty per cent of our forest industry.' For this reason, LeClair literally begged Smith to convince Cosens to appear at the association's upcoming annual meeting because, in LeClair's words, it was essential 'that the head of the provincial forestry school attend and meet personally with representatives.' Cosens did so, and even sat as the association's guest of honour.[1]

Cosens was just the man to exploit these favourable circumstances to benefit the Faculty of Forestry. He had enjoyed a long-time and intimate association with a colossus in the forest industry, Kimberly-Clark. This firm had operated a pulp and paper mill in Kapuskasing since the early 1920s, built a new pulp mill on the north shore of Lake Superior at a site it christened 'Terrace Bay' in the mid-1940s, and leased massive Crown timber limits. The University of Toronto ensured that Cosens remained on Kimberly-Clark's payroll during his term at the head of the faculty, even though it meant that he missed the occasional week of the school year 'on assignment.' The pay-off, from the faculty's perspective, was the direct link it gained with what was arguably the country's most forester-friendly company. Moreover, Kimberly-Clark was, from the moment it had arrived in Ontario, the province's darling pulp and paper maker. Not only had the Ontario government fallen over itself to meet Kimberly-Clark's requests for copious quantities of timber and water powers under favourable terms, it had permitted the firm to operate free from the suffocating 'prorationing' restrictions that it imposed on the province's other newsprint makers during the late 1920s through to the early years of the Second World War. While the policy sent Kimberly-Clark's competitors into receivers' hands, its mill in Kapuskasing operated to capacity and paid off its debt during the Great Depression. Cosens' connection to Kimberly-Clark thus represented a means of forging a tight bond between the Faculty of Forestry and both industry and government in Ontario.[2]

Cosens' ties to persons of influence extended far beyond Kimberly-Clark's corporate headquarters and Queen's Park, and he plied them to help forestry and the forest industry across Canada. He moved with the country's political and economic elite, and was easily the best-connected leader that the faculty ever had. He was on a first-name basis with the University of Toronto's presidents and the province's political leaders during these years; their doors were always open to him. Cosens' tie to Frederick K. Morrow was equally important during this period. Morrow was described at the time of his death in 1953 as 'one of Canada's leading financiers,' although little is known about the influence he wielded in the country during the 1930s and 1940s largely because, as one of his obituaries notes, 'he disliked publicity.' Morrow was appointed to the university's board of governors in 1939 at a time when he was already ensconced in the Canadian Establishment. A director or chairman of the board for a long list of companies at this time and shortly after, his association with the Canadian International Paper Company was most important to Cosens and the Faculty of Forestry. Like Kimberly-Clark, International Paper had been able to captivate governments in Canada since the mid-1920s, and had reaped immense rewards as a result. Morrow was also a major benefactor of the University of Toronto (on one occasion he donated $100,000 to its School of Nursing), and he acted as Cosens' 'unofficial' counsel during the early to mid-1940s.[3]

The brief boom the forestry movement enjoyed under Cosens' leader-ship came to an abrupt end, however, in the economic prosperity that followed the Second World War. This was not a function of an active pol-icy that sought to bury it, but rather a general indifference on the part of elected officials and the public that allowed forestry to sink to near the bottom of their priority lists. John C.W. Irwin (2T2), an outspoken critic of the Ontario government's anaemic forestry policy during the 1930s and 1940s, succinctly summed up the situation in a brief he presented to a provincial royal commission that investigated in 1946–7 Ontario's administration of timber. Irwin admitted that he could level scathing criticism at the pulp and paper and lumber industries for the manner in which they had operated over the previous decades. 'But vitally and essentially,' he pointed out, 'they do only what they are allowed to do by the contracts; and if our representatives in the Government, who are administering these resources for us, allow them to do that, and allow these things to happen, then we, the people, and the government, are responsible.'[4]

The upshot saw the Faculty of Forestry enjoy an ephemeral 'golden era.' The demands of war reduced its total enrolment to thirty-one in 1942–3, but the return of the veterans, and Ottawa's willingness to sup-port their reintegration into peacetime society, shot it up to 177 in 1945–6 and over 250 the next year. While the details regarding this dramatic turnaround will be discussed below, one of the main reasons the faculty admitted such high numbers after the war (107 in Cosens' last year as dean) was its expectation that many would fail out of the program. This did not occur, however, because the students generally brought a very high level of commitment to their studies in the wake of the war. While this created a major overcrowding problem, the good news was that the economic boom assured practically all the Faculty of Forestry's graduates jobs that many would hold throughout their working careers.[5]

As with his predecessors in the dean's office, Gordon Cosens could be an unrelenting taskmaster. He set the bar high for his students, and ex-pected them to clear it. Those who worked with him at Kimberly-Clark knew his demanding standard full well. On at least one occasion, his colleagues in Kapuskasing resented his 'foul influence' in dictating their priorities in the field.[6]

Fortunately, Cosens balanced this trait with a well-developed funny bone and a soft spot for life's simpler pleasures. Two stories illustrate this characteristic. George E. Bothwell (1T3) had been working with the

Dominion Forest Service (DFS) in Alberta when the First World War broke out. He, like nearly 25,000 other Canadians and Newfoundlanders and over 600,000 Allied soldiers, had died at the Battle of the Somme in 1916. After the war, his DFS colleagues in Alberta – led by Ernest H. Finlayson (1T2) – chose to remember him by funding a new gold medal at the faculty. Their plan was to award it, beginning in 1921–2, to the student who achieved the highest grade in the third year silviculture course. This was Cosens' métier, and in mid-1922 the faculty announced that he would be the inaugural recipient of the 'George E. Bothwell Medal for Silviculture.' The problem, however, was that Finlayson got sidetracked and never raised the money needed to pay for the award. Over twenty years later, President Cody informed Cosens that the university had granted the funds to pay for the medal that the dean had won but never received. In thanking the president of the university for taking this action, Cosens jocularly noted that he only regretted 'very much that my father did not live to see me a gold medallist. When I was first awarded it he said that times must have changed at the University when I could be a gold medallist.' Likewise, Cosens took advantage of every opportunity to indulge in his other passion, trout fishing. In the spring of 1943, he wrote a pair of his chums – 'Wes' McNutt (3T2) and 'Al' Buell (3T1), who were foresters with Kimberly-Clark's Longlac operations – to inform them that the government's local district forester had offered to fly them all 'north to some good trout fishing.' Because Cosens knew that using a government airplane for such a mission made it politically risky, he told McNutt that he hoped that 'Drew [i.e., the leader of the Opposition] doesn't hear about it.'[7]

Cosens' sense of humour would help him and the faculty confront the sullen reality that another world war brought to Canada. While the conflict's early years delivered glum news but mercifully few casualties, by the time Cosens had replaced Howe as the school's dean, hostilities were intensifying.

The impact on the Faculty of Forestry was predictable. Its tradition of instilling in its students a profound sense of public service pushed them, just as they had during the First World War, to enlist in droves. Most served in the Air Force, Artillery and Engineering Corps. 'This peculiar distribution,' Cosens explained to Cody in late 1942, was due to two factors. 'The military authorities recognise,' Cosens continued, 'that foresters have a thorough basic training in subjects particularly required by the technical branches of the armed forces, and appreciate that their field experience has developed in them initiative, leadership, and ability to deal with unusual situations. The broad scope of the course in Forestry coupled with field experience has furnished a satisfactory background for certain specialised work required by the armed services.' Although this pattern of service continued for the duration of the war, some of the faculty's students and graduates ended up applying their skills in unique ways. Two, George Tunstell (1T3) and Bertram G. Day (4T2), were involved in both tracking 'incendiary bombs carried from Japan in paper balons [sic] by the prevailing winds' and investigating their chemical composition.[8]

The faculty boasted a heavy participation rate in the conflict, but in contrast to the First World War, this time the list of the school's – and the country's – casualties was much shorter. By late 1943 its total student body had been reduced to roughly one dozen and over 30 per cent of its living graduates were on active duty. Nevertheless, only five forestry students lost their lives in the conflict, although many endured prolonged terms as prisoners of war. John E.C. Pringle (4T9) served four years with the Royal Canadian Air Force, for example, before being shot down over enemy territory and surviving three years as a POW.[9]

There were also more than a few stories of valour, some with a romantic tinge that even Hollywood could not have better scripted. William R. Parks (4T9) had been serving in the RCAF as part of a Halifax bomber crew when his aircraft went down over Denmark; he was its sole survivor. The Danish resistance reached him before the Germans did, and prepared him to be 'shipped out' ten days later. As he was making his getaway, the Germans stopped and questioned him. The fake identity card the Danish resistance had given him passed inspection by his interrogators, and they also fell for his portrayal of a deaf and dumb broommaker. In short order Parks was safely back in England. When the press reported on his experience in the late winter of 1946, it did so on the occasion of his marriage to his long-time sweetheart. The headline in the *Toronto Evening Telegram* aptly captured the storybook ending to Parks' ordeal, stating 'Dream of Wedding to Come Stilled by German's Order but Cupid Gets Final Say.'[10]

Ironically, the war years proved nearly as deadly to the faculty's graduates who were not serving in the military as it did for those who went off to battle. Working in the bush was a very dangerous activity, and Cosens was reminded of the perils of forestry employment – and life's unpleasant twists – at a time when it looked as if the war would be mercifully easy on his understudies. 'The Faculty has had no war casualties for some time now, thank Heavens,' Cosens wrote in late December 1944 to Maurice

M. 'Moose' Dixon (4T1), 'but we have had three civilian deaths. Bill Adams of '43 was drowned on the Shekak River west of Hearst while on a cruise party with Fred Wiley. Carl Stangeby of '36 who was at Red Rock accidentally shot himself with a revolver. Larry McCausland of '27, a veteran of both wars, died under very tragic circumstances.'[11]

In some respects, Cosens dealt with the war much as B.E. Fernow had a few decades earlier. Naturally, Cosens was profoundly worried about the safety of 'his boys' who were in the service and kept in touch with them as much as he could. His letters provided them with the latest news about the faculty and forestry in Canada and conveyed his fatherly concern for their well-being.[12]

But in many ways, Cosens handled the challenges that the conflict presented in a way that was diametrically different from Fernow. The latter, for instance, had exhorted his students to enlist during the First World War and had gone to great lengths to demonstrate both his efforts and the sparkling results they produced. Undoubtedly Fernow had been motivated by his determination to defend against charges that his Prussian heritage cast his loyalty into doubt, but his message had also resonated with that which other authority figures were preaching at the time. Not so during the Second World War. Leading Canadians, such as General A.G.L. McNaughton, the chief of the Canadian military, told university students – at least initially – that they could best help the cause by remaining in school. Cosens could not have agreed more, and constantly decried the degree to which the faculty's students were donning military uniforms instead of bush garb.[13]

Cosens endeavoured to do something about this situation. Unlike Fernow, he was free to do so because his 'Canadianness' was an effective aegis against aspersions of disloyalty. Cosens also enjoyed the requisite connections to the upper echelons of Canadian society. This placed him in a position from which he could exert sufficient leverage in his lobby both to protect his charges from active duty and retrieve them from the military's ranks when they landed there.

Cosens did all he could to achieve these aims. Most importantly, he waged a tireless campaign to convince the University of Toronto and Ottawa to exempt foresters from active service. In late 1942, for example, the federal government had directed the country's universities to identify which of their courses were essential to the war effort, and when Cosens learned of the inquiry, he forwarded a heartfelt appeal to H.J. Cody. After noting the disproportionately large number of the faculty's students and graduates who had enlisted, Cosens described the strategic importance of lumber and some chemical pulps in wartime. He also underscored how the export of large volumes of pulp and paper products to the Americans was integral to maintaining our 'United States credits.' The problem, he insisted, was that the enlistment of so many foresters was severely handicapping the efficient operation of these industries. It was also impairing the Ontario government's ability to protect the forest from fire, the danger of which 'has increased because of the possibility of carefully executed sabotage from incendiary fires strategically placed.' While this claim may have bordered on hyperbole, the rest of his message was evidentiary-based. 'The shortage of technically trained foresters was acute before the war,' Cosens argued, and now the situation had been exacerbated. Moreover, no personnel were available for planning 'reconstruction' forestry work. 'In light of these facts,' Cosens concluded, 'it would seem that the Faculty of Forestry should receive preferred consideration in any man power redistribution or other change contemplated in the University at the demand of the Dominion Government. It is imperative,' he added, 'that Forestry be included with and given the same consideration as other science courses.' When Cody rebuffed his initial efforts, Cosens continued to push the cause. His efforts eventually bore fruit, but as he told officials in Ottawa in early 1944, 'unfortunately this comes too late as there are only two in this years [sic] graduating class and there will be only three or four at most to graduate next year.'[14]

Cosens undeniably waged his campaign on behalf of all foresters, but he focused his energy most intensively on extricating men from the armed forces who formed Kimberly-Clark's cadre of bright young silviculturalists. Fred Nelson Wiley's (3T1) 'de-enlistment' is a case in point. Having joined Kimberly-Clark in Kapuskasing immediately after graduation, Wiley had enlisted in 1941 in the belief that he could contribute to the war effort by acting as a navigation instructor with the RCAF. The problem was that he hated flying – it made him sick, and by early 1944, it appeared he was about to be sent into active service. Before Wiley suffered through another bout of nausea, Cosens sprung into action. He pulled the appropriate strings with the Canadian military to expedite Wiley's return to Kapuskasing. 'Your red hot action,' Wiley ecstatically reported to Cosens only days after the dean had intervened in his case, had achieved its desired end. To his delight, Wiley had been informed that his work with the air force was 'fini.' Nevertheless, the requisite paperwork still had to be filled out, and the facade of propriety maintained. 'When you succeed in demobilizing,' Cosens cautioned Wiley, 'take the attitude to everybody that you were demobilized entirely at the request of Spruce Falls Power & Paper Company and make no comment to anyone … Do not make any comment about your work in the Air Force, and if you want my opinion I would praise the organization one hundred per cent.' When Wiley wrote Cosens to express his concerns about how his premature departure from the military would be perceived, Cosens tersely told Wiley that 'it did not matter a hoot how you got out of the Air Force as long as you got out.'[15]

As much as the war itself presented numerous challenges to the faculty and to universities across Canada, the cessation of fighting was a boon for them. Unlike after the First World War, when Ottawa had offered veterans practically no financial assistance, it eagerly supported them, on generous terms, during and after the Second World War. Furthermore, it had announced its intention to do so by 1941, thereby providing the universities with ample time to prepare for the day when the crush of military personnel would come home and resume their educations. The result was predictable. The University of Toronto's enrolment was seven thousand at both the beginning and end of the war, but by 1946–7 enrolment had swelled to over seventeen thousand, roughly half of whom were veterans. The University of Toronto was most affected of all Canadian schools by this phenomenon, which Robert Bothwell, Ian Drummond, and John English aptly label 'the veteran bulge.'[16]

The Faculty of Forestry shared in this meteoric growth, which both forced Cosens and his staff to take some creative steps to deal with it and opened some windows of opportunity to them. In preparation for the first major group of discharged servicemen, the university's board of governors and the Faculty of Forestry agreed, in mid-April 1945, that the forestry school would limit its admission to first year to seventy-five students for the fall term (55 of these spots would be reserved for veterans). This seemed reasonable enough. Since 1919 freshman registration had averaged seventeen students, and since the early 1930s, each year had seen the faculty receive fewer than fifty applicants. Seventy-five students thus represented roughly five times the Faculty of Forestry's average first-year enrolment. To prepare the school's facilities for this enlarged student body, the board authorized significant expenditures on building alterations, and the purchase of furniture and equipment.[17]

But it quickly became apparent that the board and the faculty had grossly underestimated the level of interest in studying forestry. By May the Faculty of Forestry had already received eighty-nine applications from servicemen and well over a hundred high school students who merited entry into the program. If the faculty admitted all ex-servicemen who qualified (or nearly so), and roughly two-dozen high school students, its first-year enrolment would be 116.[18]

Cosens discussed this potentially troublesome matter with his colleagues at the faculty and with Frederick Morrow, a member of the university's board of governors, and the upshot was a novel solution that was intended to benefit the forestry school over the long term. In reporting on the matter to the university's new president Sidney Earle Smith, Cosens indicated that the Faculty of Forestry could 'only deal with it by working two shifts in the first year and increasing the staff immediately.' To assuage any concerns that the president may have entertained about accepting such an enormous number of students into the forestry program with no assurance that they would be able to find gainful employment upon completion of the course, Cosens explained that Morrow had committed to doing his 'utmost to assist in placing them in permanent positions' after they graduated. 'As you know,' Cosens reminded Smith in reference to Morrow, 'he is a director of International Paper Company and would be of very great help.' By the time classes were set to start in the fall of 1945, the Faculty of Forestry had divided its incoming cohort of 112 into two sections, 'A' and 'B.' It would keep them separate until third year, when it was expected that attrition would allow the two groups to be amalgamated.[19]

Cosens was keen for the faculty to receive the reward – two new permanent professors – he felt it deserved for its willingness to be so accommodating, but this was easier said than done. He believed Herbert W. Beall (3T2) and John L. Farrar (3T6) were ideal candidates for the posts. Both had served for at least three years during the war and were now working as researchers with the Dominion Forest Service. Beall had ranked first in each of his four years at the Faculty of Forestry with grades that were truly extraordinary, whereas Farrar had been both an outstanding student and athlete. Smith agreed to make the two appointments, but only one of them – Beall – would be permanent immediately (the other would be temporary). To provide the teaching support necessitated by these unusual circumstances, Smith also authorized the faculty to hire a number of demonstrators and laboratory and field assistants.[20]

But Cosens was unable to sign Beall and Farrar to contracts. The former declined the offer because, just as it arrived, the DFS significantly raised his salary. Farrar, despite a resounding endorsement from the dean of Yale's School of Forestry where he had begun graduate work, was convinced that he would not make an effective lecturer (he would change his mind a decade later). Cosens then turned to Jack H. Gooden

(3T5), who was a senior forester with The Great Lakes Paper Company in Fort William, Ontario. Godden was enthralled by the prospect of teaching alongside Cosens, but felt obliged to remain at his present job. 'The matter resolved itself into the question of where I could do the most good for Forestry and the use of the forests in the public interest,' Godden's apology to Cosens read, 'and I firmly believe that I can best do that by carrying on in industry.' Cosens was thus left in the unenviable position of having won two staff positions for the faculty but being unable to fill them. Desperate to do so, Cosens turned to much more junior and inexperienced candidates. David V. Love had graduated from the University of New Brunswick's forestry school in 1941. After serving more than three years in the Royal Canadian Navy, he went to the University of Michigan's School of Forestry and Conservation to pursue a master's degree, which he was scheduled to complete in mid-1947. 'Art' Michell (4T0) had performed brilliantly as a student at the Faculty of Forestry, and worked for Ontario's Department of Lands and Forests before enlisting. Both Love and Michell joined the Faculty of Forestry's staff in time for the fall 1946 semester, the former to teach forest management and the latter harvesting and forest transportation systems.[21]

While the Faculty of Forestry was attempting to deal with its ballooning student body (in October 1946 its total enrolment stood at 263), Gordon Cosens and others felt its focus ought to be increasingly on graduate work and research; this was neither the first nor last time this suggestion would be made. Cosens knew first-hand that the faculty had a reputation for training more academically inclined foresters, and he felt moving into graduate studies in a major way was a natural step to take. For this reason, when the University of Toronto's administration raised the possibility of constructing a new building to accommodate the faculty's enormous undergraduate enrolment, Cosens recommended another use for the resources. 'I would rather see laboratory facilities provided for graduate students than new construction for undergraduates,' Cosens confided to Smith, 'as I feel the present pressure for registration in forestry is temporary.' A short while later, Cosens delivered the same message to Ontario's Royal Commission on Forestry, arguing for the faculty to become more involved in basic forest research and expand its graduate program.[22]

The Department of Lands and Forests (DLF) also wished to see the faculty carry out this mandate. W.G. Thompson, its minister, wrote to the University of Toronto's president in late 1946 to explain that the DLF's research program would require nearly fifty students each year. Thompson added that the DLF would be looking to the faculty's staff to groom students who demonstrated an aptitude for such work by providing special training in this field. The university was receptive to this invitation, explaining that its forestry school was anxious to 'develop an intensive program of research.'[23]

Long before this time, Gordon Cosens had been tapping his political contacts to further the forestry movement in Ontario in particular and Canada in general. In the end, he came within a hair of both effecting a virtual revolution in how the province managed its timber resources and resting his hands on the controls that directed its forest policy.

His tight ties to Queen's Park were clear from the leading role he played in a radical plan that was hatched in 1943. George Drew had been the leader of the Opposition Tories during most of the 1930s and early 1940s, during which time he feared the rising popularity of the Cooperative Commonwealth Federation and its socialist agenda. Drew's angst intensified during the early 1940s when the CCF suddenly appeared as a bona fide contender for office at both the national and provincial levels. Drew remained obstinate in his ideology, but pragmatism dictated that he would have to 'steal from the left' if he were going to win power. The Tories thus ran in the 1943 provincial election on the basis of a 'Twenty-two Point Program,' which promised nearly all the bells and whistles of the social welfare state for which the CCF had called in the platform it had announced a few months earlier.[24]

Drew's program had particular significance for forestry in Ontario. The CCF's campaign plans for reforming the province included a commitment to 'organize for the full use of our natural resources and the development of new industries the minute the war ends.' This sounded innocuous enough, but not when it was read in the context of the party's other electioneering pledges. The CCF also promised to 'free people from the power of the monopolies now in control of all important business and industry' (the pulp and paper industry in eastern Canada had been operating as a government-sponsored cartel for over a decade) and 'bring these monopolies under social ownership.' This idea reeked of communism, and it was anathema to both Drew and Cosens. With the CCF's growing popularity, it was a very real possibility that the party would win power in Ontario and implement its plans to nationalize papermakers like Kimberly-Clark. In a letter to a colleague at this time, Cosens confided how he feared 'the rising tide of socialism in Canada that may soon affect limit tenure and woodlands administration.'[25]

Drew sought to counter the CCF's forestry initiatives by introducing a revolutionary policy for administering the Crown forests in which Dean Cosens was front and centre. The Tories committed to creating a new body – the Ontario Forest Resources Commission – to manage the public woodlands and 'operate under long-term policies of conservation, reforestation and soil erosion.' After Drew won the 1943 election, Cosens began working out the commission's details with the premier, Minister of Lands and Forests W.G. Thompson, and Morrow. To Cosens the ideal arrangement saw him and Morrow at the commission's helm so that they could ensure the prudent management of the Crown forests and provide industry with the secure tenure to its fibre supply that it had long sought.[26]

The legislation Drew introduced in 1944 to create the commission would have achieved this end, and much more. The new body was not only authorized to do everything it deemed advisable 'to the best advantage of the forest resources of Ontario,' its creation would mark a new era in the administration of the province's Crown forests. 'The wide powers given the commission,' Drew contended in the Legislature, 'are to enable major changes to be made in the system of forest direction and control which will ultimately eliminate the Department of Lands and Forests as such, and bring all phases of its activity directly under the authority of the commission.'[27]

This news predictably set off howls of protest that were loud enough to quash the initiative. Frank A. MacDougall (2T3) had been appointed the DLF's deputy minister in 1941 and instructed to reorganize completely the department. He was naturally irate to learn of Drew's plans to replace it with a new commission. MacDougall viewed this novel departure, and specifically Cosens' role in conceiving of it, as nothing but 'a play' by industry to seize control of the province's forests. In his words, it had 'all the ear marks of a dangerous set up.' Others shared MacDougall's concerns, and it would have been politically risky for Drew to take the province that was known for its staid conservatism down such a radical path (the Department of Crown Lands predated Confederation, after all). In the end, Drew chose to form a royal commission in 1946 to investigate the state of Ontario's forests. After consulting Cosens on the matter, Drew appointed Howard Kennedy as its chairman.[28]

Notwithstanding Cosens' litany of influential political connections, he remained highly sensitive to the import of maintaining the appropriate appearance in dealing with forestry matters. The Ontario government was vulnerable to criticism in this area as it was still years away from pursuing a genuine silvicultural policy, and Cosens knew it. He was also acutely aware that political favour could be fleeting, and he should always be cultivating it.

Cosens demonstrated his sagacity in the midst of the Second World War. By this time, John C.W. Irwin (2T2) had established a reputation for being an outspoken critic of the Ontario government's lacklustre forestry policy. As a co-founder of the Clarke, Irwin and Company publishing house, he had shrewdly exploited his media savvy to carry out a series of high-profile attacks that had attracted significant public attention. He orchestrated one such incident in February 1943 when he delivered an address to the Engineering Institute of Canada (EIC). The *Globe and Mail* reported on Irwin's speech and incorrectly described him as a professor at 'the Faculty of Forestry, University of Toronto.' In Irwin's view, governments across Canada had refused to implement genuine forestry reforms because of their desire to retain control over the woodlands for patronage purposes, and he fingered Ontario as the worst offender. His general goal in speaking to the EIC, Irwin argued, was to win its support 'in the problem of preserving Canada's forests.' Considering that the report indicated that Irwin 'was roundly applauded by a big audience,' it is safe to say he had achieved his aim.[29]

Irwin's speech set off a wave of bipartisan protest, and Drew, leader of the Opposition at this time, was especially indignant. Drew immediately contacted Cosens about the article to express his anger at the unprovoked attack and inquire about Irwin's status at the faculty. Cosens explained that the *Globe and Mail* had clearly committed an egregious error in describing Irwin as a professor at the forestry school (Irwin was, however, a faculty graduate and member of the University of Toronto's senate). Drew asked Cosens to deliver the same message to C. George McCullagh, the *Globe*'s president and publisher, which Cosens did posthaste. The dean also wrote Drew and the DLF to apologize for Irwin's actions. The dean then penned a letter to Irwin, admonishing him for having delivered such an incendiary talk at this 'most inopportune time,' as the Faculty was in the 'process of negotiating with the Ontario government for an enlarged school forest area and buildings, and also a ranger course.' Drew then reached across the floor of the Legislature to join arms with Norman Hipel, the Liberal minister of lands and forests, to denounce Irwin.[30]

Cosens was determined to retain the politicians' backing because his goal during this period was to improve the state of forestry at least across the province of Ontario, if not the country. He saw the rising interest in and support for managing Canada's woodlands during the war as a double-edged sword. Granted, it would facilitate implementing effective measures, but it also created the very real possibility that government would feel compelled to bring in policies that were designed more to satisfy public opinion than lead to prudent forest stewardship. To obviate a future problem arising in this area, Cosens urged the industry to take the initiative. In a letter to one of the Canadian Pulp and Paper Association's senior officials in 1944, Cosens expressed his fear 'that unless the pulp and paper industry, and more especially the woodlands divisions, puts its house in order it will be forced so to do by measures that they will not appreciate.'[31]

To achieve this end Cosens was convinced that industry must hire a healthy complement of the country's best foresters, even if it meant luring them from government employment. The dean was so convinced of the efficacy of this approach that he even exhorted his competitors in Ontario's pulp and paper industry to adopt it. The Abitibi Power and Paper Company had suffered through an agonizing, fourteen-year receivership (1932–46), during which time its ranks of foresters had been decimated, and Cosens was anxious to see Abitibi rectify this situation. He told C.B. Davis, Abitibi's woodlands manager, in early 1947, that Abitibi was 'the most important privately owned company in the province.' As a result Cosens urged Clark to entice at least a few foresters who were working with the government 'to come over to industry. As I have repeatedly said to you,' Cosens asserted, 'I am very anxious that the Abitibi Company get a set of excellent men as your Company's very important position in the pulp and paper industry in Ontario calls for the very best we have.'[32]

Cosens thus used his influence to assist the industry in developing effective silvicultural practices. When the provincial government established the Ontario Research Council in 1946, for instance, Cosens was named chairman of its Advisory Committee on Forest Research. During the committee's first year of existence, Cosens ensured that its focus was regenerating the forest's most commercially important species.[33]

While Cosens endeavoured to further the interests of all players in the forest industry during his time as the faculty's dean, he acted first and foremost on Kimberly-Clark's behalf. It was no coincidence, for example, that these years saw the firm attract the faculty's best and brightest graduates to its workforce. Granted, this trend had begun even before Cosens had taken over the dean's office. During the Great Depression, Kimberley-Clark had bolstered its forestry staff while other pulp and paper companies had pared theirs. As it expanded its operations in

Ontario during the 1940s, the firm redoubled its efforts to hire not just any graduates but the cream of the faculty's crop each year. To achieve this aim, the firm pursued a prudent recruitment strategy. It offered the best first- and second-year students the opportunity to spend their summer internships with the firm in Kapuskasing. This afforded the company a chance to observe their performances in the field before offering them permanent employment. Furthermore, Cosens would steer the faculty's best students towards the company, as would 'Bob' Hosie (2T4), about whom more will be said below. Hosie would secretly feed mid-session exam results to George W. Phipps (2T6), Kimberly-Clark's woodlands manager in Kapuskasing, on the understanding that the information would be kept strictly confidential. This allowed Kimberly-Clark to build one of Canada's strongest forestry departments.[34]

Finally, Cosens also took steps to draw Hosie, who had been teaching at the faculty since graduation, into a much closer relationship with Kimberly-Clark. Hosie had been studying forest vegetation since his student days. While he had performed nearly all his work in this field on behalf of the public sector prior to the Second World War, during the conflict he carried out several projects for Kimberly-Clark. In 1942, for example, he oversaw a survey for the company of potential pulpwood supplies on Manitoulin Island and the Bruce Peninsula in Ontario. He continued working for Kimberly-Clark during the next few summers, and focused his attention on the silvics of the company's limits north and south of Kapuskasing. The product was a landmark report in 1945 that addressed the critical issue of regenerating spruce on these tracts, and it built on the work carried out by James B. Millar (3T1) and Ed Bonner (3T4). Hosie's opus became the cornerstone upon which Kimberly-Clark inaugurated the first sustained commercial-scale reforestation program in northern Ontario, an effort for which the company paid the entire cost. Hosie's work in this area culminated in 1953 with the publication of *Forest Regeneration in Ontario*, which chronicled all the studies in the province that had investigated how harvesting was affecting the future composition of the forest.[35]

In spite of these other ambitions, Cosens' primary goal during his years as dean was to translate his cozy relationship with the provincial government into tangible gains for the Faculty of Forestry. He made truly remarkable progress considering the very short time during which he occupied this post.

Cosens' first achievement – securing the 'University Forest' for the faculty – was one of his greatest, although credit for it was not his alone.

It was granted before he had officially been appointed dean, but his intimate ties to senior officials at Queen's Park were crucial to the process. Moreover, Cosens' connection to Morrow was also essential to the undertaking. So, too, was the desire by the DLF to acquire a practice facility where its growing ranks of employees could learn new skills and hone their existing ones. The initiative thus enjoyed the requisite political support. At the same time, Cosens, and the University of Toronto's senior administration, knew full well that his political pull may help create the forest, but paying the appropriate deference to the elected officials in Ontario would be crucial to managing it.[36]

The first sign of a breakthrough appeared as Howe was preparing to vacate the dean's office. Since the early 1900s, the faculty's backers had been arguing for it to own a tract of trees in which it could demonstrate the efficacy of forest management, provide practical instruction to its students, and undertake field research. In late 1939 Howe had reminded Cody, the university's president at the time, of the faculty's long-standing application to the provincial government for such a woodland, and noted optimistically that 'an opportunity has come sooner than we expected.' Serendipity had smiled in the form of several timber licences in Sherborne Township just below the southwest corner of Algonquin Park (this was a stone's throw from Carnarvon, the most recent location for the faculty's annual practice camps). The Royal Bank in downtown Toronto held the licences for the bankrupt Mickle, Dyment and Son Limited lumber company and was preparing to sell one of them that covered roughly 2,000 acres. Having learned of the impending sale some time before, Cosens had already begun discussing it with the bank and discovered that the asking price was $5,000. The one problem, Howe informed Cody, was that title would rest with the Crown if the faculty acquired the licence, but Howe was confident that the University of Toronto could negotiate to secure total control over the area if the faculty acquired it. Time was of the essence, Howe asserted, because other suitors would be anxiously waiting in the wings to acquire this timber in light of the period's rising demand for wood and this tract's relative accessibility.[37]

The faculty immediately investigated it and the surrounding area and opened negotiations with the DLF to obtain it, and the latter responded with uncharacteristic alacrity to the forestry school's plans. By early April 1940 Howe reported to Cody that the matter was well in hand. 'Professor Cosens and I had a very satisfactory interview [to discuss the forest] with Mr Cain, Deputy Minister of the Department of Lands and Forests, at his request,' Howe explained, and 'Mr Cain seemed anxious to do everything he could for us, in fact he suggested an extension of the area to include adjacent unoccupied Crown lands which would bring the total area up to fifty-six hundred acres.' Cain was not finished there. Although Howe and Cosens were not even going to broach the subject of tenure to the land, Howe was ecstatic when Cain 'volunteered the information that while it was a definite and long time policy of the Department not to relinquish title to any Crown lands, he [Cain] thought in this case a long time lease of one hundred years could be arranged.' The deputy minister added that he believed the Royal Bank's price was too high, 'and said he would use his influence to reduce it.'[38]

Cain was true to his word, but the government's unprecedented enthusiasm for helping the Faculty of Forestry created an unforeseen problem. In early April Cain coyly reminded the Royal Bank that the government enjoyed certain rights 'to reclaim or appropriate timber licences,' but it was committed to paying 'fair compensation' for the tract in question. Cain stated that $3,500 was a reasonable price, and the company agreed to it even though the property could easily have fetched more. The problem, Howe informed Cody in late April, was that Peter Heenan, the minister of lands and forests, had agreed to buy the licence on the faculty's behalf. This would have potentially opened a Pandora's box of problems, Howe asserted, because Heenan probably did not know of the DLF's long-standing policy regarding idle timber licences. 'In buying them up in the past,' Howe pointed out, 'the governments of the day got themselves into a peck of political trouble and so about fifteen or twenty years ago a departmental policy was established that such purchases should cease and I think it has been consistently maintained.' In order 'to maintain amicable relations and cooperation with the Ontario government regarding the forest,' Howe advised, 'we would need in particular the cordial cooperation of the Deputy Minister, a thing that perhaps we would not have if he were forced by his Minister to disrupt a long continued policy. As you know, ministers come and go but deputies go on for a life time.'

Cosens then stepped in to find a solution. He discussed the matter with Morrow, and together they arranged to have the faculty and bank engage in a contrived round of dickering over the price that the University of Toronto would pay for the tract. Not only was the deal quickly sealed, but Morrow fittingly saw to it that the board of governors charged the cost – $3,250 – to 'the Whitney bequest' (the money E.C. Whitney, the Ontario lumber baron, had left to the university). While the hoopla at the March 1941 annual banquet of the Foresters' Club was undoubtedly dampened by the sad reality that the faculty's ranks had been depleted by the call of war, Heenan's announcement that the faculty had, at long last, acquired its 'University Forest' was cause for celebration.[39]

Over the next decade or so, the University Forest grew in terms of its size and the range of activities carried out on it. What began as a 2,130-acre site in 1940 was, by mid-1945, over six times as large; it reached nearly seventeen thousand acres in the mid-1950s. Cosens, who used his own money to assist the project from time to time, knew that most of the territory that had been acquired while he was dean was 'high-graded' cutover. Nevertheless, he saw its potential, especially because the forest's size would allow for long-term, commercial-scale operations. As he wittily quipped to a fellow forester in 1944, 'quite a bit of it is little forest and a lot of acres. It will be worth more as years go on.'[40]

This was especially true in terms of drawing the DLF into a tighter relationship with the faculty. An integral part of Frank A. MacDougall's (2T3) effort to modernize the DLF was the creation of a professional cadre of field personnel whose members would receive part of their training at a 'forest ranger school.' At the January 1943 meeting of the Canadian Society of Forest Engineers, Peter McEwen (1T6), one of the DLF's regional foresters, argued that establishing such an institution was crucial to improving forest-fire protection in Ontario after the war. McEwen was soon charged with the task of planning the Ontario Forest Ranger School, developing its curriculum, and acting as its first director. From the faculty's perspective, the critical part was that the DLF built the school at Norah's Lake just south of Dorset on land within the University Forest (the main buildings were erected over the course of 1944–8 although the first 'ranger camp' was held in September 1943). The contract that the DLF entered into with the University of Toronto for the school stipulated that the former would pay for its capital costs, while the latter would rotate a number of professors through the facility over a few weeks each year to deliver the course material to the DLF's instructors and select employees. The Faculty of Forestry would also be able to utilize the facilities – at a nominal cost – when the DLF did not require them for its own training purposes.[41]

While the faculty benefited enormously from this symbiotic relationship – Cosens described the ranger school's facilities as 'second to none,' the university went out of its way to remind the Ontario government of the degree to which the latter relied on the former for the smooth functioning of the school in Dorset. This gave the Faculty of Forestry unprecedented leverage in dealing with Queen's Park, and the university's administration never tired of reminding the provincial politicians of this fact. In late September 1946, for example, W.G. Thompson, Ontario's minister of lands and forests, wrote to S.E. Smith, the university's president, to confirm that a few of the faculty's professors would be available to teach the DLF's senior staff that fall. On receiving the letter, Smith forwarded it to Cosens with a note that asked, 'what saith you?' Smith added that it would be 'easy for me to state that we should co-operate, but I know that you are anxious to do so, and I thought that I would seize the opportunity in my letter to stress the value of the co-operation.' And so Smith did. In responding to Thompson, Smith enthusiastically agreed to cooperate and explained that Cosens had arranged for two members of his staff to go up for each of the three weeks in October when the DLF's instructors would be at the ranger school. After pointing out that the university would cover all costs its professors incurred in getting to and from Dorset, Smith stressed that 'we are indeed happy to co-operate with your Department, as we have in mind the happy relations that have prevailed between it and the University.' In subsequent years, the university would repeat this message, each time underscoring the venture's 'co-operative' nature.[42]

Gordon Cosens also exploited his standing as the Faculty of Forestry's first and only 'industry' dean to generate significant financial support for it among his corporate colleagues. The school's inaugural cash prize, the Forestry Memorial Scholarship, was established in 1942 when an anonymous donor provided the seed money. Within short order, the fund had been augmented by several donations, one of which came from the White Pine Bureau. Two years later, the Canadian Lumbermen's Association established the Timber Research Fellowship in Engineering and Forestry. Then, in 1945, Cosens arranged for the Ontario and Minnesota Pulp and Paper Company to provide a series of bursaries that ran for six years and peaked at $3,000 in 1947–8. Finally, he convinced his own firm, Kimberly-Clark, to establish three fellowships (each worth $1,000 annually) at the faculty, and they were available for the first time in 1947–8.[43]

Cosens also took steps to protect his faculty's position as one of the two English-language forestry schools in eastern Canada. In the fall of 1944, Dr F.C. James, principal of McGill University, hosted a luncheon with a group of leading pulp and paper executives, including the president of the International Paper Company. During the meeting, James floated the idea of his university establishing a new forestry school. Morrow, one of Cosens' close associates, a member of the University of Toronto's board of governors, and director of International Paper, tipped Cosens off to the meeting just before it was held. This is all the time the dean needed. He informed Dr Avila Bedard, Quebec's deputy minister of lands and forests, in a confidential letter that he had compiled 'data on the [Toronto] Faculty of Forestry ... on two hours notice and passed it on by long distance telephone to the President of the International Paper Company who presented it to Dr James at the luncheon [at McGill] in Montreal.' Cosens argued that 'we here feel that in the east the requirements of forestry education are fairly well served by the three existing schools,' and Cosens reported to Bedard that his quick thinking had achieved its aim. 'Fortunately we were able to give a complete history of this Faculty,' Cosens was relieved to tell Bedard, 'and as a result temporarily stalled the introduction of another forestry school at McGill.' It is unknown what specific information Cosens passed along to his colleague in Montreal (it must have been most interesting because Bedard considered it 'very precious'), but Cosens assured Bedard that 'our negotiations with International Paper and Dr James were entirely confidential and the data has [sic] not been passed on to the University of Toronto.'[44]

Cosens was undoubtedly motivated in this instance by a desire to protect his forestry school's turf, but it appears another, more devious stimulus also prompted him to act. Cosens and Morrow were their respective firms' most prominent representatives in Canada, and both companies had been the darlings of the governments within whose jurisdictions they operated. They and the corporations that employed them thus had much to gain from working together to retain their unrivalled sway with the political authorities in eastern Canada. The creation of a new forestry school in this region represented a potential challenge to their hegemony. In heading it off, Cosens and Morrow undeniably served the interests of the Faculty of Forestry in Toronto with which they were both closely associated, an end that neatly dovetailed with protecting their private sector agendas.

Cosens' brief stint as dean of the University of Toronto's Faculty of Forestry, and the fact he served during the Second World War when its enrolment plummeted to but a handful of students (the class of 1944 consisted of one, Keith B. Lee), meant that his years at the helm produced very few graduates. Still, their backgrounds and careers reflected the trends that had begun years earlier.

This was true, for example, in the area of international connections. The Faculty of Forestry's students, for instance, continued to come from around the world. Alexander M. Kasturik (4T7) was a native of Czecho-slovakia, and Richard A. 'Dick' Shand (4T7) hailed from Barbados. A number of Cosens' graduates also became involved in international forestry work. Most notable was William G.E. Brown (4T7), who specialized in aerial photography after graduation and was the project manager for various efforts in this field in Africa during the 1960s.[45]

Pedigree also remained a major factor in determining who came through the faculty's doors. Ben Avery, one of Ontario's greatest foresters, was the father of Daniel D. Avery (4T6), and veteran Lakehead lumberman Don A. Clark saw his son, Donald C.E. Clark (4T7), become a forester. James G. Boultbee (4T7) was the brother of Richard Boultbee (2T9), and their father had a lifelong association with the Canadian lumber industry. Finally, Harold C. Larsson (4T2) graduated from the faculty five years after his brother, Oscar G. Larsson (3T7).[46]

Cosens' graduates also distinguished themselves in the forestry field. Several became senior officials in industry. Grant L. Puttock (4T7) spent his career working for Kimberly-Clark, and was the first forester to become its president in 1974. William J. Johnston (4T7) rose to become Price Brothers' vice-president of woodlands in 1965, and served in that capacity after the firm merged with Abitibi in the mid-1980s. Likewise, Duncan 'Dunc' Naysmith (4T6) spent his professional career with Abitibi and became its director of forestry in 1968. As for research, John R.F. Blais (4T5) and Kenneth 'Ken' B. Turner (4T5) distinguished themselves in the field of forest entomology. The former earned his doctorate in zoology from the University of Toronto in 1950 and spent nearly four decades working as an entomologist with the Canadian Forest Service. Turner earned his master of science in forestry, from the faculty, in 1951 and became the Ontario government's pioneer entomologist and lectured part-time at the University of Toronto in zoology and forest entomology.[47]

Two of Cosens' former students played central roles in spreading forestry education across Ontario. Harry D. Graham (4T5) helped establish the forestry technician program at Sault College in Sault Ste Marie in the late 1960s. Over the next two decades, he helped diversify it by offering streams in forest recreation and fish and wildlife, and he retired as its coordinator in 1985. Kenneth 'Ken' W. Hearnden (4T6) enjoyed a varied and colourful professional career that included reshaping the forestry education landscape in Ontario, but not before overcoming a few personal challenges. When he winced at his first taste of working in the bush as a summer intern in 1943, Cosens endeavoured to put Hearnden's melancholy into perspective: 'Are you sure your present opinion is not somewhat tempered by a combination of homesickness, flies and foul weather? You know it does affect one that way occasionally.' The dean sagaciously cautioned Hearnden against making a hasty decision, and Hearnden followed his advice. The latter went on to spend nearly two decades working his way up Abitibi's forestry ladder. In the mid-1960s, realizing that his outspokenness had cost him further advancement, he exited industry and joined the staff of 'Lakehead College' as a professor in its forestry technician program. Although much more will be said about this in the pages that follow, Hearnden distinguished himself thereafter as a major proponent of establishing a new forestry degree program in the north, a goal he achieved in 1971.[48]

Like many Faculty of Forestry alumni, several of Cosens' former students used their forestry educations as foundations for extremely successful careers in other fields. Reuben 'Rube' Schafer (4T6) was unrivalled. Like many other offspring of poor immigrants from Eastern Europe, he grew up in 'The Ward' in Toronto. He stood first in his class during each of his four years at the faculty, and his stellar record earned him the inaugural F.K. Morrow Forestry Scholarship and White Pine Bureau Scholarship. Eager to pursue graduate work in forest entomology, financial concerns forced Schafer to abandon this avenue after one year. In 1949 he set out to sell insurance, but the first six firms to which he applied rejected him. The California-based Occidental Life Insurance Company turned out to be lucky number seven. Over the next dozen years, he became its leading salesman in North America by transforming 'term' insurance into a mainstream product; over one five-year span he sold more than $26 million worth of policies. But Schafer's notoriety hardly ended there, as his outlandish behaviour continually made headlines in the 1960s. On one occasion, he invited to dinner in his home 'Popov the Clown' from the Moscow Circus when it was visiting Toronto, and Pierre Berton reported the evening's details in the *Toronto Daily Star*. On many others he pilloried feminists and marijuana smokers, conveying his attacks on billboards and in several satirical books (surprisingly, his works – *Now Show Me Your Bellybutton* and *How to Make Love to a Feminist* – never became best-sellers!). Schafer was arguably the faculty's most colourful graduate.[49]

William T. Foster (4T7) pursued a far more conservative path on the way to leaving his mark both in and outside the forestry field. He joined the DLF shortly after graduation, and worked his way up to chief of forest protection in 1968. In the early 1970s he was in charge of converting the 'Department' of Lands and Forests into the 'Ministry' of Natural Resources, after which he sat as its deputy minister. Foster retired in 1983 and then began consulting for the Canadian and foreign governments and industry on international projects from China to Chile. While serving as the chairman of the Metropolitan Toronto and Region Conservation Authority, he acted as a passionate spokesperson for the preservation of 'natural' areas within its jurisdiction and was the main impetus behind its *Greenspace Plan.*[50]

Foster's 'environmental' work reflected the degree to which the Faculty of Forestry's graduates were committed to doing much more than simply managing timber, a trend that Arthur H. Richardson also personified. After earning degrees from McMaster University during the First World War and Harvard immediately after, he worked for the DLF on its reforestation efforts in southern Ontario. In 1945 the faculty granted him one of its last 'forest engineer' degrees for his landmark report – *Ganaraska Watershed: A Study in Land Use* – that served as the rationale for establishing Ontario's network of conservation authorities.[51]

While Gordon Cosens continued to adhere to the Faculty of Forestry's long-standing refusal to admit women, he conceded that their day would come. His response to an inquiry in 1946 from Charlotte M. Cushnie, who hoped to pursue a career in forestry, reflected his view that the faculty's 'no women' policy made sense in the postwar circumstances. 'At present women are not accepted as students in the Faculty of Forestry at the University of Toronto,' he told Cushnie, but 'eventually no doubt they will be enrolled as positions in the profession develop which they could fill.' He stressed that the faculty was already pushed to its limits to find positions for returning servicemen, 'and it is hoped that the question of women taking the course will not become serious until the present and expected heavy enrolment of men from the services has become closer to normal.'[52]

Although Cosens' term as dean of the Faculty of Forestry was the shortest of any who held the office permanently (six years), the faculty made enormous strides under his leadership in several areas. Most important was the intimate bond he had forged between the faculty and the Department of Lands and Forests by inextricably weaving their interests together through ventures such as the forest ranger school. At the same time, the faculty continued to step very carefully in its dealings with the province, ever mindful of the tenuous nature of its relationship with its political master.

These dynamics were laid plain when the faculty turned to the matter
of choosing Cosens' successor. At least as early as March 1944 he had
learned that he was taking over 'limit operations' for Kimberly-Clark in
Ontario. Although he did not make this immediately known to the University of Toronto's senior administrators, the clock was ticking on his
tenure as the faculty's dean. In early 1947 Kimberly-Clark dropped an
offer on his desk that anyone would have found difficult to refuse. As
the assistant general manager of Kimberly-Clark's Woodlands Division,.
he would take charge of all the company's Canadian forestry activities.
The job came with a $20,000 annual stipend plus travel expenses, terms
that must have made his $7,000 dean's salary and $150 for expenses seem
paltry indeed. Kimberly-Clark also agreed to Cosens' request to continue 'giving scheduled lectures on logging subjects' at the faculty during
1947–8. This was Cosens' ideal job, and he submitted his resignation
letter to the president of the university shortly after receiving Kimberly-
Clark's proposal. With regret, Smith had little choice but to accept it.[53]

Smith was acutely aware that Cosens' term as dean had represented
a boon to the faculty, and he was determined to select the right successor to ensure the good times continued. In deciding on a short list of
candidates, Smith and the faculty whittled the choice down to three.
John L. Van Camp (2T2), who was teaching at Purdue University in Indiana, and Albert W. Bentley (2T1), who was working for Bowater in
Corner Brook, Newfoundland, were the 'external' candidates. Smith was
convinced, however, that it would be best to select an 'internal' one. As
Cosens explained to Van Camp in March 1947, 'the University was rather
over-emphatic about what had been accomplished in the last few years in
this Faculty and took the attitude that the present policies would probably continue if promotion were made from within. The position was
offered to Bob Hosie who refused it, and therefore Bernie Sisam is being
promoted.'[54]

The problem was that Sisam barely qualified as an internal candidate,
and this had profound implications for the more than two decades he
would serve as dean. His career will be described in much greater detail
in the next chapter, but at this point it is important to realize that since
1939 Sisam had been working as the deputy director of the Imperial
Forestry Bureau at Oxford, England. In the fall of 1944 the Faculty of
Forestry had offered him an associate professorship beginning July 1 of
the next year, but Sisam had been unable to meet this deadline. Difficulty in finding passage to Canada for him *and* his wife delayed their
departure until mid-November, which resulted in Sisam taking up his

teaching duties only in early 1946. By the time Cosens had cleared out the dean's office at the faculty, Sisam had been working in Toronto for a mere eighteen months. The problem was that there were no better 'internal' options. J.H. White was ill and had retired in 1946, and Hosie apparently did not want the job. The Faculty of Forestry's two new hires in 1946, Dave Love and 'Art' Michell (4T0), were even greener than Sisam. Consequently, the University of Toronto had achieved its aim of promoting to dean a member of the faculty, but it had chosen from a pool of relative neophytes in the field. If it expected the school's new head to enjoy the same cozy relationships as Cosens did with industry and government, and thus contribute further to the faculty's success, it was sadly mistaken.[55]

Nevertheless, in crowning the Faculty of Forestry's new dean, the university ensured that it brought the government into the discussion, and of equal importance, the government was keen to have a say in deciding the matter. Smith contacted Frank A. MacDougall (2T3), the DLF's deputy minister, about the issue in mid-March 1947, and asked to discuss it over lunch. MacDougall informed his minister of the invitation, and the import of the subject. 'The dean of this school is very close to the Department and the Government because the forests are all state forests,' MacDougall explained, 'and quite frequently the dean has to speak in public on forest matters ... [W]e have a very vital interest in who will head the school.' After their meeting, MacDougall provided his less than resounding assessment of the candidate whom Smith had endorsed. The deputy minister reported that the University of Toronto's president felt that 'the most logical man for the Deanship of the Forestry School was Mr Sisam. I believe his qualifications ... are adequate.' A few weeks later, Sisam was officially appointed the faculty's new head.[56]

Cosens resigned as a full-time member of the Faculty of Forestry in mid-1947, yet he continued to be one of its most ardent supporters until his death in 1969. In his resignation to Smith, Cosens had pledged to continue giving 'any assistance I can to the Faculty of Forestry and, of course, to this University.' He made certain that this was not simply a hollow promise. He sat as the faculty's representative in the university's senate until he moved to Neenah, Wisconsin, in 1952, and over the next few decades, he personally donated tens of thousands of dollars to the faculty.[57] He also left a large portion of his sizable estate to the forestry school to support its research efforts. With his typical modesty he insisted that his name not be 'included in the fund resulting from his bequest,' so his friends and fellow alumni came up with an ideal means of honouring his legacy. They named it the 'Rob Roy Fund in Support of Forestry Development' because, as Sisam explained to the University of Toronto's president at the time, membership in the fishing club of the same name was 'one of his [Cosens'] keen interests and great pleasures.' Furthermore, Cosens prodded his fellow Faculty of Forestry alumni and his employer, Kimberly-Clark, to follow his lead in opening their wallets to the faculty. In fact, Cosens never missed an occasion on which to foster support for his alma mater. Stuck in Cochrane overnight in 1950 because of 'rotten train service' from Kapuskasing to Toronto, he took the opportunity to discuss the matter of facilitating research into spruce regeneration with A.E. Wicks, the head of a local lumber concern. The upshot, Cosens joyfully reported to Sisam, was a $200 donation from Wicks (to be followed by a like amount for each of the next four years) that was dedicated towards pursuing this field of study that was so close to Cosens' heart and to which he devoted much of his professional career. It stands as a fitting synopsis of his profound and lifelong dedication to both forestry and the faculty from which he graduated.[58]

Chapter 6

'Today It Is Not always Ranked Professionally as First,' 1947–1957

When in 1947 John William Bernard 'Bernie' Sisam was ushered in as the dean of the Faculty of Forestry at the University of Toronto, he came as an accomplished researcher but a novice in many other domains that were critical to succeeding in his new position. A native of Nova Scotia and graduate of the University of New Brunswick and Yale, Sisam had begun working as a silvicultural researcher for the Dominion Forest Service in the early 1930s. His outstanding work had earned him a job in the spring of 1939 as deputy director of the Imperial Forestry Bureau at Oxford, England, where he was exposed to forestry developments throughout the world. While this was certainly a potential strength, when he came to Toronto in late 1945 he had never taught in a university, had no first-hand understanding of the political milieu in Ontario, and he had no connections whatsoever to the province's forest industry. Despite these lacunae on his curriculum vitae and the fact that he had served less than two full years as an assistant professor at the University of Toronto, the university promoted Sisam to full professor when he became dean in mid-1947 in an apparent attempt to improve his stature.[1]

Nevertheless, 'Bernie' Sisam had arrived in this post at what seemed like a most auspicious time. Canada's postwar economy was firing on all cylinders, and this was especially true of the forest industry. The country's universities were bursting at the seams from the postwar demand from both veterans and high school students for admission, and governments at both the national and provincial levels were generally inclined to provide the funding to accommodate them. Upon completing their post-secondary studies, most of these graduates were virtually assured of landing well-paid jobs, many of which would be theirs for life. In this epoch of good times, many of the issues university deans faced seem truly farcical in retrospect. In June 1951, for instance, Sisam and his staff decided that a fine of $7 per 'man' was sufficient punishment 'for misconduct of the group of Forestry students responsible for the water-fight at the Ranger School during the spring camp.'[2]

But underneath this veneer, Sisam – and foresters across the country – faced myriad challenges. Despite the substantial tax revenues that Canada's dramatic economic growth generated, its citizenry concomitantly called for a level of social services that placed ever-greater demands on public funds. In the ensuing competition for financial resources, forestry still ranked near the bottom. Unfortunately for Sisam, the university's administration shared this view and placed forestry far behind other priorities.

For Sisam and the Faculty of Forestry, other circumstances contributed to the challenges they faced, especially at the undergraduate level. In response to the unprecedented demand for post-secondary education, the University of Toronto grew during the late 1940s and early 1950s. Thereafter, it began planning for a massive expansion to accommodate the projected spike in enrolment that would occur when the baby boomers began applying for admission in the early 1960s. Unfortunately for the Faculty of Forestry, its student body began shrinking dramatically during these years. After peaking at 302 students in 1948–9, it was down to seventy-nine by 1956–7 (the incoming classes for these years were 61 and 19 respectively). With the foresters declining precipitously in relative importance on campus at this critical juncture in the University of Toronto's history, their claim to greater support from the administration carried little weight. At the graduate level, Sisam was able to lay a firm foundation for the faculty's program between 1947 and 1957, but he did so largely through his own efforts. The university only extended a helping hand to him in this area after Sisam resorted to shrewd means of pressing it to do so.

Sisam's term as dean began amidst a sea of hope for forestry in Canada. At the federal level, Ottawa had responded to provincial pressure in the 1940s by inching towards offering support for forest management initiatives. At a conference on resource conservation in 1945, the federal government endorsed the concept of managing the country's woodlands on a sustained yield basis and agreed to help fund provincial programs that aimed at compiling inventories of their forest resources. Four years later, Ottawa enacted what came to be known as the Canada Forestry Act. It called for the establishment of national forests and forest experimental stations, and created a framework for the federal government to share the costs for provincial projects in fields such as forest protection, silviculture, and forest inventories.[3]

At the provincial level, the signs were even more promising during the mid- to late 1940s. The province's forest industry was enjoying a boom period, as new mills appeared and existing ones augmented their facilities. The Ontario government introduced legislation requiring the pulp and paper companies to sign agreements that obliged them to submit, within five years, an estimate of their Crown timber holdings with maps that indicated their annual and long-range cutting plans. In addition, Ontario's Department of Lands and Forests (DLF) initiated in 1946 – for the first time in the province's history – a comprehensive forest resource inventory (FRI) of its commercial woodlands. These initiatives created enormous employment opportunities for foresters in both industry and government. Finally, Ontario's high-profile Royal Commission on Forestry (1946–7), chaired by Major-General Howard Kennedy, produced a report that called for the province to adopt measures that would see it operate on a sustained yield basis.[4]

Over the next few years, it seemed as though the Ontario government was genuinely committed to realizing this aim. As the data from the FRI rolled in the DLF's senior foresters – almost all of whom were faculty alumni – frequently met to discuss how best to manage, and not simply harvest, the Crown forest. Harold Robinson Scott, minister of lands and forests (1946–52), ostensibly supported this approach. On a visit through northern Ontario in 1949, he insisted that 'only a province-wide program of planned reforestation can save Ontario's billion-dollar pulp and paper industry.' He thus announced that 'by 1953, Ontario hopes to have in effect a reforestation scheme where every pulp operator will replant as many trees as he cuts down.' The DLF had already established its first northern nursery at the Lakehead in 1946, largely at the insistence of Peter Addison (2T9), and it seemed as though this would simply be but one small piece in a much larger puzzle.[5]

Sisam's initial involvement in these activities involved research. After he succeeded Cosens as chairman of the Ontario Research Council's Advisory Committee on Forest Research, he ensured that it continued to focus on how best to regenerate the most valuable species in the province's forests. It was at his insistence that the council asked 'Bob' Hosie, Sisam's colleague at the faculty, to review and report on all the regeneration surveys that had been carried out in Ontario, the results of which were published in 1953. Sisam also played a role in facilitating the cryptically named 'RC-17,' an unprecedented example of government and industry cooperation in studying the impact that logging was having on the forest's productivity (it was based near present-day Manitouwadge, Ontario).[6]

Soon enough Sisam moved into the political realm of the province's forestry movement through his chairmanship of Ontario's Advisory Committee on Lands and Forests. One of Kennedy's recommendations in 1947 was the creation of a body – composed largely of business representatives – to check the government's arbitrary power over the Crown forests and ensure policy continuity. With the Opposition in the Ontario Legislature continually harping on the Conservative government's dilatory approach to implementing Kennedy's prescription for what ailed Ontario's forests, Premier Leslie M. Frost decided to call its bluff in late 1950. He appointed the committee to advise the minister, and named Sisam its first (and only) chairman.[7]

From the outset, Sisam and Frost had diametrically opposed understandings of the committee's raison d'être. The dean saw it as a bona fide opportunity to apply his expertise to manage Ontario's Crown forests as well as possible. Sisam's perspective resonated perfectly with his predecessors' – deans Fernow and Howe and almost-dean Judson F. Clark – who had been asked to counsel the provincial state on this matter. Conversely, Frost saw the body's mandate as one that entailed instructing him in the art of maximizing the revenue generated from harvesting wood from public lands. Moreover, Frost viewed the creation of the committee as a strategic political move, one that would redound to his benefit in a variety of ways. As he candidly commented years later, establishing the committee represented a 'chance to put his enemies on the spot; to create a buffer for himself; and to get rid of repeated allegations that Departmental decisions were "motivated by politics."'[8]

For a while at least, it appeared that Sisam's view of the committee would prevail. Over the course of its first few years, he plunged headfirst into its activities. He was front and centre in consolidating the existing separate pieces of legislation regarding the administration of Crown forests into the Crown Timber Act of 1952. At the same time, the committee demonstrated its autonomy by issuing its own press releases without the DLF's approval. Furthermore, in August 1951, when Scott (the minister) sent an important directive to the major players in Ontario's forest industry without consulting the committee, it immediately reprimanded him for his summary actions and convinced Scott's successor, Welland S. Gemmell, to withdraw the edict the next year.[9]

But by mid-1952, if not earlier, Sisam was beginning to see that the promise of 'a new day' in forestry was still just that. Slowly but surely, the government, specifically the minister and premier, marginalized the committee. In late October 1953, for example, Frost dropped a major bombshell on it when he declared that the DLF would thereafter be responsible to a cabinet, not the Advisory, committee. Moreover, the government insisted that one of its officials, John A. Brodie (2T3), begin attending every meeting of the Advisory Committee, a move Sisam and his fellow members adamantly opposed. Then in 1954 the government refused to allow the committee to influence in a meaningful way the content of its White Paper, the first ever put before the Ontario Legislature. Considering that the document was a general statement regarding the province's approach to managing Crown lands, the members of the committee were justified in interpreting this act as a major umbrage.[10]

The chasm that separated the committee from the government was the latter's intransigence on the matter of reinvesting timber revenues in forest management. Almost from the time the premier had first struck the committee, its members had agreed that the government should pay for silvicultural measures because it owned the land on which the trees grew. The committee still felt this way even after the government, in 1952, formally made industry responsible for regenerating the forest.[11] Nevertheless, the government made its position on the matter perfectly clear when it increased stumpage dues explicitly to generate a pool of money to pay for forestry work but then refused to allocate the funds for this purpose. Like every previous government in Ontario since at least the turn of the century, the Tories of the 1950s felt it was far more expedient to allocate precious resources to politically popular activities such as road and school construction instead of forest regeneration. The government repeatedly delivered this message to the committee, and was especially blunt on the occasion of one of its gatherings in late 1955. 'The Provincial Government still depends on income from stumpage charges as a source of revenue to help pay for social services and some of the costs of other non-revenue producing departments,' the minutes from the meeting explained.[12]

It was thus hardly surprising that the Advisory Committee, although it continued to function until 1961, drifted into obscurity by the mid-1950s. The successive ministers retained it simply because, as Minister Clare Mapledoram remarked, it was still 'a sort of buffer between myself and the companies.' Sisam grew increasingly vexed by the committee's impotence, and pressed Frost to rectify the situation. The premier held

Sisam at bay with a healthy dose of platitudes regarding the committee's importance, and instead bent Howard Kennedy's ear when he wished to be advised on matters affecting Crown forests. Years later, Sisam lamented how, in reference to the committee, it seemed pointless to 'go on giving advice [with] nobody really paying much attention.'[13]

Sisam's travails with the Advisory Committee accurately reflected the attitude of both the provincial and federal governments, and the electorate, towards forest management. In the postwar era of rapidly expanding social programs and public expenditures on them, silviculture was barely on the radar. In his 1947 report Kennedy had presciently predicted that, 'unless the public is willing to spend large sums of money on forestry in the next quarter-century, efforts towards improvement, or even maintenance, of the present forest conditions, will continue to be little better than a gesture.' Sadly, even the gesture was practically inconspicuous.[14]

The story was much better for the Faculty of Forestry in so far as making progress with the Ontario government between 1947 and 1957 was concerned. Sisam earned a major victory with regard to the faculty's involvement with the Ontario Forest Ranger School. Under the terms of the original agreement with the DLF, the faculty had agreed to provide the teaching staff to lecture in their areas of expertise to the government's instructors and senior personnel. To carry out this obligation, the faculty had been forced to shuttle professors for several weeks at a time north to the Haliburton region. This had become an insufferable burden in the late 1940s for Sisam's teaching staff, as it was already overtaxed with the responsibilities involved in teaching nearly three hundred students on campus. In the fall of 1947 Sisam convinced Frank A. MacDougall (2T3), the DLF's deputy minister, to resolve the problem. Henceforth, the government and the University of Toronto would equally share the cost of hiring a faculty member whose full-time job would be to run the ranger school and manage the surrounding University Forest. For more than the next decade, Quimby Hess (4T0), Alexander D. 'Dal' Hall (4T8), and Norman L. Kissick (4T8) successively acted in this capacity.[15]

Sisam was also able to gain the DLF's cooperation in administering the University Forest. Because the property grew to roughly seventeen thousand acres during these years, the Faculty of Forestry was able to plan and carry out commercial harvesting operations, which it initiated in the mid-1950s. But under the terms of the agreement the University of Toronto had signed with the DLF for control over the forest, proceeds from the sale of the timber that was cut would first be used to reimburse the DLF for its costs associated with the property. Sisam scorned this provision because he felt revenue generated from the forest ought to be reinvested in it in the form of research and development. He approached MacDougall about the matter, and although the deputy minister was worried that such an arrangement could raise eyebrows with the government's timber licensees, MacDougall agreed to abide by Sisam's wishes. 'This, I feel,' Sisam told the University of Toronto's president in late 1954, 'is a step in the right direction.'[16]

Sisam was unsuccessful, however, in his campaign to convince the government to expand the accommodation available at its ranger camp so that the Faculty of Forestry could use the site more often during the school year. Despite Sisam's intense lobbying, he failed in this effort because neither the university nor Queen's Park would pay to augment the facility.[17]

In dealing with the University of Toronto during his first decade as dean, Sisam repeatedly learned that forestry was essentially an afterthought for the administration. This was a period of dramatic growth for many of the university's programs. The Faculty of Applied Science and Engineering, for instance, mushroomed. In the context of the Cold War, governments saw engineers – and the research they carried out – as integral parts of a successful strategy to combat communism, which translated directly into significant public funding for this field of study.[18]

Sisam understood that forestry hardly qualified as a matter of 'national security,' but he found his running dialogue with the university to be more frustrating than its lack of support for the faculty. Time and time again, the university's senior administrators asked him for detailed statements on his plans for strengthening the faculty, and the dean dutifully invested significant time and effort into drafting them. After submitting his designs, however, the university would seldom offer a prompt appraisal of the chances of – let alone a timetable for – realizing them; on a few occasions the administration did not even acknowledge his submissions. When the university actually met the faculty's needs, it did so through merely temporary arrangements that it made evident would come undone the minute it found a better use for the resources it had allocated to the foresters.

From the moment Sisam became the Faculty of Forestry's dean, his goals were to sustain its undergraduate stream and, more importantly, develop its nascent research and graduate program. He made this clear over the course of 1947 in response to one of the university's earliest requests for his faculty's projected financial needs. By late in the year, it was apparent that his undergraduate enrolment would remain high because unprecedented numbers of high school students (roughly 80) were applying to the faculty. Even if Sisam limited admission to sixty after 1948–9, as was his desire, his undergraduate student body would remain above two hundred for the foreseeable future (and well above that total in the short term). Since enrolment at this level was more than three times the faculty's historical average, he rightly pointed out that its existing building and staff were grossly inadequate to accommodate the undergraduates, let alone graduate students and research work. As far as the latter was concerned, Sisam felt that this should be the faculty's focus. He stressed that developing these programs would 'involve expenditure of money over and above that available to the Faculty under its present appropriation.' In particular, funds were needed to hire at least two new professors, one in the field of 'biological forestry' and another to plan and direct research and graduate work. In addition, the faculty would require new research equipment and laboratories (including a greenhouse) both on campus and at the University Forest, and money for travel allowances and fellowships for both staff and students. The overarching goal of this new work, Sisam concluded, would be to help 'solve some of the important fundamental problems of forest management in this Province.' He added that this matter was especially pressing because the four Canadian forestry schools were being criticized by industry and government officials for their meagre or non-existent research facilities and graduate programs, and these deficiencies were seen as major hurdles to managing prudently the country's woodlands.[19]

The University of Toronto offered Sisam limited support in grappling with his burgeoning enrolment, and went out of its way to look for signs that the Faculty of Forestry was withering. The university allocated $3,500 in the faculty's 1948–9 budget to hire the specialist in 'biological forestry' that Sisam had requested, but experts in this field could not be had for this salary. When the university only agreed to pay a new appointee a higher salary if Sisam found the money from within his budget, he balked and continued to operate short-handed. Over the next few years, the faculty's undergraduate enrolment began a rapid descent as the veterans completed their studies. After a peak of just over three hundred students in 1948–9, enrolment fell to roughly half that number in 1951–2. Still, this level of enrolment was double the pre-war average. Sisam felt that this attested to the popularity of his field of study, but the university's administration saw things differently. In the fall of 1951

Sidney E. Smith, president of the University of Toronto, brought the sudden contraction in the size of the Faculty of Forestry's registration to Sisam's attention. 'Do these figures indicate that the interest of young men in Forestry is waning?,' Smith asked Sisam. The dean jumped to his school's defence by explaining that the previous year's graduating class was one of the last of those that had been dominated by veterans. In contrast, he pointed out that 'the freshman class that entered the Faculty this September are 100 per cent High School students, and represent a 50 per cent increase over the largest entrance class in any year before the war.' Sisam added that the demand for foresters was still very strong and declared defiantly that, 'without reservation … interest in forestry is increasing rapidly in many directions including, of course, the young men in our High Schools.'[20]

On the graduate and research fronts, Sisam waged an intense battle to win resources from an administration that clearly doubted the logic behind supporting him. In pushing ahead in these areas over the next few years, he was hampered by more than merely the university's reluctance to assist him. The postwar boom had created such a demand for foresters that graduates were reluctant to continue their studies, and Canadian forestry students had traditionally pursued graduate work in the United States because there had been so few opportunities in this field in Canada. It would thus take time, Sisam argued, to break this pattern and convince them that they could and should earn post-graduate degrees north of the border. In Sisam's eyes, the time was ripe to do so because of Ottawa's decision to implement in 1949 the Canada Forestry Act, which promised funding for forestry research. The key to his campaign, he insisted, was demonstrating that his faculty could accommodate such work, specifically by building a greenhouse and laboratory (and the ancillary 'support' space) on the roof of the Faculty of Forestry's existing building on St George Street. The university's administrators rejected this idea, however, arguing it would be too expensive and aesthetically unpleasing. More importantly, the university's officials were at a loss to explain how Sisam could possibly believe the facilities were 'urgently needed' when the faculty had only three students in the School of Graduate Studies.[21]

Smith came up with a novel approach to meeting Sisam's request, and it reflected the university's feeling that space on campus was far too precious to allocate to the faculty. In the president's notes on the matter, he recorded how he 'told JWBS no money' and that the penthouse option was not viable. He also scribbled a dozen-line letter that represented at least faint hope for the faculty: 'how about ER Wood?' it asked.[22]

Judson F. Clark was Ontario's first Provincial Forester (1904–1906) and the best dean the Faculty never had. Here he is sitting on a white pine stump during a visit to a hardwood stand in Lincoln County in southern Ontario (ca. 1905). Clark declined the offer to be the Faculty's first dean after realizing that the provincial government was not interested in implementing meaningful forestry reforms. (Archives of Ontario, RG1-448-1, Box 110, Photo 828)

Even before the Faculty's establishment, its advocates had recommended it be given its own practice forest in which students could gain field experience. Because the wish was not granted, the Faculty turned to nearby Rondeau Provincial Park on the north shore of Lake Erie as a suitable site on which to carry out these activities. Its experience there became a harbinger of things to come. When locals learned of the Faculty's plans to do some selection cutting in the area, they pressured the government to cancel the project, which it did. Nevertheless, in the spring of 1908, the Faculty's first cohort undertook a number of forestry projects in Rondeau, including treeplanting. Pictured here are (l to r) T.W. Dwight (1T0), F.M. Mitchell (did not complete the programme), Isaac Gardiner (Rondeau's Superintendent), Lecturer A.H.D. Ross, Dean Fernow, and J.H. White (0T9). Both Dwight and White would later join the Faculty's teaching ranks. (AO, RG1-448-1, Box 110, Photo 3004.A)

Springs and summers spent working in the field as interns provided the Faculty's earliest graduates with a wide range of 'bush' experiences. The Dominion Forest Service employed Edward Beverly Prowd (1T5) as a timber cruiser in northern Manitoba in 1914, and here he returns to camp proudly holding what he described as 'our supper.' (UTA, A1972-0025/006[P006.094])

The First World War exacted an enormous toll on Fernow's 'boys.' In addition to those who returned with mental and physical injuries, the Faculty lost fifteen past and present students. This was an astounding total considering Forestry had annually graduated an average of fewer than five students during its first dozen years of existence (1907–1919). (UTA, A1972-0025/006[P.006.101])

Faculty graduates were on the vanguard of new developments in forestry. As Ontario's Assistant Provincial Forester after the First World War, C.R. 'Charlie' Mills (4T3), to whom the Faculty officially conferred his degree thirty-four years after he had first enrolled, played a central role in implementing the government's controversial policy of giving primacy to aircraft in forest fire protection. This 1926 photo of an HS2L Flying Boat, the biplane that was the workhorse of the government's air service at the time, was taken on Lake Ramsey in Sudbury. Mills is seated in the foreground to the right of William H. Finlayson, Ontario's Minister of Lands and Forests (1926–1934); two pioneering bush pilots, W. Roy Maxwell and C.J. 'Doc' Clayton, are seated left to right on the wing. Note the open cockpit in the aircraft's nose; it provided a bird's eye – albeit frightening – vantage point for fire detection, timber sketching, and aerial photography. (AO, RG1-448-1, Box 105)

Locating a forestry school in one of the country's largest urban centres often created some stark juxtapositions. Here, the Faculty's incoming cohort in the fall of 1926 is heading out on a dendrology field trip. While the students' formal attire was hardly unusual for those attending university at the time, the fact that several of them are sporting work boots with their Sunday best was. Moreover, they were probably the only forestry students in Canada who lined up to take public transit to do field work. While they may have been chagrined by this reality, they must have been thankful for the opportunity it afforded them to load up on the wares offered by the store pictured behind them; pipes were so prevalent on such trips that they seemed to have been mandatory equipment! l to r: A.P. MacBean, W.D. Start, J. Raeburn, G.R. Sonley, A.W. Leman, E.E. Grainger, W.O. Faber, F.N. Wiley, G.S. Andrews, A.B. Wheatley. (UTA, A1972-0025/003[66])

Despite making up but a relatively tiny part of the University of Toronto, the Faculty had a long history of demonstrating a school spirit that transcended its small size. As a result, it won a disproportionately large number of interfaculty awards, including the 1930 men's basketball crown. Back row, l to r: J.P. Townson (3T3), L.E. Simpson (3T1), F. Leslie (3T1), D.W. Gray (3T0); front row, l to r: J.B. Millar – Manager (2T9), W.D. Start (3T0), E.E. Grainger (3T0), F.N. Wiley (3T1), J. Raeburn (3T0). (UTA, A1972-0025/001 [P.003-090])

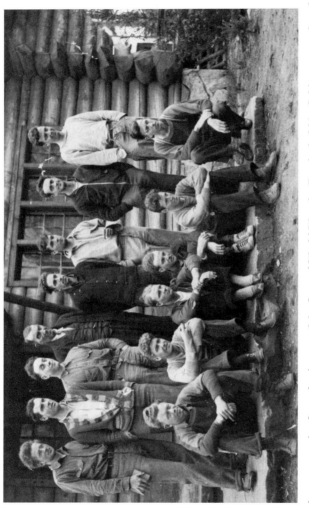

Still lacking its own practice forest by the eve of the Second World War, the Faculty held its field camps and workshops in a number of different locations. Although logging companies often hosted the forestry students, in the mid-1930s Forestry made use of Sherwood Forest Camp (near Carnarvon, Ontario), which was owned by alumnus Cecil H. Irwin (2T2). This is a photo of students at Sherwood in 1937 during the first scaling course offered in Ontario. Back row, l to r: R. Silversides (3T9), F. Sider (3T8), R. Hyslop (3T7), Mr. Duquette, G. Murchison (3T9), G.H. Bayly (3T9), R. Sexsmith (3T8), P. Ward (3T8); front row, l to r: J. Reynolds (4T0), Jim Barron (3T8), L. McConnell (3T8), F. Hick (3T9), T. Powell (3T7), R. Grinnell (4T0). (AO, RG1-448-1, Box 134)

The Faculty's swelled ranks after the Second World War galvanized its already renowned esprit de corps. Attesting to its enthusiasm was its prize-winning float in the All Varsity Parade on Home-Coming Weekend in mid-November 1948. Most of these students were members of the class of 1950. Front row, l to r: W.G. Dowsett (partially hidden), J.D. Stone, J.R.M. Williams, D.L. Lindsay, E.K. Dreyer; second row, l to r: C.H. Burton, J.G. Wiskin, W.A. Bartlett, R.H. Banks, W.E. Jonas, J.R Prendergast, E.J. Winters, V.G. Smith; top group, l to r: W.D. Ross, W. H. Hilborn, G.F. Coyne (standing), L.W. Booth. (UTA, A1972-0025/002[32])

It was only after Dean Gordon Cosens began exerting significant influence within the Faculty in the late 1930s that it was able to secure its own forest and first-rate accommodations for field work, the latter in the form of the Forest Ranger School at Dorset. Both are pictured here in 1953. (UTA, A1972-0025/003[07])

The Faculty made good use of the facilities at Dorset from the moment it acquired them. At the spring camp there in May 1949, Professor T. W. Dwight discusses the growth of white pine (visible in the background) with second year students (l to r) Kenneth A. Armson (5T1), John M. Anderson (5T1), and Michael Adam (5T2). Armson later joined the Faculty staff (1952–1978) and served as Ontario's third – and last – Provincial Forester (1986–1989), and Anderson enjoyed an illustrious career that included serving as President of the University of New Brunswick. (Photo Courtesy of Ken Armson)

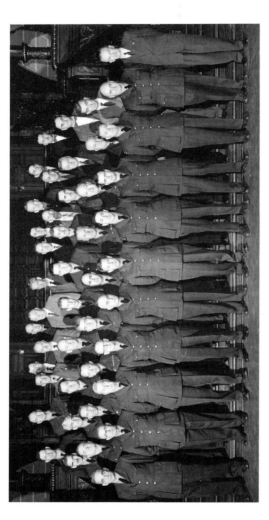

Throughout its existence, the Faculty's graduates rose to positions of prominence. After the Second World War, they came to dominate the upper ranks of Ontario's Department of Lands and Forests (even if they were unable to implement substantial forestry reforms). When the participants in the Department's Regional and District Foresters Conference on 'Land Planning for Recreation Use' posed for this picture in January 1954, eighteen of the twenty-three District Foresters and all six Regional Foresters were Faculty graduates; they are denoted by their years of graduation. Front row, 1 to r: I.C. Marritt (2T2), F.S. Newman (1T3), F.L. Simmons (1T5), W.E. Gimby (2T7), W.B. Greenwood (2T5), A.S. Bray (3T1), A.J. Herridge (4T9), G. Hamilton, J. Taylor, G.F. Meyer (3T2), C.M. Clucas; second row, 1 to r: A.B. Wheatley(3T0), A. Leman (3T0), D.N. Omand, W.E. Steele (2T7), H. Middleton (4T0), G. Phillips, L. Ringham (4T9), G. Ponsford; third row, 1 to r: R.D.K. Acheson (3T3), A. Crealock (3T2), J. Barron (3T8), P. Addison (2T9), J.M. Whalen, R.H. Hambly (4T7), F.L. Hall (4T2), R.D.L. Snow (2T8), F.E. Sider (3T8), R. N. Johnston, P. O Rhynas; fourth row, 1 to r: R. Boultbee (2T9), Q. Hess (4T0), R.V. Whelan, T.E. Mackey (2T6), J.F. Sharpe (2T2), F.W. Beatty, W.J.K. Harkness, J.S. McMillen, T.S. Yoerger, A.R. Carey; fifth row, 1 to r: A. Doerr, G. Delahey, J.M. Main, E.L. Ward (2T7), G. H. Bayly (3T9). (AO, RG1-448-1, Box 10)

Dr. Carl C. Heimburger (2T8) personified many of the themes that defined the Faculty's graduates, namely in terms of international connections, the achievement of excellence, and the diversification of their endeavours after the Second World War. Born in Russia, he earned undergraduate and graduate degrees in Europe and North America, including his BScF from the U of T. He then went to work as a research scientist for the Ontario government, specializing in tree breeding, particularly hybrid poplars. In this mid-1950s photo, he is tying bags over experimental pine grafts. (AO, RG1-448-1, Box 29)

Described by the *Globe and Mail* as one of the 'biggest moving jobs ever carried out in Toronto,' the U of T nudged – inch by inch – the Faculty's building a few hundred feet north over the summer of 1958; note the large square timbers that facilitated the process. While the Faculty believed this created a golden opportunity for it to expand its scope of operations, the university saw it simply as a means of making room for the engineers' new Galbraith Building. (UTA, A1975-0027/P)

One of the few benefits that accrued to the Faculty from its move 'up St. George Street' was the extension of its basement to include a forest products laboratory over the course of 1958–9. In this photo from the late 1960s, newly-hired Wood Science Professor Dr. John Balatinecz demonstrates the strength properties of wood with the aid of the Faculty's Tinius Olsen testing machine. (Faculty of Forestry, #526)

Erik Jorgensen came to the Faculty in 1959, but his profile rose dramatically in the early 1960s when he led the fight against the Dutch Elm disease and then 'invented' the field of urban forestry. The arrival of Vidar Nordin as the new dean in 1971 was a change that Jorgensen did not appreciate, and he departed shortly thereafter. He is pictured here with a log from an elm tree that was afflicted with the disease. The fungus that caused it is passed from tree to tree by elm bark beetles, and the holes that they made in this log have been marked with pins. (Courtesy of Erik Jorgensen)

Great optimism greeted the Faculty's move to its new home in the Earth Sciences Centre in 1989. At the official 'log-cutting' ceremony to mark the occasion are Adam Zimmerman (left end of the saw); and Bernd K. Koken, Chairman of the Board, Abitibi-Price Inc. (right end of the saw), Eric Turk (Vice President of the Foresters' Club) and R. C. 'Cam' Lewis (President of the Foresters' Club) respectively hold the far and near end of the logs (both graduated in 1991); George Connell (U of T's president) is between Turk and Lewis; and Rod Carrow (the Faculty's dean) is smiling over Lewis's right shoulder while watching the proceedings. (Faculty of Forestry, #347)

To commemorate the 'death' of the Faculty's undergraduate program in early 1993, the Faculty's students initiated a plan to plant a memorial tree. Deepa Tolia (9T4) is overseeing the ceremony, with Dean Rod Carrow the first person in the ring of onlookers to her left (the tree later died). (Faculty of Forestry, #355)

Rorke Bryan engineered an about face in the Faculty's direction, particularly in terms of creating the new Masters in Forest Conservation [MFC] program. He is pictured here (front row, fifth from the left) on the MFC field course in Sweden in 2000; this was the fourth and largest cohort in the program's history. Shashi Khan, Rorke's colleague and the Faculty's forest economist, is kneeling two spots to Rorke's left. (Courtesy of Rorke Bryan)

Bryan was integral to cementing new international connections for the Faculty. He is pictured here at the Nanjing Forestry University in China in 2000 shaking hands with Dr. Shinyuan Yu after signing a partnership agreement that included a major CIDA-funded project on carbon sequestration and forest management in China. Yu is also a double alumnus of the Faculty, having earning both his M.Sc.F and Ph.D. there. (Courtesy of Rorke Bryan)

Although the Faculty no longer controls its own practice forest, it has entered into a fruitful relationship with the Haliburton Forest, which is under the auspices of Dr. Peter Schleifenbaum. The site is now host for the Faculty's annual fall camp, as well as a host of other teaching activities and staff research projects. Here, MFC students David Richardson and Kristi Faccer enjoy Haliburton's Canopy Walk in September 2001. (Faculty of Forestry, #661)

In many respects, the termination of the undergraduate program took the wind out of the Faculty's sails, but it gradually bounced back. Over the course of 2006–7, two of its students, Adam Kuprevicius and Tylor Peet, decided both to restart the U of T Timbersports Club, which provides an outlet for students who are keen to participate in traditional lumberjack activities, and to enter a team in events in which forestry students from across North America battle. Here is the Club's squad for the Fleming Competition in Lindsay, Ontario, in November 2007. (Photograph by Jorge Solorzano Filho and courtesy of Jessica Kaknevicius)

The Faculty has grown in many ways during its first century of existence. After excluding women for nearly six decades, they are now a major component of both its student body and professorial ranks; it also offers programs that touch all corners of the planet. In 2007, Dr. Sandy Smith, the Faculty's entomologist, helped lead the new Tropical Forest Conservation course that the Faculty now offers during reading week to qualified undergraduate students at the U of T. Smith is pictured with Elvis Stedman at the field station in Dominica. Dr. Sean Thomas (the Faculty's Canada Research Chair, Forests and Environmental Change) is to the left and PhD student Adam Martin is on the right. (Faculty of Forestry, #47)

Events held in connection with the Faculty's Centenary celebration in 2007 facilitated members of the 'Old Guard' reconnecting with their alma mater. As they always had, each graduating class remained tight knit. Here are some of the surviving members – and their spouses – from the 'Class of 1951' photographed in Hart House's Great Hall at the spring reunion in 2006. Front row, l to r: William Cleaveley, Nan Collict (wife of the late Fred Collict), and Louis J. Nozzolillo; back row, l to r: John Larke, Paul Masterson, and Robert Bourchier. (Faculty of Forestry, #852)

Edward Rogers Wood was a member of the legendary clique of Canadian entrepreneurs who had ridden the economic boom at the turn of the twentieth century to the highest rung of the country's social ladder. He and his wife, Agnes Euphemia, took up residence in 1902 on a two-acre property in downtown Toronto – which they named Wymilwood and which was renamed Falconer Hall when they vacated it – that they leased from the University of Toronto. During the early 1920s, they began laying out and building an estate on a glen on the west branch of the Don River at a time when this area was on Toronto's farthest outskirts. They named it Glendon Hall (or Glendon for short), a title that was, and still is, used to describe their property and the stately manor they built on it. During these early years and after, the Woods turned these grounds into a mixture of woodlands and parklands that were recognized then and now for their natural splendour. E.R. Wood died in 1941, and eight years later, 'Pheme' amended her will in order to bequeath the property to the University of Toronto's governors.[23]

When Mrs Wood died a short while later, the university's attention turned to the wording of her bequest, which was so vague that it gave the university enormous latitude in handling the land. Her will described how it was her wish that the university use the property 'in connection with the work of the Department of Botany of the University of Toronto, or in such way as it may think best for the purposes of the University, its teaching staff and students.' If it became 'impossible, inadvisable or impractical' to do so, the document explained that the board of governors should exercise its discretion in using the land 'for the best advantage of the University for other purposes consonant with the spirit and intention of this gift.'[24]

Smith informed Sisam in December 1949 of the possibility that the Faculty of Foresty could use Glendon, but the dean was certainly not the only party at the university to eye the property. For several reasons, most importantly the wording of Mrs Wood's will, the Department of Botany felt that Glendon should be its domain. Since 1928 botany professor Robert Boyd Thomson had been working in the area with the goal of creating a 'scientific garden and arboretum.' There was, however, also the Department of Zoology to consider, as it envisioned Glendon as a site on which to showcase its research activities. Likewise, the School of Graduate Studies felt that Glendon was an idyllic spot on which to accommodate its exploding ranks of staff and students. Finally, the Department of Civil Engineering (in the Faculty of Applied Science and Engineering) had already begun using Glendon to teach its students surveying and geodesy (considering the engineers stashed small caches of radioactive waste at Glendon in the mid-1950s, it was probably a good thing they did not end up with control of the site!).[25]

Outside interests also thought Glendon would fulfil their needs. Health officials supported building a new hospital there, while the Ontario College of Art saw Glendon as a great spot on which to build its new home (the college was already leasing the main house). And finally, the Ontario government wished to replace the Toronto Normal School by constructing a new College of Education on the property.

Forestry rode botany's coattails to gain permission to use Glendon at this time, but the University of Toronto saw the Faculty of Forestry's claim to the property – and the botanists' for that matter – as being tenuous at best. Botany definitely had the strongest claim on Glendon, but the property was considered far too large for only one department to use it. Moreover, its greenhouses, woodlands, and tract of level land atop the valley slope (ideal for a laboratory and nursery) made it logical for the university to allow forestry to share in its use. Not only would this satisfy Sisam's long-standing request for these facilities, but it would do so at a fraction of what it would have cost the university to provide them downtown. As a result, over the protests of the vanquished suitors, Simcoe Hall (i.e., the university's senior administrators) decided in the fall of 1950 to give the Department of Botany and the Faculty of Forestry the opportunity to use the grounds and asked each to prepare short- and long-term plans for them. In discussing the matter internally, however, the university's administrators stressed that the two were granted use of the property only on a 'tentative basis.'[26]

Forestry and botany immediately began despatching proposals to the university's administrators and turning Glendon into a hive of activity. Within a few months, they had spent the small grant (under $5,000) that Smith had provided in late November to make the necessary repairs and alterations to the greenhouses there. Even before these improvements had been made, two of forestry's graduate students had begun working at Glendon, and in February 1951, the Faculty of Forestry and the botany department submitted a joint strategy for their 'permanent' occupation of the site. They estimated it would require $24,000 annually to operate the facility, a figure they contended would allow for 'a scale of operation that is adequate but not ample.'[27]

For Sisam, acquiring Glendon was sine qua non to developing his graduate and research programs. Certainly, the Faculty of Forestry had opportunities to carry out these activities at the University Forest and in the laboratory at the ranger school at Dorset, but these sites were remote from Toronto, the hub of the province's power wheel. In Sisam's mind, industry and government would be far more willing to support the faculty's research and graduate work if it was known that these efforts were *already* under way within range of the provincial capital. He was thus convinced that Glendon represented the missing and anxiously awaited 'calling card' that he required to solicit outside funding for his initiative in this domain.[28]

And so, Sisam immediately began publicizing the Faculty of Forestry's good fortune in securing this site. He informed the Ontario Forest Industries Association about this breakthrough in January 1951, and five months later, he excitedly described to S.E. Smith the new 'University of Toronto Forestry Series' of booklets he was set to begin publishing. It would report on the faculty's research projects at Glendon, and the initial volume stressed the seminal role this property was playing in these efforts.[29]

By February 1952 Sisam had mapped out a course of action for converting news of the faculty's work at Glendon into financial support from industry, one that the university backed. His top priority was to secure funding from the province's pulp and paper industry for a chair or professorship in 'biological forestry research,' and he was advised by the industry's leaders to pursue the matter through the OFIA's entire membership. In doing so, he shrewdly noted that the industry in New Brunswick had recently pledged its support for a chair in logging at that province's forestry school. Sisam estimated that $10,000 per year would be needed to fulfil this need, and that the position should be guaranteed for at least one decade to give it adequate security. Smith wholeheartedly encouraged Sisam to 'make the Glendon Hall story known' if it would assist in attracting research dollars. When the Faculty of Forestry published in February 1952 its first *Bulletin*, which contained Hosie's summary of regeneration surveys in Ontario, Smith congratulated the dean on the accomplishment, and poked fun at his own limited grasp of forestry nomenclature. 'I learned,' Smith informed Sisam, 'that "stocking" has nothing to do with wearing apparel.'[30]

Smith's jocularity belied, however, his and the university's true perspective on Glendon. The university's president was already actively entertaining – even soliciting – ideas for alternative uses for the property. In January 1951, for example, Smith had informed Harold A. Innis, dean of Graduate Studies, that the university was 'not committed' to a use for Glendon and that he favoured 'it as a centre, residential and otherwise, for the School of Graduate Studies.' By mid-1952 the University of Toronto had reached an understanding with the Ontario government regarding leasing the latter a large section of the property for its new Ontario College of Education (OCE). Later that year, W.E. Phillips, chairman of the board of governors, quietly asked Hamilton Cassells, the University of Toronto's solicitor, for advice on the matter. Cassells provided a pithy analysis that argued that the university would be wise to sell the land on the open market and consider using 'some of the funds for the work of the Department of Botany having regard to Mrs Wood's wishes as expressed in her will. If the property were retained for University purposes or leased for a long term of years,' he cautioned Phillips, 'I think constant trouble would have to be faced in the years ahead.'[31]

Once Sisam and his colleagues in the botany department discovered the University of Toronto's covert schemes to dump Glendon, they dug in their heels to defend what they now felt was their turf. In particular, they pleaded with Simcoe Hall not to assign part of the property to the OCE as it would sound the death knell for the work they were carrying out there. Their fear, and considering the evidence it was well founded, was that they would be 'gradually but for practical purposes completely displaced' over time. For this reason, they asked that another location be found for the OCE, one that would 'not involve the impairment, not to say the destruction, of one educational and scientific institution in the cause of establishing another.'[32]

It was during this pitched battle that Sisam was driven to lambaste the University of Toronto for the meagre and conditional support it had offered the faculty since he had been appointed dean. When the president directed Sisam in early 1952 to make some deep cuts to the Faculty of Forestry's budget, slashing that Sisam felt would have a fundamental impact 'on the future development of the Faculty,' the dean sent the administration a stinging missive. In Sisam's eyes, the university's fiscal stringency in dealing with his faculty was causing it to lose its pre-eminent place in Canadian forestry education. 'I do not view any of them with much favour,' Sisam declared defiantly to Smith in reference to the cutbacks he was forced to make at the time. 'In a sense,' he continued,

each of these proposals is a negation of the principle of progressive development on which it was hoped the future policy of the Faculty might be based. In this connection and having certain pertinent factors in mind, I further suggest that the time has come to raise our sights with regard to the financing of forestry education in this Province, and that for the future, rather than relating the forestry budget to the minimum requirements, every effort should be made to increase the appropriation of this Faculty so that it may meet its legitimate requirements for expansion and be worthy of its place in the University of Toronto. Here I might point out that although ours was the first forestry school in point of time to be established in this country, today it is not always ranked professionally as first. You will realize that this is a matter of some concern to me. And I may say that it is one which I would like to remedy as quickly as possible.

Sisam then delineated the specific reasons for his discontent. The Faculty of Forestry's undergraduate program was now averaging 150 students compared with forty-four prior to the war. Nevertheless, Sisam pointed out that 'the number of senior members [on staff] was five in 1930, and has remained at that total throughout.' Moreover, whereas the faculty had done relatively little in the way of graduate or research work before the conflict, he described how, 'at present, there is considerable activity in both these directions.' This exacerbated his understaffing problem, and he pleaded for a permanent professor 'to assist with the biological subjects.'[33]

Sisam's tirade – and a touch of serendipity – undoubtedly contributed to him achieving a few goals at this time. To buttress his teaching corps, the university agreed to hire Kenneth A. 'Ken' Armson as a lecturer. More importantly, when negotiations between the university and the Ontario government regarding building the OCE at Glendon broke down in early 1953, the University of Toronto granted the forestry faculty and the botany department $15,000 to support their work there during the next academic year. Although it was much less than the annual grant of $24,000 for which they had asked, they greatly appreciated the university's largesse. Then in mid-June, Smith informed botany and forestry that he could now 'express assurance with respect to the continuity of our retention of Glendon Hall. Therefore the Faculty of Forestry and the Department of Botany are justified in planning long-term development of research undertakings.'[34]

Even with this good news, Sisam was still unable to entice industry into opening up its wallet to his faculty. After his appeal to the OFIA in early 1952, he met intermittently with its directors to lobby them for financial support. 'But membership in this association,' Sisam explained disappointedly to Smith in the spring of 1953, 'includes many small operators and wood exporters, who are interested mainly in short-term forest exploitation with little concern in the problems affecting long-term management.' This was a major problem because this group represented by far the OFIA's largest block of members. For this reason, Sisam believed it would be best 'to go back to the original scheme of working through a few large companies.' Although these firms periodically gave him reason to believe a breakthrough was in the offing over the course of 1954–5, he came away empty-handed.[35]

The matter became all the more pressing for Sisam in early 1955. At that time, the Ontario government announced that it was terminating the Ontario Research Council as of 31 March. As chairman of the council's Advisory Committee on Forest Research practically from its inception, Sisam was devastated by this news. He immediately protested to the premier about this summary decision, especially because the council had handed out over $10,000 annually to several faculty members to support their silvicultural studies.[36]

With his faculty's fortunes at a low ebb, Sisam was in a position that his predecessors as dean had known full well and for which they had developed a particularly effective strategy that had never failed to gain the undivided attention of the university's administration and the provincial government. Frustration over their litany of unanswered requests for help had driven these erstwhile deans to produce, at the critical moment, an offer of employment from an outside party. In comparison with their paltry academic stipend, the extracurricular salary and benefits always appeared practically irresistible. Perhaps more importantly, the very notion that one of the University of Toronto's own would abandon it for greener pastures was a thought that repulsed its senior administrators. The upshot was inevitably an effusive – albeit temporary – display of affection and support for the Faculty of Forestry. Feeling like he was in dire straits, Sisam desperately needed just such an offer that he could wave in front of his bosses' noses. Fortunately, one came just in time.

In the fall of 1955, just as Sisam was at wit's end with the University of Toronto, the province's forest industry, and the Ontario government, lady luck shone on him, and he took full advantage of the opportunity presented. Lincoln R. Theismeyer, president of the Pulp and Paper Research Institute of Canada, apparently approached Sisam with a job offer that included a salary that dwarfed that which the University of Toronto was paying him. Sisam ensured that news of this offer reached the university's administration; he apparently personally delivered a copy of it to the president! Whether Sisam had a bona fide interest in leaving Toronto is unclear, but he certainly was not the first dean to use this ruse to pry from the university and the provincial government the support that he sought.

And support he got. After discussing the matter with officials at Queen's Park, they agreed to provide $5,000 annually for the faculty's research and reinvigorate the Advisory Committee of which Sisam was chairman. When the matter was raised with the university's board of governors, it approved increasing his salary in the new year by nearly 30 per cent (from $11,000 to $14,000) and his expense allowance by 200 per cent (from $400 to $1,200). Sisam also received a promise from Phillips, chairman of the board of governors, to tap the latter's 'associates' and 'endeavour to raise money from the forest industry for research within your Faculty. I know that he has in mind,' Smith told Sisam, 'that he will seek forthwith a contribution from a company in which he has an interest; that should be an example to other companies.' Smith also promised that the University of Toronto would endeavour 'to solicit from the Provincial Government, particularly through the Ontario Research Foundation [which had replaced the Ontario Research Council] and through the Department of Lands and Forests, and to solicit also from the Federal Government, grants for research.'[37]

Sisam's thank-you note to Phillips and the board was significant, for it stressed the metamorphosis he was endeavouring to effect in the Faculty of Forestry. 'Inevitably,' his letter began, 'there are growing pains associated with the progress of a relatively small unit of a large university from what has been almost exclusively an undergraduate program to the higher and more specialized levels of university work. Requirements of staff, equipment, accommodation, and sustained financial backing are all involved,' he continued, adding that 'there is, however, an urgent need that greater attention be given to these aspects of forestry education in our university.'[38]

Phillips' commitment to push his business contacts to cough up research funding for Toronto's forestry school was crucial to achieving this end. In the context of postwar Canada, Phillips was as well connected as any businessman. He had married the daughter of R.S. McLaughlin (whose family's carriage-manufacturing business in Oshawa was the forerunner to General Motors of Canada), and was a remarkably prosperous industrialist in his own right before and during the Second World War. Thereafter, he had teamed up with E.P. Taylor, who was already well on his way to amassing an awesome stable of companies. In association with a select few other leading Canadian entrepreneurs, they founded the Argus Corporation, Canada's prototype 'dynamic growth' holding company. In the buoyant years after the war, 'Argus Corp' was, as Michael Bliss puts it, 'a fabulous success.' Furthermore, during the 1950s it acquired control over Dominion Tar and Chemical Company (now Domtar) and several major pulp and paper mills in Quebec. Phillips was thus a Canadian tycoon with a vested interest in and intimate familiarity with the country's forest industry.[39]

By the time Smith had told Sisam about Phillips' plans, Phillips had already called on his associates to the Faculty of Forestry's advantage. Back in early November, he had spoken to Leslie M. Frost, the premier, about the possibility of the faculty losing Sisam, and convinced Frost to provide the $5,000 annual grant to fund the Faculty of Forestry's research program. Phillips had also advised M.W. McCutcheon, the vice-president of Argus Corporation, to avail himself of his contacts at the Canadian Pulp and Paper Association (CPPA) to generate additional research funding for the forestry school in Toronto. McCutcheon soon reported that he had obtained 'an undertaking [from the CPPA] to contribute $10,000 a year to the University for the advancement of the research program in the Faculty of Forestry.'[40]

In late November Phillips set his sights on and secured a similar donation from a more local source. He met with Douglas W. Ambridge, the august president of Abitibi (Ontario's largest pulp and paper maker) who was already keen to support practical research related to his firm's activities. Not only did Ambridge agree to donate $10,000 annually over ten years to establish a research chair in forestry at the faculty, he made it clear that 'his Company would not desire to control in any way the activities or work of the incumbent.' Two days later, Ambridge informed Sisam of Abitibi's generosity.[41]

And so, over the course of a few weeks in November 1955, the fortunes of the faculty had taken an about-turn. No sooner had Sisam dangled the prospect of leaving Toronto in front of the Ontario government and the University of Toronto's senior administration than they had swooped in to revive the languishing spirits of the dean and his forestry school. At Queen's Park, he had drummed up a commitment to a major annual research fund grant and more respect and authority as a member of the province's Advisory Committee. At the university, he had received a huge raise at a time of budgetary restraint. Moreover, the university had promised – and promptly fulfilled its pledge – to obtain for the faculty one of the fundamental building blocks to a graduate and research program that it had long sought: long-term, substantial funding. Finally, the University of Toronto had committed – in writing no less! – to solicit more funds for the faculty from the Ontario and federal governments. When the dust had settled, Sisam could hardly contain his exuberance over these recent developments. 'I am much looking forward to the next few years in the Faculty of Forestry,' he told Smith at the time, a sanguine outlook that the president apparently shared.[42]

As Sisam was basking in the glow of the good news over the course of late 1955 and early 1956, he learned that some of the promises that had been made to him and the faculty were just that. The Ontario government had pledged and did resuscitate the Advisory Committee to the Minister of Lands and Forests for a while in early 1956, but just as quickly it withered away again and was beset by rigor mortis. As far as the CPPA's $10,000 annual grant went, Sisam had decided that the money could be best used to establish a program, and not simply a chair, in forest economics. No sooner had he made this decision than he learned that the funding had never existed in the first place. McCutcheon informed the University of Toronto in February that, indeed, he had spoken to senior members of the Canadian Pulp and Paper Association about the possibility of contributing to the faculty's research program, and they had supported the idea. But, he explained, 'the Association has never to my knowledge made any commitment in this regard.' And it would not.[43]

Dean Sisam was disappointed but not deflated by this turn of events, and even though his attempts to find alternative funding for the forest economics program proved abortive, things were still looking rosy. In early 1956 the university received the DLF's first instalment of its annual $5,000 contribution to the faculty's research program. In expressing his gratitude, Smith contended that he was 'confident that we can build within the University of Toronto a Faculty of Forestry second to none' and that the Ontario government's support was an essential part in the 'writing of a new chapter in the Faculty of Forestry.' Around the same time, Abitibi's directors officially approved the $100,000 grant to the faculty ($10,000 annually for ten years). Writing Ambridge to thank him for Abitibi's munificence, Smith underscored its significance to the forestry school. 'You and your associates could hardly estimate what a "shot in the arm" you have given to the Faculty of Forestry in the University of Toronto ... We hope to establish one of the best schools of Forestry on the North American continent. Dean Sisam is delighted by the developments, and I know that he has no misgivings about the wisdom of his decision to continue to give leadership to this field, which is of such importance to our provincial and, indeed, national welfare.'[44]

In hiring the first – and, as it would turn out, the only – Abitibi Chair of Forest Biology, the faculty selected wisely. J.L. 'Jack' Farrar (3T6) had earned his master's and his doctorate in Forestry at Yale University and spent his entire working career at the federal government's Petawawa Forest Experiment Station. He was its chief research officer when the faculty lured him to Toronto in 1956. He was also a renowned perfectionist, and a demanding mentor who expected his understudies at the Faculty of Forestry to aspire to his high standard (this probably explains the 'Dr Death' moniker he earned while working in Toronto!). Farrar had turned down Cosens' offer to join the Faculty of Forestry roughly a decade earlier because he felt he was a poor lecturer, but he had no second thoughts about accepting the Abitibi Chair that seemed tailor-made for him.[45]

Although Sisam was clearly taking the Faculty of Forestry in a new direction during his first decade as dean, the same themes that had marked its earlier days still defined its development. Its international connections, for instance, which had begun with the appointment of a 'foreigner' as its first dean, continued. This was especially true of the faculty's student body during the 1947–57 period. The faculty still boasted the usual complement of those who had been born in the United States, the United Kingdom, India, the British West Indies, China, and Norway. But the roll call grew to include natives of Thailand and Holland, and a variety of political refugees from Europe. Many of them were former Displaced Persons whose countries of origin were now behind what Winston Churchill had famously called the 'Iron Curtain.' Considering that the forestry heritage in their native lands dwarfed Canada's, it is hardly surprising that so many émigrés from Estonia, Hungary, Poland, Czechoslovakia, Lithuania, and Latvia ended up studying at the faculty.[46]

Two students went against this trend of leaving a communist world for a capitalist one, and made landmark contributions to forestry in their homeland. Chung-Li Huang and Tung-Kung Yuan came to Toronto in 1947 to pursue master's degrees in forestry (at the time they constituted two-thirds of the faculty's graduate program). Although China was still nominally under Chiang Kai-shek and the Nationalists, over the next few years Mao-Tse Tung and the Communists gradually established total control over the country. With the Nationalists' world collapsing around them, in September 1948 they recalled the students whose advanced studies they were supporting in North America. Huang and Yuan not only defied the edict and completed their programs in 1949 and 1950 respectively, they chose to return to Maoist China and begin careers as forestry professors, careers that would span nearly four decades. In 1986 Zhang Xiangping, an aspiring forestry student from Beijing, wrote the faculty to inquire about its graduate program and explain his inspiration for doing so. 'I long to study in Toronto,' Xiangping's passionate letter explained, 'with deep feelings for my teacher Huang Zhongli (or Chung-Li Huang) who had been head of the Department of Forest Management in the Chinese Academy of Forestry and died in 1983 and had gotten the master [sic] degree in the University of Toronto early in 1950s and his fellow Zhu Zhao-hua got the Man of the Trees Award in 1985.'[47]

Other students from this era became involved in forestry the world over. Soon after graduating, Alan R. Turner (4T9) went to work on aerial surveys on Africa's Gold Coast. Likewise, John H. Hewetson (5T1) began managing the Firestone Rubber Company's plantations in Liberia in 1953 and remained there until well into the next decade. Other alumni waited until late in their professional careers to venture overseas. Robert M. 'Bob' Dixon (5T0) was instrumental in developing a simple and cheap sampling method for the Ontario government's inaugural FRI. The Food and Agricultural Organization of the United Nations and the Canadian International Development Agency later utilized Dixon's expertise in this area to help establish forestry programs in Chile, Peru, and Guyana. Likewise, Terence G. Honer (5T5), Victor G. Merritt (5T1), and Douglas A. Skeates (5T3) assisted with projects in tropical forests around the globe.[48]

Another trend that continued at the Faculty of Forestry during these years was the tendency to attract students whose fathers were either graduates or engaged in some aspect of the forest industry. Thomas A. 'Tom' Buell (5T6), John D. Irwin (5T1), Donald G. Parsons (5T4), and Norman J. Turnbull (4T9) were all the second generation in their families to earn their bachelor of science in forestry degrees from Toronto. John R.T. Andrews (4T8), William G.L. Cleaveley (5T1), and Robert W. Morison (5T4) were sons of foresters or forest rangers, whereas Benjamin F. 'Ben' Merwin Jr (5T2) and John T. Somerville (4T9) were sons of lumbermen. The fathers of Edwin B. Ayer (4T9) and Thomas P. McElhanney (4T9) were respectively a forest entomologist and superintendent of the Forest Products Laboratory. Finally, Douglas P. 'Doug' Drysdale (5T5) came from a family that owned the Ballantrae Tree Farm north of Toronto, which continues to be one of southern Ontario's most popular 'cut your own' Christmas tree operations.[49]

Just like their predecessors, the Faculty of Forestry's graduates from the 1947–57 period were at the forefront of the movement to manage prudently our natural heritage. Many of them took up this cause at universities across North America. Michael Chubb (5T5), for example, pursued graduate work and later taught at Michigan State University and specialized in protecting that state's unique sand dunes. Lawrence S. 'Larry' Hamilton (4T8) became an expert in the development of water resources and the founding head of a centre devoted to this field of research at Cornell University during the 1960s (more will be said about this later). Many foresters were also only able to devote their full-time attention to natural resource management in the twilight of their careers. Paul G. Masterson (5T1) and Donald G.E. Harris (5T4) both worked for industry for roughly three decades before branching out in the 1980s. The former became the supervisor of the Kortright Centre for Conservation (in Kleinburg, Ontario) and director of the Metropolitan Toronto and Region Conservation Authority Foundation, whereas the latter joined Parks Canada as its chief of natural resource conservation in the Atlantic region after having served as its district superintendent on Prince Edward Island.[50]

One graduate from these years stands alone as far as environmental stewardship is concerned. Orie L. Loucks (5T3) completed his master of forestry in Toronto in 1955, his doctorate in botany five years later at the University of Wisconsin-Madison, and he began teaching there in 1962. Over the next four decades, he distinguished himself as a leading scholar in the field of applied ecology, focusing on the impact of human development on the environment. During these years, Loucks served as a member of numerous environmental boards and committees, and was an outspoken friend of the environment in 'his own backyard.' While teaching in Madison, for example, he fought for the preservation of rare or scientifically important natural areas, and his presentation during the case that decided whether to ban the use of DDT in Wisconsin was critical to the court ruling in favour of doing so. In recognition of his dedication to these causes, in 1971 the Madison Newswriters' Guild awarded Loucks a 'Page One Citation.' In presenting the distinction, it described him as 'a tireless and articulate crusader for environmental sanity long before the words "ecology" and "environment" had crept into the national vocabulary.' Roughly three decades later, the (American) National Wildlife Federation honoured Loucks with a National Conservation Achievement Award, which recognizes those on the vanguard of grassroots conservation. 'A forester, ecologist and eminent professor,' the federation's description of his accomplishments reads, 'Orie Loucks is considered "one of the intellectual giants of the environmental movement."'[51]

Like Loucks and many of his predecessors at the Faculty of Forestry, numerous graduates from this period joined the professorial ranks or became scientists. In addition to those whose careers have already been cited, this list includes John M. Anderson (5T1). After earning his doctorate in zoology in 1958, he joined the Department of Forestry at the University of New Brunswick and then served as UNB's president for six years (1973–9). John H. Blair (4T8) worked for the Ontario and Minnesota Paper Company for more than a decade before joining the forestry school at the Lakehead in 1971. While on staff there, his innovative forestry initiative with the local Gull Bay First Nation led to it designating him as an honorary chief in recognition of his work. Similarly, Alexander T. Cringan (4T8), Arthur W. Ghent (5T0), William H. Hilborn (4T9), Archibald R.C. 'Archie' Jones (5T0), Ronald W. Stark (4T8), and Gordon F. Weetman (5T5) joined post-secondary institutions across the continent.[52]

Many forestry graduates chose to pursue their academic careers at their alma mater. Ken Armson (5T1) excelled as an undergraduate, did graduate work at Oxford, and returned to teach at the faculty for the next two decades. In 1975 the Ontario government seconded him to review its administration of Crown timber. The next year Armson produced a landmark report that forever changed – for the better – how the province's forests were managed. The government accepted his major recommendation, namely, that the Crown sponsor and supervise forest management activities while the timber companies carry out the work. In 1986 Armson was appointed Ontario's third, and final, provincial forester. David C.F. 'Dave' Fayle (5T7) had also been a superb student at Toronto (he won the CIF's gold medal) before earning a graduate degree from Oxford, a doctorate from the University of Toronto, and a job at the faculty. In the meantime, he had become deeply involved in preserving and restoring 'heritage' homes, work for which he has received notable acclaim. Other alumni from this period who returned to teach at the faculty included A.D. 'Dal' Hall (4T8), Philip J. Pointing (4T9), and Victor G. 'Vic' Smith (4T9).

A handful of Faculty of Forestry graduates from these years earned their doctorates and then dedicated their lives to research. Most were employed with the Canadian Forest Service or FORINTEK (the successor to the Canadian Forest Products Laboratory), including Robert J. Bourchier (5T1), Clayton T. Keith (5T5), and Leo Sayn-Wittgenstein (5T7). Samuel N. Linzon (4T8) was especially noteworthy among this period's scientists. Born in 'The Ward' in Toronto, the city's principal receiving area for a potpourri of poor immigrants, Linzon entered the faculty after his plans for pursuing dentistry went awry. Although he was a mediocre undergraduate student, he pursued graduate studies in botany even though he had failed this subject while at the faculty. After completing his doctorate in this field in 1964, Linzon spent the next several decades establishing an international reputation in phytotoxicology, specializing in the impact that acid rain was having on Ontario's fauna.[53]

Some graduates simply ended up in interesting lines of work. Ewart J.K. Bagg (4T9) was a baseball scout for the Baltimore Orioles in the mid-1980s. Roger Y. Edwards (4T8) earned a master of arts degree at the University of British Columbia two years after leaving Toronto. Thereafter, he developed and designed B.C.'s parks interpretation program, and then moved on to perform the same function for the Canadian Wildlife Service in the late 1960s. In 1974 he returned to British Columbia as the director of the Provincial Museum (now the Royal B.C. Museum), a position he retained until he retired in 1984. Joseph L. Mennill (4T8) survived three years as a prisoner of war in Germany during the war and took a job with the Department of Lands and Forests immediately upon graduation. In 1962 he took advantage of a unique offer that came his way when he became the first head of the Ontario College for Police.[54]

As for non-traditional lives, no graduate from this period tops George C. Wilkes (4T9). He was a career civil servant in Ottawa (mostly in what is now the Competition Bureau, Industry Canada), and earned his master of science in forestry degree from the faculty in 1952 (his landmark report on forest taxation was published as a faculty *Bulletin*). Wilkes' life was altered forever by the implementation of the War Measures Act during the October Crisis in the fall of 1970; the suspension of habeas corpus inspired Wilkes to devote his energy to fighting for human rights. He sponsored a plethora of foster children from the developing world, and became an activist in Sandy Hill, his Ottawa neighbourhood. He successfully fought to preserve this 'heritage' section of the nation's capital and led the movement to gain a local health centre. Wilkes later spearheaded a campaign to build a public monument to human rights in Ottawa, which the Dalai Lama of Tibet unveiled in 1990. Ten years later, the Office of the Governor General awarded Wilkes the Meritorious Service Medal, which is bestowed 'to recognize a deed or activity performed in a highly professional manner or according to a very high standard that brings benefit or honour to Canada.'[55]

Unlike Wilkes, most foresters who graduated between 1947 and 1957 followed more traditional career paths, and several worked their way up to senior management positions at their firms. Tom A. Buell (5T6), for example, spent nearly two decades after graduation with Weldwood of Canada, and was appointed its chairman of the board in 1979. William G. Harris (5T5) was named president of Northern Plywoods Limited at the Lakehead in the 1960s. He then moved into the world of pharmaceuticals, and by the late 1980s was a vice-president with multinational CIBA-Geigy. Donald Penna (5T0) and Arthur E. Davis (5T2) rose to the rank of vice-president with Kimberly-Clark, and John E. 'Ted' Bothwell (5T0) served as Weyerhaeuser's timberlands manager during the 1970s.[56]

The Faculty of Forestry also continued to produce graduates who rose to senior positions within the civil service, and this period proved especially fruitful. Doug Drysdale (5T5), Arthur J. 'Art' Herridge (4T9), and Donald R. Wilson (4T9) became supervisors or chiefs of their branches within Ontario's Department of Lands and Forests/Ministry of Natural Resources. James W. Keenan (5T2) ascended the same ladder to become an assistant deputy minister by 1977. In 1982 he was appointed assistant secretary of management board, and three years later deputy minister of correctional services. Likewise, John W. Giles (4T8) moved up the ranks of the provincial civil service and reached assistant deputy minister by 1972, a rank he held in several departments over the next few decades.[57]

Among the faculty's graduates who went on to become civil servants, Robert D. 'Bob' Carman (5T4) blazed a trail that few would duplicate. As a student, he finished first in his class every year and won a handful of scholarships, including the Canadian Institute of Forestry's coveted gold medal. He pursued his interest in silviculture first with the Department of Lands and Forests and then in Alberta with Des Crossley on the latter's pioneering forest management effort at Hinton. Carman returned to Ontario in the mid-1970s to rejoin the provincial civil service, specifically as deputy minister of community and social services. Seven years later, he was named secretary to cabinet, which is one of the most senior non-elected posts in government, and in 1989 he became secretary of the cabinet and clerk of the executive council. In 1995 'Bob' Carman was appointed a member of the Order of Canada, in recognition of his lifelong dedication to public service.[58]

The Faculty of Forestry at the University of Toronto celebrated its golden anniversary in style in the fall of 1957. It hosted the Canadian Institute of Forestry's annual meeting, and nearly half of the four hundred foresters who had graduated from the faculty returned to share in the festivities. To mark the occasion, the University of Toronto agreed to confer honorary degrees on six legendary officials – including alumni R.W. 'Bob' Lyons (1T6) and Wallace A. Delahey (1T5) – who had made major contributions to the country's forestry movement.[59]

The fact that nearly all of the conferees were from industry was significant, as it reflected the Faculty of Forestry's relatively narrow approach to forestry at a time when dramatic changes in Canada were beginning to render this perspective anachronistic. The quarter-century of prosperity that followed the Second World War produced a society that was much wealthier and more urbanized, and blessed with unprecedented amounts of leisure time and mobility (due to the now-ubiquitous automobile). These forces created ever-stronger calls for accessible and affordable recreational areas in terms of camping opportunities and ownership of weekend and summer retreats. The Ontario government responded aggressively to these new demands by expanding the provincial park system from eight parks in 1954 to ninety just over ten years later. In addition, it placed increasing emphasis on 'managing' wildlife and fish for the sake of the growing tourism industry.[60]

These forces also made the public more critical of the forest industry's operations. The sensational battles pitting loggers against environmentalists were not occurring yet, but companies operating in locations where recreationists felt they had a solitary claim to the local woods were facing major challenges. In addition, with less and less first-hand contact with the 'bush,' the perception of the forest and how it ought to be managed became increasingly romanticized and disconnected from a factual understanding of the woods' ecology. Increasingly, foresters were being seen as the enemy of those who sought tranquility in the forest, a perspective that the faculty's graduates were confronting more frequently in their work.[61]

The incipient signs of this dynamic were already apparent in Ontario by the mid-1950s, but the faculty was doing a poor job of reading them. To be sure, there was still a strong demand for 'traditional' foresters, and forestry was still as much about danger and 'roughing it in the bush' as it had ever been. This brutal reality is evinced by the handful of students and graduates from this period who either lost their lives or were seriously injured carrying out their duties. The former list includes Douglas C.G. Robertson (4T7), who died at Longlac barely a year after graduation when he was pinned by a tractor, and George E. Heasman (5T3), who was fatally injured in a trucking accident while working for Abitibi near White River on his summer placement in 1950. But an increasing number of foresters were employed in 'new' fields such as wildlife conservation and park management. In fact, as early as the mid-1930s, hunting organizations in Ontario had recommended that the faculty create courses in wildlife conservation and management, but little came of it. In the main, Sisam steered the Faculty of Forestry in the same direction as his predecessors, where the focus was almost entirely on forest, instead of land, management. It did a poor job of incorporating into its thinking the various factors – such as preserving recreational values – that were gaining unprecedented political attention.[62]

In contrast, a few of the country's other forestry schools were doing a much better job of sensing the changing winds. The Faculty of Forestry at the University of New Brunswick, for example, had fundamentally diversified its course offerings after the Second World War. UNB had wholeheartedly supported this initiative by hiring a handful of new professors in fields like entomology and promoting the 'department' to faculty status in 1946. UNB's forestry school had also begun teaching wildlife management in a meaningful and significant way, if only at the graduate level. In 1947, with the backing of Ducks Unlimited and the Wildlife Management Institute of Washington, DC, a wildlife research station was built on campus (duck-banding stations would be erected later in Labrador and on the north shore of the Gulf of St Lawrence).[63] Sisam had indeed revisited his faculty's curriculum in the late 1940s and early 1950s, and made some changes as a result, but collectively they constituted fine-tuning an engine that was soon to be in need of an overhaul. To the Faculty of Forestry's credit, it had dropped the outdated 'forest engineer' degree, and raised its admission requirements to keep out students who had chosen forestry only after receiving their rejection letters from faculties with higher standards. In terms of curriculum, Sisam – and many of those who offered him counsel on the subject at this time – was focused on achieving one main aim: preparing students for graduate studies in silviculture. In an effort to realize this goal, starting in the early 1950s third-year students intensively examined selected areas of the University Forest during their time at spring camp, and then 'worked up' the data during their last year at school. In addition, the faculty placed greater emphasis on subjects of 'general educational importance' (such as modern world history) and offered fourth-year students a short list of course options from which they could choose a 'speciality.' But there were precious few major new initiatives.[64]

Compounding the problem was the university's neglect of the Faculty of Forestry's undergraduate program, a condition that Sisam had repeatedly brought to Simcoe Hall's attention but which the latter had generally refused to acknowledge or correct. Attesting to the faculty's decline was the brain drain that was afflicting it. By the mid-1950s droves of high school students from Ontario's hinterland were heading 'down east' (to UNB) instead of 'down south' (to the University of Toronto) to study forestry. Sisam was distraught over this exodus of potential students, and brought it to Smith's attention in the spring of 1956. The dean pointed out to the president that this migration was due, inter alia, to UNB's lower tuition, and that the University of Toronto's recent decision to increase its fees was only going to exacerbate this situation. Smith responded to Sisam's concern by defending the University of Toronto's policy. Pointing out that 'increases in fees are in the air from the Atlantic to the Pacific,' Smith reassured Sisam that the faculty's strengths would compensate for the higher cost of attending such a first-class institution. 'With the excellent staff you have,' Smith assured Sisam, 'and the new programs ... and with the research work that will be carried on, I am confident that (as in the Ivy League universities, which have always had higher fees than the state institutions) the marketing principle of going to the place where you can get the "mostest" for your money will be appreciated.' Although some departments and faculties at the University of Toronto could certainly boast that they offered their students the 'mostest,' the Faculty of Forestry's undergraduate stream was not one of them. This deficiency would become more noticeable over the rest of Sisam's tenure as the faculty's dean.[65]

Chapter 7

'Forestry Has Suffered Its Share of Frustrations,' 1957–1971

The hullabaloo that marked the Faculty of Forestry's Golden Jubilee in the fall of 1957 could not mask the challenges it faced at the time and over the rest of Bernie Sisam's tenure as dean (1957–71). At the University of Toronto the problem was manifold. Although the mid- to late 1950s saw Simcoe Hall plan a major expansion to accommodate the baby boomers who would soon begin arriving, forestry did not figure prominently, if at all, in it. In some respects, this was understandable. The faculty's enrolment had peaked at just over three hundred in 1948–9 and then begun a precipitous decline, bottoming out in 1957–8 at seventy-three. It rebounded slightly thereafter, and was around a hundred by the mid-1960s. The figures for its graduate enrolment also followed this pattern. As a result, at the very time when the University of Toronto was mapping its long-term growth strategy, forestry's significance was shrinking to a level that called into question not only its relative importance but even its viability. This reality, and the faculty's mistaken belief that the administration shared its view that it was a cornerstone in the august university, caused relations between forestry and the administration to deteriorate during these years. The faculty was at least partly to blame for the university's approach. In the face of rapidly changing circumstances both within and outside the university, it demonstrated an obstinacy that acted as a major hurdle to achieving its aspirations. Even if the University of Toronto had offered the faculty strong support, Dean Sisam probably would not have moved it in the direction it required.[1]

The problems that confronted the faculty reflected forces that extended far beyond Simcoe Hall. One of the greatest new threats was the advent of a rival forestry school in Ontario and the provincial government's willingness to facilitate its establishment. An equally imposing cause for concern was the degree to which forestry remained a low priority for Canadians. From 1957 to 1971 the Ontario government took merely nominal steps towards applying silvicultural principles to managing Crown forests.[2] In Ottawa the story was only slightly better. Forestry enjoyed a renaissance in the early 1960s, but by the end of the decade, its health was in steep decline. Dal Hall (4T8), executive secretary of the Canadian Forestry Association, poignantly summarized the root cause of forestry's plight across the country in the fall of 1966. After Mitchell Sharp, the federal finance minister, announced that the government was terminating its limited support for provincial silvicultural initiatives, Hall wrote to Sisam to lament this situation. 'I think it is characteristic of the position of forestry in Canada,' Hall despaired, 'that there has not yet been recorded in *Hansard* any comments regarding Sharp's statements concerning the discontinuance of Federal-Provincial Forestry Agreements. On the other hand,' he continued, 'there has been all sorts of mention made of the proposed delay in Medicare which is much closer to the political needs of the members of the House. Until we are able to really impress both members of parliament and the people of Canada that forest resources do indeed affect all of them individually then we shall continue to get short changed in terms of legislation.'[3]

These factors undoubtedly created trying times for the Faculty of Forestry, but its story between 1957 and 1971 was not all doom and gloom. It enjoyed a few major triumphs, particularly in terms of creating several new fields of expertise. Nevertheless, these years exacted a heavy toll on the faculty. In fact, its undergraduate program come within a whisker of falling by the wayside, and its future hardly looked assured when Sisam stepped down as dean after nearly a quarter of a century at the helm.

It was in the throes of planning the faculty's fiftieth birthday party in mid-1957 that Sisam received word that his forestry school was about to undergo a major move – literally. During the late 1950s the University of Toronto constructed new buildings to house divisions whose dramatic expansion could no longer be contained within their existing edifices. Although one of the central projects in this effort was a new home, the Galbraith Building, for the burgeoning Faculty of Applied Science and Engineering, forestry's building occupied space that stood in the path of realizing this plan. To make way for the engineers, the university decided to transplant the foresters a few hundred feet north.[4]

The faculty's relocation was the perfect metaphor for all that was wrong in its relationship with Simcoe Hall. The latter saw the forestry building as something it could summarily move to facilitate the expansion of 'more important' departments. In contrast, Sisam and his colleagues viewed their relocation as moving 'up' St George Street in a manner that presented them with a golden opportunity. At the new locale, their building was abutted on three sides by undeveloped space into which they could grow. In this period of unprecedented construction on campus, they saw it as being well within the realm of possibility that they would also participate in this process.

Sensing that the University of Toronto was at a crossroads in its long-term planning, Sisam sent Simcoe Hall a lengthy memorandum in an effort to ensure that his faculty received due consideration. It began by listing what he saw as the forestry school's major deficiencies, the most important of which was its facilities. Although the faculty had done as much as it could to adapt its building for research and graduate work, the structure was grossly inadequate for these purposes. Sisam also stressed that he had long been lobbying for the university to support the faculty in strengthening itself in four fields: forest biology and soils, forest economics, forest pathology, and wood technology. The university had helped fill the first lacuna through such acts as hiring Ken Armson (5T1), but the demand from graduate students to work in this field was now exceeding the faculty's ability to accommodate them. 'Other forestry schools are expanding and developing to meet the needs of today, which are so different in many respects from those of 25 years ago,' Sisam frustratingly declared, and the only way to meet the faculty's 'minimum requirements' was to address all the aforementioned issues.[5]

The university's response typified the paltry support it was prepared to offer the faculty. It paid to renovate partly the faculty's relocated building and excavate a large basement that extended beyond the building's walls. This was an integral part of Sisam's strategy for expanding the faculty, as he envisioned extending each floor of the forestry building over the enlarged substructure in the near future. In the meantime, however, the renovations created nearly as many problems as they solved. The basement was plagued by major flooding, seepage, and peeling walls, and the forty-year-old building was still in dire need of a major facelift. Likewise, Sisam was ecstatic in 1958–9 when he learned that the administration was finally granting him the long-awaited appointment in forest pathology. The dean promptly went out and lured Erik Jorgensen from the Forest Biology Division of the Canadian Forest Service (CFS). Sisam was embarrassed, however, by the sad reality that his inadequate facilities left him nowhere to house his new professor.[6]

Unfortunately for Sisam, he unwittingly undermined his own campaign to win more support from the University of Toronto. He and his staff had decided that they should decrease the size of their student body in light of the faculty's space limitations. Whereas they had previously restricted first-year enrolment to fifty students, they reduced this figure to forty beginning in the fall of 1961. While this was prudent under the circumstances, it created the appearance that the faculty's physical facilities were sufficient to meet the school's needs. It also dramatically reduced the visible signs of overcrowding that would have provided valuable fodder for Sisam in his lobby for new quarters.[7]

At this time Sisam was especially concerned about his inadequate accommodations, particularly for his graduate and research programs, because the federal government had recently announced that it was dramatically raising forestry's profile in Ottawa. In July 1960 John Diefenbaker's Conservative government both elevated the CFS from branch to departmental status and greatly expanded its scope of activities. Sisam soon learned that this move would create an unprecedented demand for 'foresters with post-graduate degrees.' This expectation was heightened with the implementation one year later of the Agricultural Rehabilitation and Development Act (one of its main thrusts was reforesting substandard farmland) and the convening of the Resources for Tomorrow Conference in 1962 (its goal was to foster stronger federal-provincial cooperation in the stewardship of renewable resources). Together, these signs seemed to portend a golden era for forestry graduate studies and research, and Sisam was determined to ensure that his faculty was well positioned to share in this bounty.[8]

Sisam was also sensitive to how society's changing attitude towards 'the woods' was reshaping the forestry field. His colleagues south of the border had long ago recognized that working in the bush was quickly becoming an activity that involved far more than simply harvesting trees, and they reoriented their programs accordingly. The University of Michigan's Department of Forestry, for example, had evolved into the School of Forestry and Conservation in 1927; four years later it offered three distinct program streams, one of which was forest zoology. Although Cornell University was a little slower off the mark, its forestry school metamorphosed into the Department of Conservation in 1948. Even the forestry school at the University of New Brunswick, which was situated in a province whose vitality was still largely timber-dependent, had begun diversifying its course offerings after the Second World War. The pressure to follow this lead was even greater in Ontario. The forest industry was still integral to sustaining the province's prosperity, but its import was slipping – some would argue it had disappeared long ago – from the public consciousness. Its operations were generally confined to northern Ontario, far from most of the province's residents, who increasingly perceived the woodlands in terms of recreational and ecological, not fibre, values. Furthermore, whereas in 1900 Crown timber revenues constituted roughly 30 per cent of the Ontario government's total income, sixty years later the figure was just over 1 per cent. Finally, the Faculty of Forestry at the University of Toronto was situated both in the centre of the urban environment whose inhabitants were taking an increasingly dim view of the work its 'typical' graduates were performing in the field, and on the campus of a university that perceived it as a tangential operation. In this context, it was imperative for the faculty to adapt.[9]

From the late 1950s to the mid-1960s, then, Sisam repeatedly recognized that the faculty ought to branch out in new directions, but he allowed this to occur only within tightly prescribed parameters. 'In recent years,' he noted in mid-1961 to a colleague and former graduate, 'there has been an increasing awareness of the need for a stronger integration in the various phases of management of natural resources. For this to be accomplished, it is important that the specialists in any one field of activity be knowledgeable, in general terms at least, about the aims of those working in other fields, the implications of those aims, and the means being undertaken to attain them.' Driven by this impetus, he and his colleagues at the faculty repeatedly revisited their undergraduate curriculum. To assist with this exercise, they followed the lead of a few other faculties and departments at the University of Toronto, each of which had created a small 'advisory committee' to which they had appointed leaders in their respective fields. The men whom Sisam chose to sit on the Faculty of Forestry's committee reflected his aversion to taking the faculty in a new direction. In the main they represented forestry's 'traditional' interests that harvested timber for conversion into paper and lumber.[10]

And when the faculty began implementing its 'revised' undergraduate curriculum in the mid-1960s, nearly all the changes were superficial. There was a greater emphasis on improving the students' ability to communicate, and the faculty began offering – but only in a limited way – an opportunity for fourth-year students to specialize in an 'area of concentration.' But overall, Sisam and his cohorts had decided to leave their core curriculum intact. Their attitude towards changing the name of their faculty spoke to their intransigence to overhauling their program. A leader in forestry education in North America had advised Sisam that 'the word "forestry" might be a detriment rather than an asset in the name of a University school or faculty.' Consequently, the faculty and its advisory committee agreed to append the suffix 'and Resource Management' to the school's title, a change that was intended to be accompanied by a broadening out of the faculty's curriculum. But the talk was not converted into action. Although Sisam introduced a new 'wood sciences and forest products' stream, which students could enter after first year, the initiative had nothing to do with taking a new approach to forest stewardship.[11]

Two aspects of this situation are noteworthy. After the mid-1960s Sisam operated under the impression that his faculty had 'thoroughly reviewed' its undergraduate curriculum and made 'many changes' when this was simply not the case. Second, the paradigm for a new direction in which he could have taken the faculty was right before him. By this time Ontario's Department of Lands and Forests (DLF) had begun considering recreational, environmental, and other values in managing Crown lands. It was thus keen to find a means by which it could offer its staff, specifically its foresters, an opportunity 'to gain a broader background of knowledge and understanding of the various disciplines involved' in carrying out their work. To meet this need, Sisam teamed up with Kenneth Clark Fisher, the chair of the zoology department, to create a one-year, post-baccalaureate diploma course (jointly sponsored by the Faculty of Forestry and the Department of Zoology) in resource management. The 'DipRM' was officially launched in September 1961. Its early success compelled Sisam to open a dialogue with Fisher regarding the possibility of establishing a joint undergraduate program in 'Fish and Wildlife Management,' but tellingly, nothing tangible came of these discussions.[12]

This did not mean that the faculty lacked successful initiatives during these years, but these must be considered in their proper context. 'Urban forestry,' for instance, quickly became one of the Faculty of Forestry's best-known programs, but its prosperity was a function of circumstance, not Sisam's efforts on its behalf. The development of this field also attested to the University of Toronto's determination to circumscribe its support for its forestry school.

An illness that afflicted and often killed a species of tree that city-dwellers in eastern North America had grown to love gave the faculty a most unexpected and much-needed boost in the early 1960s. The American elm's large, handsome, and graceful qualities had enraptured city officials and homeowners and made it a species of choice for urban landscapers. A fungus introduced from Europe around the beginning of the Great Depression, however, and spread by elm bark beetles thereafter, had begun ravaging the stately trees. This set off a mad search for an elixir to defend against the 'Dutch Elm Disease,' as it swept through successive neighbourhoods. Its arrival in Ontario after the Second World War signalled the beginning of this phenomenon in the province, and Eric Phillips, chairman of the University of Toronto's board of governors (1945–64), was acutely concerned about the situation. When his own elms were attacked, he called Erik Jorgensen, the Faculty of Forestry's recently appointed professor of forest pathology, to ask what could be done to counter the onslaught. Phillips also wondered why the university was not doing anything to prevent its wonderful collection of elms (by far the most prevalent tree species on campus) from being converted into an above-ground graveyard of leafless branches and debarked trunks.[13]

No sooner had Phillips wondered than the University of Toronto acted, and it became the central command post for the province's campaign against the disease. J.F. Westhead, the head of the university's grounds-keeping crew in the superintendent's office, was directed to address the matter post-haste. Jorgensen was instrumental in this cause, and Westhead became chairman of the Dutch Elm Disease Control Committee for Metro-Toronto and Region (which in 1964 would become a province-wide organization and be renamed the Ontario Shade Tree Council). In September 1962 the university established a Shade Tree Research Laboratory in the superintendent's building on Huron Street which became the control committee's headquarters, and early the next year the University of Toronto hosted a heavily attended symposium on the disease. The university – not the Faculty of Forestry – was thus the hub of the effort to counter the disease, and Westhead was its figurehead.[14]

This arrangement worried the University of Toronto's senior administrators because they feared it would obtrude on their strategic goals, and so they slyly thrust the torch for the effort into the faculty's hands. The matter came to a head when Westhead was invited to become the president of a formal organization to orchestrate the fight against the disease. John H. 'Jack' Sword, the university's vice-provost, was dead set against Westhead taking on this task for several reasons and forcefully expressed them to the university's president in May 1963. Sword was particularly concerned about 'the appearance of another University of Toronto sponsored drive for funds – an appeal which some of us think is not properly a University of Toronto responsibility.' He thus convinced Westhead to decline the offer. Sword added that stepping back in this manner would also permit the university's administration to terminate its commitment to fund the laboratory that was integral to the shade tree disease research program. Instead, Sword pointed out that 'full responsibility and authority for making the case for the development of the laboratory would rest with Dean Sisam and his Forestry colleagues.' By mid-1963 the university had pledged to provide short-term funding to the research laboratory, which it relocated to the Borden Building.[15]

Jorgensen and his lab flourished throughout the rest of the decade and into the early 1970s. This research laboratory owed its genesis to the concerns over the Dutch elm disease, but Jorgensen ensured that its mandate was stretched to investigate all shade trees. It passed a major milestone in 1967 when it became home to the Superior Shade Tree Program for Ontario, which was devoted to selecting and propagating native trees for landscaping in the province. The product was a series of cutting-edge studies, publications, and discoveries that brought Jorgensen and his lab international acclaim. Among them was the creation of a fungicide (Lignasan) to fight Dutch elm disease, which Divenda N. Roy helped develop after he joined Jorgensen's team in 1967. Moreover, Jorgensen was the pioneer in Canada in the domain of 'urban forestry.' Not only did he practically invent and define the term and the field, in 1965 he introduced a master of science (forestry) in urban forestry at the faculty and, four years later, a fourth-year undergraduate course in the same area (both were firsts in Canada).[16]

While these developments were a bright light for the Faculty of Forestry, tension soon developed between it and Jorgensen's operation. Not only did he work in his own building, but he also had his own budget and staff, and the liberty to direct research however he pleased. More importantly, Jorgensen wanted it that way, even though Sisam kept pleading with the university's administration to expand the Faculty of Forestry's building so that Jorgensen's work could be brought – both figuratively and literally – under the dean's roof. 'To ensure the continued development of a coordinated programme of work within the Shade Tree Research Laboratory,' Jorgensen told the vice-provost in mid-1964, 'it is believed essential that the Laboratory is provided with the greatest form of autonomy by placing the responsibility for staff and project developments in the hands of the Head of the Laboratory [i.e., Jorgensen].' This approach created more than a hint of jealousy towards him from among his fellow staff members.[17]

Just as Jorgensen was beginning to establish the Shade Tree Research Laboratory in the early 1960s, an unprecedented challenge to the faculty appeared. Other Canadian forestry schools faced the same issue as well, but Sisam's forestry school in Toronto would be more directly affected than the others.

The bête noire was the proliferation of forestry education in Canada. The postwar boom in the forest industry had created a need for a new cadre of woodsman whose members were schooled in the 'technical' side of forestry but not to the extent that was required by those who completed a four-year, undergraduate degree. The Faculty of Forestry at the University of Toronto had recognized the need for such a program when it had first opened its doors in 1907, but had terminated this stream for lack of interest in it. In addition, after Ontario's Department of Lands and Forests had created its forest ranger school in Haliburton at the end of the Second World War, representatives from northwestern Ontario lobbied for a similar institution in their bailiwick. By 1948 the Lakehead Technical Institute (LTI) in Port Arthur (now part of Thunder Bay) was offering a two-year diploma program in 'forest technology,' and Sisam immediately recognized the potential danger it posed to forestry schools in general. He feared, as he told his university's president in the fall of 1948, 'the competition of graduates from that school for certain positions in industry that would otherwise be available to graduate professional foresters as a first appointment or "interne period," as it were preparatory to taking up more advanced work.'[18]

Over the next roughly fifteen years, forestry education grew steadily at the Lakehead, and it did so with little official connection to the Faculty of Forestry in Toronto. By 1962 the LTI had become a degree-granting institution known as Lakehead University, although it still offered only a two-year 'tech' program in forestry. For students in this program who wished to earn their bachelor of science in forestry degree, attending Sisam's school was rarely an option because Lakehead's forestry department was still operating as a college. This meant that most students enrolled there after completing Grade 12. In contrast, admission to universities in Ontario demanded completion of Grade 13 (the University of Toronto also required a minimum average of 60%). Lakehead thus entered into a special arrangement with the forestry school at the University of New Brunswick, whereby the latter would accept the former's graduates but require them to complete three years of university to earn their professional forestry degrees.[19]

Shortly after Lakehead earned its degree-granting status in the early

1960s, a movement arose for it to acquire its own 'professional' forestry school. J.W. 'Jack' Haggerty, who joined Lakehead's staff in 1951 and after whom its 'university forest' is named, led the charge. The push also drew support from a number of local forestry officials and businessmen, although the extent of this broader appeal is moot.[20]

Sisam was determined to head off this potential rival, and in doing so he was aided by several forces. A number of the Faculty of Forestry's well-placed and loyal alumni acted as 'moles' at the head of the lakes and kept Sisam informed on Haggerty's activities there. They also did their best to defend the faculty's interests, although their actions spoke as much to their profound desire to do what was best for the province as a whole. A.L. Ken Switzer (3T4) was most notable in this respect. He was a senior forester in the Thunder Bay District and chairman of the forestry committee that Lakehead's board of governors had struck. When Haggerty began pressing Lakehead to support his plan to establish a professional forestry school in 1962, Switzer derailed the effort by pointing out that Ontario's most urgent need at the post-secondary level was more arts and science faculties.[21]

Over the next few years, Haggerty battled to achieve his end, one that included firing salvos at Sisam's faculty in areas in which it was highly vulnerable. After Haggerty converted Douglas Fisher, the member of Parliament for Port Arthur, to the cause, they raised the matter with the federal government. Sisam successfully countered this move by convincing officials in Ottawa that any funds put towards forestry education would be better spent on the existing Faculty of Forestry in Toronto. Fisher and Haggerty then used the local media to publicize data that highlighted Sisam's Achilles' heel. Their figures indicated that, during the 1962–3 school year, seventy-nine of UNB's 159 forestry students had come from Ontario, and other Ontarians were taking their forestry training at the University of Michigan. In a letter to the editor of the local daily, Fisher stressed that there was nothing wrong with the province's students studying outside Ontario, but that 'the numbers suggest that Ontario has not the [forestry] facilities of the right kind at present and the Lakehead College should consider providing them.' Likewise, when John P. Robarts, the premier, toured the north with his cabinet in early 1964, Haggerty handed the entourage a brief outlining Lakehead's case. It was only logical to locate a forestry school in the middle of Ontario's largest 'woodlands community,' the document argued, and offering a five-year course with an entrance requirement of only Grade 12 would also allow legions of young Ontarians to be trained in the province. The brief closed by pointing out that 'U. of T. has a poor attitude towards Forestry,' and that 'forestry could be the key course on which the reputation of the Lakehead College could be built,' a situation that was in diametric contrast to the one Sisam faced in Toronto. The *Port Arthur News Chronicle* published an editorial that outlined the main issues Haggerty had raised – its headline screamed 'Lakehead College Merits Forestry Degree Course' – but it prudently observed that the effort's success hinged on winning over 'those in Toronto who control university purse strings.'[22]

Sisam's allies at the head of the lakes did their best to thwart this campaign, but they recognized the sensitive nerve Haggerty and Fisher had touched. Sisam's policy of limiting first-year enrolment to forty precluded the faculty from adequately addressing the province's need for undergraduate forestry education in terms of numbers, let alone quality. Moreover, at this time the University of Toronto was undertaking a major push to become Canada's centre of graduate studies. It had already begun leaning on its smaller faculties to drop their undergraduate programs and focus instead on post-graduate studies; the School of Social Work had followed this directive in 1965. In late February 1965 Switzer wondered with uncanny prescience whether this would ultimately be forestry's fate. 'It appears that the U. of T. is tending towards becoming a graduate school much as I believe Yale and Harvard are,' he commented to Sisam. 'Is this likely to happen to the Faculty of Forestry in the foreseeable future?' Sisam categorically rejected this suggestion, but the dean must have been acutely aware that his perspective was incongruent with the university's at this time.[23]

Sisam did sense that it would be politic for him to extend an olive branch northward at the same time as he defended his monopoly on granting forestry degrees in Ontario. 'From what I have read,' Sisam confided to Bill Harris (5T5) in March 1964, 'it would seem to me that there are considerable pressures, personal and political, being brought to bear on the establishment of a professional forestry course at the Lakehead College.' After a meeting with Lakehead's senior administrators, Sisam agreed to consider admitting at least some of its forestry graduates into his faculty. He concomitantly pressed his university's president, Claude Bissell, to demand that the province not create a new northern forestry school before both strengthening Toronto's Faculty of Forestry and studying the entire matter of forestry education in Ontario. Bissell was also chairman of the recently established Committee of the Presidents of the Provincially Assisted Universities in Ontario. He saw to it that the committee became part of a movement to compel the Canadian Institute of Forestry to carry out a survey of forestry education in Canada to determine its status and recommend its future development. Later in the year Art Herridge (4T9), president of the Ontario Professional Foresters Association, delivered the same message to Bissell and Premier Robarts. Within months the CIF was preparing to undertake the study under the direction of George A. Garratt, the former dean of Yale University's School of Forestry.[24]

Sisam sensed the rising tide of support for Lakehead – more than a few provincial politicians felt that he ought to be taken down a notch – and events in 1965 left him no choice but to cooperate with it in a way that laid the foundation for the upstart university to gain eventually its own degree-granting program. When the University of New Brunswick terminated its arrangement with Lakehead whereby graduates from the latter's two-year forest technology program could earn their professional degrees at the former, Sisam stepped into the breach. Within a year Lakehead University began offering essentially the same first- and second-year forestry courses as the Faculty of Forestry in Toronto, and Sisam welcomed into third year those students who successfully completed this program. While Sisam saw this as a means by which his faculty would retain its hegemony in Ontario, Lakehead viewed it as the penultimate step towards gaining its own degree-granting forestry school.[25]

Sisam's battle with Simcoe Hall for improved facilities became even more pressing in light of this new threat to his forestry school. His pleas for help in the early 1960s, including one to the university's influential presidential committee on accommodation and facilities in late 1962, fell on deaf ears, however.[26]

Fearing that the presidential committee was ignoring his forestry school's needs just as the committee was completing its long-range planning exercise, Sisam intensified his campaign. In mid-1965 he sent Bissell a lengthy summary of his vision for the faculty's development. After commenting dejectedly to the president of the university that he felt 'that the Faculty of Forestry has little significance in the University scheme of things at the present time,' he stressed that the public's growing awareness of the need to manage woodlands prudently imbued the faculty with a weight that transcended its relatively small size. Although some may question the suitability of teaching forestry in the heart of a major metropolis, Sisam asserted that many of the world's most successful forestry schools were situated in urban centres, and it was essential that his remained at a university that boasted strong science departments. At the same time he delivered a quasi-ultimatum to the university's administrators. If the University of Toronto was 'to continue to offer professional and postgraduate education in forestry and resource management,' he insisted that 'steps must be taken in the immediate future to improve and expand facilities and staff in the Faculty of Forestry.' It was the toughest language to which the dean had resorted in pushing his cause with the administration, and his demands were more numerous and complex than ever before. They included a new building and equipment, permanent research quarters, and the resources to inaugurate new undergraduate and graduate programs in forest hydrology, meteorology, forest genetics, and resource economics.[27]

While Sisam's appeals were heartfelt and, in many ways, logical, they reflected a fundamental disconnect from the University of Toronto's thinking at the time. The mid-1960s saw the university's two largest professional faculties – medicine and engineering – enjoy exceptional growth in terms of space and staff; the latter saw its professorial ranks double in size (from 75 to 150 members) and shrink in terms of teaching load. This augmentation reflected the import Ontario's electorate, and its elected officials, placed on training engineers and physicians. Naturally, this influenced where the university located them in its strategic planning. The opposite was true for forestry and the Faculty of Forestry, and there was good reason to think the latter would soon fall victim to the University of Toronto's strategy of converting its small professional faculties into graduate programs. Finally, and most importantly, the university simply did not share Sisam's view that the Faculty of Forestry's significance transcended its relative size, especially after 1967. That year the Ontario government introduced its 'funding formula' to university education that reinforced the adage that, in fact, size does matter. Thereafter, the amount of money a university received was based on enrolment, with graduate students receiving double the public funding of undergraduates. In this context, the University of Toronto could not help but view the faculty's relative puniness as a major handicap.[28]

The mid-1960s continued to deliver reminders to Sisam of this grim reality, but the dean saw a few rays of sunshine on the horizon and hoped to bask in them. He suffered a major setback in 1965 when R.W. Kennedy, whom Sisam had hired only a few years earlier to develop the forestry school's nascent wood science program, resigned to join the Forest Products Laboratory in Vancouver and teach graduate courses in the University of British Columbia's Faculty of Forestry. Although Kennedy explained he was leaving because he had 'reached a plateau beyond which I cannot make significant contributions unless provided with infinitely more facilities for research,' and the university's administration was clearly displeased by the defection, Simcoe Hall typically took no action to reverse the Faculty of Forestry's languishing condition. This was especially troublesome because Sisam's colleagues were overwhelmed by a nearly 50 per cent increase in the number of graduate students at this time. Their spirits were buoyed, however, when the federal government announced that, beginning in 1967–8, it would make $40,000 available in block funding to each of the country's four forestry schools for their graduate and research programs. This breakthrough resulted from Sisam and the deans of the other forestry schools lobbying the Canadian Forest Service for discretionary grants to help them build their research programs to the point where their staffs would become more competitive with the research-granting agencies. In an attempt to solicit greater support, Sisam outlined his ambitious plans for expanding this realm of the Faculty of Forestry's work. He projected that he would soon need to sustain roughly '40 candidates at the Master's level and 15 at the Ph.D. level at any one time,' and this would require the appointment of at least twelve more full-time staff and the expenditure of an additional $300,000 on capital projects and $360,000 on annual operating costs.[29]

From the mid-1960s practically until Sisam retired in 1971, the dean became increasingly desperate over the University of Toronto's indifference towards his faculty. When the university suggested that the foresters resolve their issue of inadequate accommodation by moving into the mining building in early 1966, for instance, Sisam cynically asked whether the offer was bona fide; it was not.[30]

Just over one year later he was utterly contemptuous of the university for having the gall to ask him for yet another long-term plan for the Faculty of Forestry. Sisam expressed his disdain in a lengthy letter to the university's acting president, 'Jack' Sword, in August 1967. 'It is presumed,' Sisam's dispatch asserted, 'that planning for the further development of the St George Campus is going on continuously, and so far as I am aware without any proper attention being given to the needs of the Faculty.' Moreover, Sisam was deeply worried by the myriad stories he was hearing about his forestry school's uncertain future and the impact they were having on his colleagues. 'From time to time,' the dean explained, 'suggestions and rumours circulate within the University community regarding the future of the Faculty, some of which are speculative and most of which suggest misinformation or misapprehension. While, no doubt, one shouldn't pay too much attention to this sort of thing, nevertheless it does tend to undermine staff morale.' In his view, if the University of Toronto continued to include the Faculty of Forestry in its operations, it behoved the university to support the forestry school 'as long as the latter is providing the services implicit in the terms of its establishment and there is a continuing demand for those services.'

For Sisam the crux of the matter was the University of Toronto's continual refusal to ameliorate the Faculty of Forestry's abysmal accommodations, despite his constant pleas for it to do so. He felt that the university had implied at the time it had moved his faculty's building, in 1957, to make way for the Galbraith Building 'that Forestry could expand in the future as the need arose.' Over the next decade he had repeatedly responded to the university's requests for long-term plans with proposals that included new quarters for the faculty. Yet the university had rarely even responded to his submissions, let alone granted his principal request. Now he had learned that Simcoe Hall was about to expand into the space adjacent to the forestry building, thereby permanently hemming it in.

Sisam pointed to the University of Toronto's recent callous attitude towards the Faculty of Forestry's work at Glendon as an object lesson in its habitual neglect of the foresters. The Faculty of Forestry and the Department of Botany continued to use the property even though the university had promised it in 1960 to the new York University. In the mid-1960s the University of Toronto authorized York to begin developing the site – without informing Sisam of this decision – an oversight that he felt was inexcusable. The University of Toronto was 'so little concerned with the needs of Botany and Forestry,' Sisam chastened its senior administrators, 'that it was only through Dr Howarth, who at that time was acting on behalf of York University, that I learned that the ground on which our greenhouse and laboratory unit was then standing would be subject to excavation within a matter of weeks. Fortunately, it was possible to arrange for an alternate site [on the property] and have the unit moved, though there had been no prior planning for this.'[31]

The university took steps to assuage Sisam's anxiety, but made it clear that forestry was not an important part of the university. The University of Toronto's senior officials, most notably Moffat St A. Woodside, its provost, met with the dean to discuss his concerns but avoided any promise to address them. 'There is no doubt at all that Sisam is exceedingly worried,' Woodside reported later, and his 'worry is even more apparent in our conversation than it is in the attached letter.' But, the provost declared candidly, 'it is the casual belief around the University that the Faculty of Forestry is a very tiny operation which is not really expanding.' Ultimately, Woodside invited Sisam to address the 'President's Council' about the future of this faculty.[32]

Address it Sisam did, in October 1967, and this time he delivered an unequivocal ultimatum. He raised many of the issues he had before, and supplemented his case with data that indicated that the demand for foresters far outstripped the supply at both the undergraduate and graduate levels. In 1965, for example, first-year enrolment had been twenty-six, whereas two years later it had risen to fifty-eight (the faculty had raised its first-year cap from 40 to 60 students). In addition, in a few years the Faculty of Forestry would begin accepting twenty to thirty third-year students who had completed their first two years of training at Lakehead University. The problem was that the faculty could only accommodate a maximum of forty students in each year. Sisam closed by presenting, with unprecedented candour, the University of Toronto with a choice. If the university did not wish to support forestry, he proclaimed, 'then an alternative arrangement should be sought.'[33]

Despite Sisam's stern warning, the administration made it clear to him that forestry's future at the University of Toronto was tenuous. With York University itching to spread its wings over the entire property at Glendon, the Faculty of Forestry and the botany department were cooperating to find a new site for their expanding research activities. The university rejected each suggestion they presented, and added that, because 'the future needs [of Forestry] are still very uncertain ... to commit our own lands may not be the best plan.'[34]

Increasingly despondent over the situation, Sisam was ready to embrace novel solutions to the faculty's inadequate facilities both at Glendon and downtown. After discussing the issue with the administration through late 1967 and early 1968, he agreed that Erindale, the University of Toronto's new suburban campus west of Toronto, might represent the panacea for what ailed his faculty. He recognized that there were practical problems associated with moving there, such as inadequate student housing and distancing the faculty from the 'support departments' on the St George campus on which it relied. 'Nevertheless,' he predicted confidently, 'it is believed that with the support of the University these difficulties can be overcome and that in the long run the advantages could be highly significant.' Sisam thus urged the University of Toronto to investigate this option and assess its strengths and weaknesses.[35]

But the university did nothing of the sort. Instead, it considered Sisam's encouraging response as all that it needed to deem the foresters' migration from downtown to Mississauga as a fait accompli, and published this news in an intra-university newspaper in early December 1968. The administration's story on the subject indicated that the Faculty of Forestry had discussed and 'expressed enthusiasm for a move to Erindale college.' This observation was inaccurate, however, as the opposite was true. Erik Jorgensen, for example, despatched a terse letter to the university's president after reading the story, and his communiqué reflected the faculty's generally dim view of relocating to the suburbs.[36]

Sisam saw this latest crisis as marking the nadir in his relations with the University of Toronto, and he wrote Claude Bissell to convey this message. The dean suggested to the president of the university that the rapport between Simcoe Hall and his Faculty of Forestry had been degenerating for some time because there was 'a certain lack of enthusiasm and understanding on the part of the University for the purpose and function of this sector. The result has been a growing sense of frustration on the part of the Faculty Staff which in turn can only lead to more difficulties,' he reported sadly. 'There is an urgent need to resolve this basic problem as quickly as possible.' He pointed out that the most recent misunderstanding involving Erindale attested to this malaise, as did the administration's decision to ignore his long-term designs for the faculty. Most importantly, Sisam lashed out at the administration for its plan to develop the land around the forestry building in a manner that would 'effectively establish its obsolescence by cutting off any possibility of lateral expansion.'[37]

Sisam and his colleagues had every right to complain about the University of Toronto's systemic insensitivity and neglect, but on this occasion, their behaviour contributed to the very problems they were confronting. When it came time to consider the 'Erindale option,' neither Sisam nor most of his staff gave the matter a fair hearing. Having been left bobbing in the water for so long, it was almost as if they could not recognize a potential lifesaver when one was thrown to them. After discussing the matter, Sisam reported that his colleagues opposed the move because, in their eyes, the faculty's success hinged on remaining downtown amid the university's science departments and crucial government and industry offices. But it was easy enough to read between the lines and discover the issues that truly lay at the heart of the matter. Prominent among them were the prestige attached to being at the downtown campus and the stigma attached to being located at a campus in what was still considered 'the boonies.' So, too, was the intransigence on the part of senior staff to trade their lifelong pattern of travelling a few minutes a day to their downtown offices for undertaking a genuine commute to the suburbs. Finally, it appears that many in the Faculty of Forestry, even the distraught dean, naively clung to the view that the University of Toronto would eventually come to its senses and offer the faculty its wholehearted support. The upshot was that the Faculty of Forestry hastily closed a door of opportunity that the university had opened to it. Moving to Erindale undeniably had its drawbacks, but it could have served as an immediate means of meeting practically all the faculty's needs and reverse its predicament of being a small fish in a big pond by becoming the marquee attraction at the suburban campus.[38]

Sisam did little to improve the situation a short while later. In the wake of the Erindale debacle, the university had asked him to prepare an updated precis of the situation. In it the dean was completely unrealistic. In insisting that the best location for the Faculty of Forestry's new home was on the St George campus, he called for the university's administrators to decide the issue in a vacuum. Just this once, he urged, those who were formulating education policy in Ontario in general and at the University of Toronto in particular ought to divorce their thinking from reality. 'It is most important that decisions on these matters be made solely in the interests of forestry education and research to serve the future needs of this province and country,' he argued, 'and not in terms of political expediency or administrative strategy, and further that they be made with reference to the special needs of the Faculty, if necessary as an exception to the general trend in University development.'[39]

Toronto's Faculty of Forestry then received a much-needed expression of support. In late 1968 Lakehead University launched another offensive to land its own professional forestry school. This attack was the fillip that spurred the Ontario Professional Foresters Association to come to Sisam's defence. At its annual general meeting in November, it passed a resolution that stressed the glaring need to manage more effectively the province's woodlands, the integral role foresters would play in achieving this goal, and the gross inadequacy of the Faculty of Forestry's facilities in terms of both space and quality. 'Practically every other faculty or school connected with the University of Toronto,' the document pointed out, 'has been accorded new or greatly expanded and improved facilities since 1945, mostly at the expense of the Provincial government.' The resolution thus declared that the OPFA would establish a committee to approach the university, the forest industry, and Queen's Park 'to ask that accommodation adequate to the teaching requirements and appropriate to the dignity of the profession and its importance in the provincial and federal economies, be provided as soon as possible.' When Ken Armson (5T1), a veteran faculty member and past president of the OPFA, reported to his colleagues on the matter at a staff meeting, he was far more forthright, declaring that 'the facilities of this Faculty were deplorable.'[40]

At the same time, the University of Toronto looked to Queen's Park for direction in dealing with the faculty. Lakehead was pressing the Ontario government to investigate the province's future needs in terms of forestry education, and William G. 'Bill' Davis, the minister of university affairs, was corresponding with Claude Bissell about the situation. Bissell was adamant that Davis ought to state the government's position vis-à-vis the Faculty of Forestry in Toronto. The president of the university told the minister that, with apparently no pun intended, 'I think it is wise if we have a clear-cut policy at this University' regarding forestry. The matter was a burning one for the University of Toronto for several reasons. Sisam was scheduled to retire in mid-1971, and as Sword (the university's provost) put it, 'acceptance of responsibility for recommending a successor raises in turn the question of the future of the Faculty of Forestry in the University of Toronto.' Furthermore, the faculty's undergraduate and graduate enrolment was now surging, making the issue of where to house the foresters a vital one. Finally, Sword stressed that the university owed it to the faculty to clear the air. 'Forestry desperately needs the assurance to plan and Forestry has suffered its share of frustrations,' he openly admitted to a colleague in January 1970.[41]

By this time forestry's fortunes across Canada were plummeting. The federal government had reduced its profile at the national level by demoting it from departmental to branch status, dramatically cutting its budget, and discontinuing the support offered to provincial forestry initiatives. Compounding the forestry profession's problems was the rise of radical 'environmentalism.' Spearheaded by a cabal of young zealots, its goal was to paint issues surrounding timber harvesting in black-and-white terms (foresters bad, canoeists good) by pandering to the emotions of urbanites whose increasing spatial separation from the woods made them highly vulnerable to such appeals. This was especially true in Ontario, where the 'tree-huggers' turned Algonquin Park into their most famous (or infamous) battleground. Boasting that they were void of self-interest in battling to 'save' the forest, they – and the mainstream media that proved remarkably receptive to their message – vilified those who depended on harvesting trees for a living.[42]

In the midst of these swirling crises, the OPFA was pushing ahead with its own inquiry. At its annual meeting in November 1968, the OPFA had struck a committee to study the state of forestry education in Ontario. The committee consisted of chairman Ken Hearnden (4T6), Art Herridge (4T9), and Leo Sayn-Wittgenstein (5T7). When Hearnden took a job with Lakehead University the following August, he resigned his position (also in 1969 he succeeded Herridge as the OPFA's president). Herridge thus became head of the OPFA's professional education committee.

Herridge was the driving force behind the committee, and in the course of carrying out his investigation, he learned some disquieting news about Toronto's Faculty of Forestry. He held two crucial in camera tête-à-têtes with Bissell, in April and November 1969, during which the university's president told Herridge that the Faculty of Forestry faced bleak prospects at the University of Toronto. 'He was quite candid,' Hearnden later recounted in reference to Bissell, 'in discussing the problems of providing adequate financial support to small professional schools such as forestry, when confronted by the needs of engineering, medicine, arts, and other schools having a greater public impact and claim upon the university purse. It was made apparent to the [OPFA's] committee that the prospects for the provision of "adequate accommodation to the teaching requirements and appropriate to the dignity of the profession" were decidedly remote, and that the Toronto school could expect to receive no more than maintenance support from the University coffers.'[43]

This news spurred Herridge to go out on a limb by presenting a radical report. It outlined two major recommendations to the OPFA's council: undergraduate forestry education in Ontario should remain generalist in nature, and the Faculty of Forestry ought to be moved from the University of Toronto as long as this would not undermine the quality of forestry education in the province (the implication was that Lakehead's program would be merged with Toronto's at the new location). Although the latter directive naturally created quite a stir in the OPFA's inner circle, the association endorsed Herridge's two recommendations and investigated the most desirable site at which to relocate the undergraduate degree program in Ontario. By May it had determined that Laurentian University in Sudbury best met its criteria, followed closely by Guelph, Trent, and Waterloo, and it forwarded its findings to the University of Toronto at this time.[44]

Herridge's actions were not the work of an impulsive renegade, for there was a definite method to his madness. He was making a calculated gamble, namely, that the mere suggestion of moving the Faculty of Forestry – steeped in decades of tradition and esteemed for being the first school of its kind in Canada – would spur the University of Toronto to rush to embrace it like a long-lost brother. Although Herridge appeared as a turncoat who was acting to undermine both Sisam and the Faculty of Forestry, the opposite was true. He was simply carrying the ultimatums that the dean had been delivering to the administration over the previous few years – either support the faculty or it will move – to a new level that was intended to resolve the matter once and for all. In fact, Sisam respected Herridge's ploy, and continued to hold him in high regard.[45] Moreover, Herridge was a veteran senior civil servant who knew all the ins and outs involved in paving the way for new initiatives. If he had considered transplanting the faculty from Toronto to Laurentian a bona fide option, undoubtedly he would have contacted officials in Sudbury and laid the groundwork for such a move, which he did not do.[46]

Thereafter, the situation became so plagued by indecision and delay that it would have been farcical had the Faculty of Forestry's future not hung in the balance. Lakehead pushed its cause with – and the University of Toronto sought guidance from – the Ontario government, the OPFA presented its report to the Committee on University Affairs, and all parties anxiously awaited the CIF's study of forestry education in Canada, which was now long overdue owing to the illness of its author, George Garratt. In the meantime, several stimuli drove Sisam to plead with Bissell to take a stand on the issue. The number of applications to the Faculty of Forestry was far outstripping the spaces available at a time when the University of Toronto was forcing the faculty to make budget cuts as part of the university's general austerity policy. More importantly, Sisam was worried that the OPFA's recent recommendation to move the faculty from Toronto might become a self-fulfilling prophecy. His fear was heightened by the letters he began receiving from well-placed foresters who supported the OPFA's diagnosis and prognosis of the problem. Gordon Godwin, an executive vice-president with the Ontario Paper Company, for example, told the dean at this time that there was 'quite a strong body of opinion that such a move would be desirable, principally because of the feeling that the University of Toronto holds forestry in low regard.'[47]

With the issue at an impasse, the University of Toronto began considering fresh approaches to resolving it. In September 1970 Sword, the provost, discussed the matter with Sisam and made it clear that the Faculty of Forestry 'had no prospect of further capital assistance before 1975 and that a long list of priorities is already before us [i.e., the University of Toronto] for use of the first capital funds that become available.' For this reason, Sword suggested that forestry eliminate its undergraduate program to free space for research and graduate work and become part of 'some kind of inter-disciplinary centre or institute dealing with Forestry and Resource Management ... In other words,' the university's provost later reported, 'we talked about radical solutions to the problem of the constraints that are now demoralizing and destroying the health of the Faculty of Forestry.' The upshot was that Sisam would prepare, yet another, report on the Faculty of Forestry's spatial requirements for the next decade and other possible means of extricating his school from this quagmire. He also agreed to extend his tenure as dean for another year in light of 'the murky future that Forestry appears to face.'[48]

Over the fall of 1970 Sisam returned to his desk to draw up another long-term policy paper for the Faculty of Forestry, and he received some extra fodder for the effort. The dean had learned of two major recommendations that would be coming out in the long-overdue Garratt Report, and he forwarded them to Simcoe Hall. One called for a moratorium on undergraduate forestry programs in Canada, and the other insisted that the universities – and the provincial governments – that hosted the country's four existing forestry schools dramatically increase the funding for them. By the end of October Sisam had polished off a ten-year sketch of his faculty's requirements. In it he contended that the Faculty of Forestry would need to accommodate three hundred undergraduates (the number of students for whom Garratt had projected space would be needed) and forty graduates; this would require greatly enlarged facilities and five new appointees. Just as he was despatching this plan to the university's administration, the Science Council of Canada produced a report that was scathingly critical of the conditions with which Sisam and his fellow forestry deans had been dealing. 'Canada's four faculties of forestry are understaffed, underfinanced and their facilities are inadequate,' the document declared emphatically. It called for one decade of sustained support from the universities and provincial governments to repair the damage.[49]

And then, just as Sisam was being bombarded with more bad news, suddenly the ground shifted beneath his feet. Claude Bissell, the university's president, was committed to making the difficult decisions that the period's fiscal reality demanded, and his statement to the Committee on University Affairs, which advised the Ontario government, in late November 1970 explained the implications of his approach for Sisam and the Faculty of Forestry. 'With the imminence of full enrolment, and the consequent freezing of the budget, we shall be faced with the problem of re-assessing our professional commitments,' Bissell declared matter of factly. 'If we are to have continued development in certain schools under the formulae, then we must either phase out other schools or gradually starve them into inactivity. Forestry and Law are two pressing examples of special needs: Forestry if it is to survive; Law if it is to sustain its momentum.' But no sooner was the university's president reporting on the Faculty of Forestry's imminent demise than it became clear that reports of its death were grossly exaggerated. In early December the University of Toronto's position swung 180 degrees when it decided to continue to support the Faculty of Forestry, including its undergraduate stream and a new master's program in resource management, which was under consideration in the School of Graduate Studies. Sword summarized the situation for Bissell at this time and pointed out that it had consequences in two realms. Internally, it meant 'an immediate search for a Dean, a willingness to put additional financial resources into Forestry and a first priority effort to provide improved physical accommodation for the Faculty.' Externally, the provost continued, it 'probably means reaching an agreement which should not, I think, be difficult, with the Committee on University Affairs and with Lakehead University, that Toronto will continue to be the Forestry Faculty in the Province and that it will continue to be willing to accept for advanced standing, students taking preliminary training at the Lakehead University.'[50]

The University of Toronto's about-face was a result of several factors. By this time Garratt had submitted his opus on forestry education in Canada, in which he had admonished the University of Toronto for its dereliction of duty vis-à-vis the Faculty of Forestry. Undoubtedly, the university was keen to portray itself as actively addressing this embarrassing situation. In addition, despite the Faculty of Forestry's small size, its alumni were some of the university's most faithful and generous supporters (especially in terms of money). Coincidentally, at this very moment these patrons of the University of Toronto were informing its development office that they would no longer donate to the university because of its disregard for the faculty. In an era when post-secondary institutions were turning more and more to alumni for succour, the university needed to retain their confidence. This was doubly true because the University of Toronto was concomitantly learning of the perils involved in ignoring alumni from small departments. The university's recent and abrupt decision to close the Faculty of Food Sciences had triggered a well-orchestrated and powerful response from its alumni. The result was a highly publicized battle that cast the University of Toronto as coldly taking a Goliath-like attitude towards the Davids in its family. The last thing the university needed was another ugly spectacle, especially one that involved the Faculty of Forestry, whose activist alumni already felt alienated. Finally, the University of Toronto had also apparently learned that the provincial government favoured it retaining its monopoly on undergraduate forestry education in Ontario and would presumably offer it the attendant support.[51]

Simcoe Hall had definitely extended its hand to Sisam and the forestry school, but it had done so in a careful and calculated manner. It had reassured Sisam and his faculty's supporters that forestry would continue to operate at the university, expressed its 'willingness' to increase its funding for forestry, and to make it a priority both to find the foresters a new home and choose Sisam's successor. But that was it. The university had deliberately refrained from providing the faculty with a concrete promise to increase its level of funding. In fact, the University of Toronto made it clear that the Faculty of Forestry's future vitality depended on generating *external* financial assistance. As Bissell told an anxious Robert B. 'Bob' Loughlan (4T9) in January 1971, 'I can assure you that it will be our policy to retain and develop the Faculty of Forestry, and to pursue additional sources of support. These additional sources of support will be necessary if the Faculty is to carry out in an effective way its increasingly broad and complex responsibilities.'[52]

One aspect to this situation would prove immensely problematic to the Faculty of Forestry. The university had based its decision to reach out to forestry on a mistaken assumption, namely, that 'Toronto will continue to be the Forestry Faculty in the Province.' Within months, that premise was shown to be faulty, and this would be the critical turning point in the faculty's history in terms of casting the fate of its undergraduate program.[53]

The deciding influence was Lakehead's renewed, and ultimately successful, push to gain its own professional forestry program. The University of Toronto's long-standing refusal to provide sufficient accommodation for the Faculty of Forestry had forced Sisam to admit only sixty students to first year. The attrition rate in forestry freed up a sufficient number of spaces by the time the students had reached third year such that Toronto could accept all of Lakehead's students. This arrangement worked smoothly as long as Lakehead required spots for only two dozen or so third-year students in Toronto. But Lakehead's forestry school was being deluged with applicants, and it was more than happy to force the issue by admitting far more students into first year than the University of Toronto would be able to handle two years later. In 1970, for example, Lakehead received requests for admission from 181 applicants and admitted fifty-two of them. Over the course of the first part of 1971, Lakehead conveyed its growing disquiet about the looming predicament to the Committee on University Affairs. So, too, did its second-year students, half of whom would face the bleak prospect of not being able to complete their schooling in Ontario. The local MPP added his weight to their cause, and in chorus, they all urged the government to grant Lakehead University a professional forestry school. The government pleaded for patience, as it was awaiting the 'overall review' of the situation that the Committee of Presidents of Provincially Assisted Universities of Ontario was carrying out.[54]

The issue was decided when the University of Toronto inadvertently played right into Lakehead's hands. In a cunning move, W.D. Bohm, Lakehead's director of admissions, wrote to his counterpart at the University of Toronto in early 1971 to confirm the number of third-year students Toronto would be able to accept from Thunder Bay. He promptly received word that the Faculty of Forestry had room for only twenty-five to thirty Lakehead students, and the number would decrease in the future as the faculty in Toronto increased the number of students it admitted into first year. W.G. Tamblyn, Lakehead's president, forwarded this critical information to Committee on University Affairs Deputy Minister E.E. Stewart. 'We might be beyond this point now [i.e., Lakehead had more than 30 students in second year] depending upon the failure rate in our second year and yet this is the first indication of a possible limitation that we have received from Toronto. As I indicated to you previously,' Tamblyn bemoaned to Stewart, 'this is our primary concern, and the concern of our students. We do have the physical capacity for larger numbers and it makes it difficult to turn down students on the basis that Toronto does not have sufficient capacity in their third and fourth years.'[55]

In the end, the crisis Lakehead had shrewdly fostered gave its campaign the necessary boost to go over the top. It had been able to foment a potential public relations disaster for the government, namely, the inability of a province that boasted of its vast natural resources to graduate five or six dozen foresters each year. The upshot saw the Committee on University Affairs decide, in early February, to 'recommend to the Minis-

ter of University Affairs that Lakehead University be allowed to proceed
with the development of the degree program in Forestry. This recommendation,' Stewart noted, 'will be made independent of any immediate
consideration of the existing degree program in Forestry at the University of Toronto or of an ultimate long-term plan for Forestry education
in this Province. The recommendation was motivated by a number of
factors, the most important of which may have been an indication by the
University of Toronto that it would not be able to accommodate all of the
students who were currently undertaking the first two years of the degree
program at Lakehead University.'[56]

In taking this step, the government ignored the seemingly sage advice
of some of the Faculty of Forestry's well-placed friends within the provincial civil service. George H. Bayly (3T9), the deputy minister of lands
and forests, was the most outspoken. He pleaded with the Committee on
University Affairs to refrain from making a hasty decision and first undertake 'a thorough analysis of the Ontario situation ... Our concern,'
Bayly asserted, 'is that funds already appear to be limited in providing
an adequate basis for forestry education at the University of Toronto. To
provide degree grading [sic] status at Lakehead University will undoubtedly involve additional staff and funds. If the current facilities at the University of Toronto are inadequate and continue to meet the needs of the
students from the Lakehead, it would seem desirable to direct additional
funds to the University of Toronto so that the required expansion could
take place there.' Ultimately, Bayly's entreaty read, 'it would seem pref-
erable to ensure that we have one strong Faculty rather than two inadequate Faculties.'[57]

Nevertheless, Lakehead University had finally achieved the end towards which it had been striving for the better part of a decade. Sisam
and his colleagues should not shoulder the blame for this defeat, as
the measly support the University of Toronto had been giving them left
them little choice but to restrict enrolment and conserve resources. But
Lakehead's success in this instance spoke to the potential value of creating a crisis in order to highlight one's plight. Instead of pursuing this
avenue, Sisam had chosen to manage the faculty in a way that obviated
such disasters. Maybe, just maybe, it would have been wiser to allow them
to happen and hope that their occurrence would have driven the University of Toronto and the provincial government to come to the Faculty
of Forestry's aid in a meaningful way.

The tumultuous events that marked the second part of Sisam's reign as dean of the Faculty of Forestry at the University of Toronto could not overshadow its graduates' remarkable achievements and successes. Those who earned their degrees from the forestry school between 1957 and 1971 continued to follow several patterns that had begun a half-century earlier.

One was the series of international connections that wove their way through the faculty. Its student body included strong representation from across the globe, especially Europe. The same was true of its staff. Erik Jorgensen, for example, was from Denmark, Divenda Roy and Jagdish C. Nautiyal were from India, and Jaroslav Vlcek from Czechoslovakia.

John J. Balatinecz had taken a circuitous route to joining the Faculty of Forestry in 1970. A native of Hungary, he had been a student in the Faculty of Forestry at Sopron when the entire school – its dean, staff, students, and their families – had fled during the uprising against the Soviets in 1956. The Canadian government, the University of British Columbia, and the Powell River Company Limited provided the forestry faculty-in-exile with a refuge, and it became a division of UBC's forestry school the following year.[58]

The graduates from this period also continued to leave their mark on international forestry. An increasing number applied their expertise in 'the developing world.' Various parts of Africa benefited from the work performed by, among others, Michael S. Allen (7T0), John E. Ambrose (6T6), Lloyd O.W. Burridge (6T6), and Kevin P. Campbell (6T1). Several other faculty alumni from these years established themselves in the southern hemisphere. Ronald N. 'Butch' O'Reilly (7T0) earned his master of forestry degree from the University of Canterbury's School of Forestry in Christchurch, New Zealand, and spent over a decade teaching there. Brian J. Myers (7T0) pursued graduate work at the Australian National University and worked as an experimental scientist for that country before opening his own forest tree nursery in 'Oz.'[59]

Familial ties to the faculty and the forest were still forces that drew many students to the forestry school in Toronto. The father of John D. Walker (6T7) was an employee of Ontario's Department of Lands and Forests, and William D. 'Bill' Addison (6T3), John D. Brodie (6T1), and James F.K. 'Fred' McNutt (6T1) were the second generation in their families to graduate from the University of Toronto's Faculty of Forestry. So, too, was Grant D. 'Dave' Puttock (7T1), who went on to earn his doctorate, and join its professorial ranks in 1980.[60]

Puttock was only one of many from this period who entered the hallowed halls of academia. Paul A. Cooper (6T9) and Robert J. 'Bob' Fessenden (6T5) followed in Puttock's footsteps by earning their doctorates and joining the faculty. Others, like Brodie, Peter A. Murtha (6T1), Crandall A. Benson (6T6), and Douglas J. Gerrard (6T0) taught forestry but not in Toronto. Juris Dreifelds (6T5), Walter S. Good (6T4), Lawrence T. Kirby (6T3), Edward B. MacDougall (6T1), and Hendrick P.B. Moens (5T9) became university professors but not in forestry, while Ronald E. 'Ron' Chopowick (6T2) and Glenn N. Crombie (5T9) played important administrative roles at Seneca College in Toronto and Cambrian College in Sudbury respectively.[61]

Like their predecessors, many of the Faculty of Forestry's graduates from the 1957–71 period enjoyed brilliant careers in their chosen fields. Several rose to senior ranks in the forest industry and businesses closely related to it. Michael R. 'Mike' Innes (6T5) earned his master of science in forestry two years after graduating, and went to work for Ontario's Department of Lands and Forests. He then joined Abitibi, the country's newsprint colossus, and climbed its corporate ranks. Over the next four decades Innes held titles that ranged from the firm's manager of forestry to its vice-president, environment. Like Innes, Richard D. 'Dick' Fry (6T5) worked with the provincial government before being hired in the late 1970s by the pulp mill in Marathon, Ontario. By 1979 he was the local chief forester, a position he held for nearly a quarter of a century. Fred McNutt (6T1) earned a master of business administration degree from the University of Toronto, went to work for his family's lumber firm, William Milne and Sons, in Temagami, Ontario, and served for decades as its president and general manager. Jack M. Taylor (6T9) earned his master of forestry from Yale University two years after graduation and began working as a forest economist with the Western Forest Products Laboratory (now FORINTEK) in British Columbia. After a stint with industry, Taylor worked in Vancouver for the Canadian Imperial Bank of Commerce's corporate division as its manager of forestry.[62]

Many alumni from these years continued the tradition of leaving their marks as both scientists and senior managers in the public sector, particularly in a manner that reflected their concern for environmental stewardship. While Neil W. Foster (6T8), Frederick J. 'Fred' Hutcheson (6T8), Edward S. Kondo (6T4), and Ian K. Morrison (6T2) fit under the first rubric; those in the second were far more numerous. John W. Ebbs (6T8), Ian H.H. 'Tony' Jennings (6T3), Larry S. Lambert (7T0), Harold J. McGonigal (6T0), and Edward G. Wilson (5T9) rose to executive positions in the Ontario government. So, too, did Richard M. 'Rick' Monzon (6T3). He joined the Ontario government the same year he graduated and never looked back. He rose through the ranks of the DLF/Ministry of Natural Resources, and served successively as its director of policy and planning and as assistant deputy minister. In 1988 Monzon was appointed to the Management Board of Cabinet as one of its assistant deputy ministers. William D. Tieman (5T8) also reached the rank of assistant deputy minister in Ontario's Ministry of Northern Development and Mines.[63]

As far as exceptional graduates from these years is concerned, Diether E. 'Doug' Buck (6T3) stands out. Born in Indonesia, he studied agriculture in Germany before earning his degree from Toronto's Faculty of Forestry, and thereafter began working for the Ontario government in the Temagami area. While Buck retained the post for the rest of his career, he took time out in early 1989 to take part in a unique expedition that climbed Mount Kilimanjaro. Having received a heart transplant years earlier, he participated in the adventure with two other organ recipients in an effort to raise awareness of their cause.[64]

Finally, two major trends during this period also deserve attention. There was nothing new about foresters entering the teaching profession when they had been unable to find work in their field, but this proclivity became especially pronounced as forestry opportunities grew scarce beginning in the mid-1950s. The result was that, by 1970, over sixty Faculty of Forestry graduates were instructing in classrooms across Ontario. When the idea of developing an alumni directory was raised around this time, James R.M. 'Mack' Williams (5T0) jokingly suggested that 'a good way to track down alumni was through the Ontario Secondary Schools Teachers' Federation.' Some like Tieman (5T8) lamented this situation. 'It seems to me to be a considerable waste of forestry training,' Tieman commented dejectedly to 'Miss' Harman, the dean's motherly secretary. Harman characteristically invited Tieman to keep his chin up and recognize the silver lining in this storm cloud. 'What would appear to be a loss to forestry surely must help to raise the standards of teaching,' Harman pointed out, 'and it does provide an opportunity for missionary work on behalf of forestry among the rising generation where it is most needed.'[65]

The other trend that began between 1957 and 1971 was most definitely new. In early 1961 Rose Marie Rauter, a graduate of Jarvis Collegiate in Toronto, applied to the Faculty of Forestry because she found the city confining and loved the outdoors. What made this decision so noteworthy was the fact that the Faculty of Forestry had heretofore only admitted men, and it was hardly inclined to change its policy of its own volition.

While the press could not resist the temptation to sensationalize Rauter's breakthrough as the faculty's first female student – headlines patronizingly heralded the arrival of 'Rosie the Lumberjack' and the 'Babe in the Woods' – the explanation for how she gained entrance into the country's oldest forestry school hardly merited the commotion. Until this time the University of Toronto had left it up to the faculty to determine who among its annual applicants warranted admission. Women would be members of this group, but Sisam would encourage them to discuss their candidacy with him. Although he recognized their academic ability, he would discourage them from pursuing forestry by describing the many difficulties they could expect to face in the course of studying in this field (having men share tents with women was simply unthinkable!). To overcome this practical problem, Sisam would suggest that they apply with at least several other female friends, something that he knew full well would not occur. In the early 1960s, however, Simcoe Hall began vetting applications for admission to the university, and sent offers to those who merited entry without considering their sex. When the Faculty of Forestry's staff learned in September that Rauter had been admitted, 'Bob' Hosie, its grizzled veteran, announced to Sisam that he would resign if the dean permitted her to show up for class. Not only did Hosie withdraw his threat – the fact that Rauter was a superb student and went on to distinguish herself as a forest geneticist no doubt assuaged his concerns, ironically, he became one of her greatest boosters.[66]

Rauter's presence hardly unleashed a gender revolution in the faculty. She was the only woman to attend the forestry school during the 1960s, and it was not until 1970 that another woman was admitted. Even then, women remained a rare exception at the faculty for the next few years. As a result, these individuals continued to draw the media's attention, largely because the press seemed obsessed with tracking the progress of another anomalous 'Miniskirted Forester.'[67]

But the Faculty of Forestry had no time for such frivolous distractions in early 1971. With the University of Toronto having offered a limited commitment to sustain the faculty, the Ontario government having authorized Lakehead University to establish its own professional forestry program, and Sisam's term as dean drawing to a close at the end of June, the forestry school had its hands full.

The latter issue was the most pressing, and the manner in which the task of choosing a new dean was addressed reinforced the view held both inside and outside the faculty that it needed to move in a radically new direction. The faculty had undeniably undertaken new initiatives in the 1960s, such as urban forestry and wood science, but in the main it had steadfastly clung to its old ways.[68] Not surprisingly, when the non-faculty members of the decanal search committee (including James Ham, the University of Toronto's future president) reported to Bissell on their initial discussions, they stressed that appointing a new head of forestry was simply part of a larger plan to launch it into a new orbit. The critical issue, they agreed, was to ensure that the new dean 'would be conscious of and prepared to work at the evolution of suitable relations between Forestry as a school for training professional foresters and evolving theories of Environmental Studies.'[69]

The faculty's advisory committee delivered the same message far more forcefully. In early May 1971, L.A. 'Bud' Smithers, its chairman and Ontario director of the federal government's Department of Fisheries and Forestry, forwarded to the search committee the qualifications it felt the new dean ought to possess 'based on the major problems facing the Faculty.' The person must be capable of leading the fundraising campaign needed to support an expanded and 'up-to-date forestry facility' and come with 'an established reputation as a leader in forestry in its broad sense.' In general, Smithers stressed the advisory committee felt that

> the next ten years will be critical ones for both the forestry profession and forestry educators in Ontario. Foremost in our minds is the thought that the forestry profession needs a new public image that identifies it more clearly with the people-oriented demands of the future. These demands will emphasize a balanced and compatible production of recreation opportunity, wood production and enhanced environmental quality of forested wildlands. This image must come, not only through curriculum development, but also from the public image of the Faculty members and in particular, the new Dean of the Faculty ... We must have a strong leader who ... has the confidence of all and can spearhead our message to the public.

Smithers provided details as to the ranks from which such a candidate should be plucked, and a few examples to illustrate the advisory committee's thinking on the matter. The person should come from, at a minimum, the 'second level of management' in a provincial or federal department concerned with renewable resources, a comparable level of the forest industry, or the 'first level' of responsibility in academia. As far as category one was concerned, Smithers indicated that Art Herridge (4T9), who was director of the DLF's Division of Resource Management, 'is an excellent example of the type of person we need.' The committee felt Des Crossley (3T5), chief forester in Hinton, Alberta, represented an industry official who had 'the national reputation and drive to fill this assignment.' Finally, I.C.M. Place, CFS director for the Maritimes region, was seen to possess 'all the appropriate qualifications.'[70]

But when all was said and done, yet again the Faculty of Forestry would not get its man. Those who possessed the requisite credentials were uninterested in the post. And although the new dean would bring an unprecedented vivacity to the forestry school and a profound commitment to improving it, his qualifications fell short of those for which the faculty's advisory committee had been aiming.

Chapter 8

'Rebuilding a Neglected and Deplorably Weak Faculty,' 1971–1985

In the middle of 1971 Vidar John Nordin was selected to succeed 'Bernie' Sisam as dean of the Faculty of Forestry at the University of Toronto. A native of Sweden, Nordin had earned undergraduate degrees from the University of British Columbia and his doctorate in forestry from the University of Toronto in 1951. Over the next two decades he had worked his way up the ranks of the Canadian Forest Service's research branch, specializing in forest protection, management, and education. By the time he was hired to be the Faculty of Forestry's dean, Nordin was research manager in the CFS's program directorate and also chairman of the International Union of Forestry Research Organization's subject group.[1]

Although Nordin came to the Faculty of Forestry with an exemplary publication record in tow, his biggest asset was his dynamic personality. His boundless energy and flare for working a room on any occasion infused the dean's office with an elan it had never seen before. This spirit fuelled his seemingly indomitable optimism about his chances for success in rebuilding the faculty. In Nordin's world the wine glass was always half-full, never half-empty. On one occasion when the university informed him that his faculty would receive none of the many things for which he had asked, a senior administrator noted astonishingly, 'of course, all I had for him was bad news but he bounces back easily!'[2]

To be sure, Nordin had every reason to be upbeat during the first few years of his tenure as dean, as this period saw him enjoy a prolonged honeymoon and the faculty's fortunes rise to a new high-water mark. With the University of Toronto having agreed to rekindle its support for the faculty, Nordin made carpe diem his mantra. Toronto's Faculty of Forestry doubled the size of the space it occupied on campus, undertook a wholesale curriculum review, hired new professors, saw its funding for research and graduate work skyrocket, and forged ties with foresters across the globe. It was hardly surprising that, during these heady days, the faculty's undergraduate enrolment swelled from just over two hundred at the time of Nordin's arrival to over 260 four years later, and its graduate program enjoyed steady growth (it was soon averaging about ten incoming master's and doctoral students annually).

But still, all was not well with the Faculty of Forestry. Its facilities were dispersed over a half-dozen locations in and outside downtown Toronto. Nursery work and soils research were conducted at Glendon College, some distance north of campus; the Shade Tree Research Laboratory operated out of the Borden Building on Spadina Avenue; and the faculty used the original forestry building at 45 St George Street, the top floor of a building across the street, and laboratory space in the civil engineering building. This set-up was obviously not conducive to fostering unity among staff and students.

The problems went much deeper than that, however. Certainly, the University of Toronto's decision to end its benevolence towards the faculty in the mid-1970s did not help the situation. But while Nordin and some of his colleagues continually blamed the administration for their deficiencies, the real problems were rooted within the faculty itself. This was an era during which environmental issues gained unprecedented international attention. Not only did the United Nations hold two major conferences to discuss related topics (the first in Stockholm in 1972 on the environment and the second in Nairobi five years later on desertification), the Canadian and American governments executed landmark legislation to deal with pressing matters such as acid rain and pollution in the Great Lakes. Nevertheless, the Faculty of Forestry was unable to convert its grandiose plans for reinventing its role at the University of Toronto into tangible action that tapped this newfound interest in nature. More importantly, by the late 1970s many observers were expressing serious doubts about the quality of the faculty's program, and the faculty was missing its enrolment targets by wide margins even though its admission standard was arguably the lowest on campus. When these forces combined to imperil – yet again – the Faculty of Forestry's future at the University of Toronto, Nordin and his staff fell into the same trap as had their predecessors in the late 1960s. They dug in their heels and short-sightedly refused to consider options that might have delivered the manna that they so desperately needed. The upshot saw Nordin take the faculty on a journey that led to dizzying new heights, but also left its undergraduate program closer than ever to the chopping block.

Vidar Nordin agreed to come to Toronto after he received, as he put it, a commitment from the University of Toronto to develop a faculty 'of excellence,' and the first step in this process involved identifying the ailments that both he and the administration believed afflicted the forestry school. Certainly, it boasted a 'fine nucleus' of professors, Nordin commented soon after taking over as dean, but that was the long and short of its strengths. 'The major aspects of the Faculty operation,' he described, 'have been bad – not enough of the right people, poor housing and facilities, inadequate operational and research funding, and a failure to effectively exploit existing resources. This reflects decades of neglect and subsequent serious lack of capacity to be responsive to rapid changes in technology and demands of society.'[3]

Nordin's first priority was seeing to it that the University of Toronto increased its financial support for his faculty even though this was a period during which university budgets in Ontario were being slashed. During his first year as dean, for instance, the Faculty of Forestry saw its budget grow by nearly $100,000, and the next year he succeeded in winning the astounding 17 per cent increase in funding that he had requested. In fact, the University of Toronto's munificence towards forestry was so out of step with its stringent fiscal policy that it fostered deep resentment among some members of other academic divisions who were painfully aware that their pockets were being picked in order to fill Nordin's.[4]

For obvious reasons, Nordin's other high priority was addressing the Faculty of Forestry's 'extremely inadequate' accommodation. He insisted that the university take 'minimum emergency measures' to address this 'absolutely critical' situation for the short term, for he felt it was 'hampering the total teaching and research program and contributing to an unfavourable atmosphere amongst Faculty and students.' The construction of a new office building at the corner of Beverley and College streets proved timely. Within one year of Nordin becoming dean, the University of Toronto had agreed to lease its fourth and fifth floors for the Faculty of Forestry (this additional 10,000 square feet of space was equal to the size of forestry's original building on St George) even though the rent was expensive (over $50,000 per month). Nordin still complained to Simcoe Hall that the faculty's total space remained inadequate and diffused, but the foresters had much more breathing room.[5]

This additional space allowed Nordin to begin increasing the faculty's undergraduate enrolment. No doubt this move was also a function of the funding formula for post-secondary education, a system in which a larger student body translated directly into greater provincial funding (basic income units, or BIUs). In Nordin's case, he annually raised the faculty's first-year admission quota by ten-student increments, so by 1974 enrolment was at ninety.

At the same time Nordin believed he must dramatically broaden the Faculty of Forestry's activities. After his first year in Toronto, he laid before Donald Frederick Forster and Jack H. Sword, respectively the University of Toronto's provost and acting president, the ambitious new course he was charting and to convince them that pursuing it was indeed desirable. 'The Faculty of Forestry has been too narrow in the scope of its teaching, graduate, and research programs and has not exerted the necessary leadership in natural resources,' Dean Nordin explained. He added that it also had an unequalled opportunity to seize the reins in this area because of its 'many strong sister departments' at the University of Toronto. For the new dean, the Faculty of Forestry had only one option as far as its future direction was concerned. It could no longer accept the status quo, which in Nordin's view was synonymous with '"lumberjack" production.' Neither could it grow into an '"Umbrella" Faculty of the Environment' because inter alia Nordin recognized that the expertise in the environmental sciences already existed at the University of Toronto at both the undergraduate and graduate levels. Instead, it must follow the lead of the Western world's major forestry schools by becoming a 'Faculty of Natural Resources and Environmental Management,' whose name indicated the new comprehensive role it would assume in land stewardship. Moreover, Nordin assured the administration that he could undertake this radical departure by cooperating with existing faculties and thereby cause only 'minimal structural adjustments' at the university.

Nordin provided the university with considerable detail regarding his restructuring plans. They included absorbing the Division of Landscape Architecture because of its 'close affiliation with natural resources and management.' Naturally, the accreted faculty would continue granting the bachelor of science in forestry and bachelor of landscape architecture degrees, but it would also award a string of new degrees at both the undergraduate and post-graduate levels. The former would include a catch-all bachelor of science that was designed as 'a liberal environmentally-oriented bachelor's degree' for students seeking a 'broad, flexible program' but who had 'no specific career goal.' For master's students,

Nordin envisioned his faculty granting both an 'MSc' in a host of 'scientific aspects of natural resources for graduates in forestry, agriculture, science and applied science,' and the country's first 'MLA' to students pursuing postgraduate studies in landscape architecture. Finally, he rhymed off a seemingly infinite number of 'collaborative programs with other Divisions' at both the master's and doctoral levels. These included 'water resources management, water resources science, ecology, urban and regional development, resource economics, [and] air pollution.' In Nordin's eyes, 'a very simple machinery,' such as a steering committee in the School of Graduate Studies, could encourage the development of these 'interdisciplinary graduate and research programmes in environmental sciences which we do not presently enjoy.'

While this was the ultimate goal, in the meantime Nordin set out to modernize the faculty's curriculum, particularly at the undergraduate level, in terms of both course offerings and pedagogy. The upshot was the adoption of a semestered school year (as other faculties and departments had already done), and the introduction of first-year courses in 'orientation,' effective written and oral communication, and 'interpersonal dynamics.' These were intended to assist graduates in countering the attacks by 'environmentalists' on foresters and the work they were doing. Furthermore, over the next few years Nordin introduced other new courses and discarded redundant ones.[6]

To expand the range of subjects that the Faculty of Forestry offered and the research it conducted, and revitalize its staff, Nordin used several avenues to attract a host of new professors. The university hired a number of tenure-stream professors in areas such as recreation and ecology. At the same time Nordin fostered ties with senior officials within both the Ontario and federal governments, particularly the Department of Environment in Ottawa and the Ministry of Natural Resources in Toronto (formerly the Department of Lands and Forests). These connections led to the faculty gaining the services of a handful of adjunct professors who were bona fide leaders in fields such as genetics and tree breeding. These efforts also bore fruit in the form of research grants and funding for specific teaching positions. Finally, Nordin lobbied for support from other groups that were connected to 'resource management.' His biggest coup on this front was his successful effort to have the Canadian National Sportsmen's Show underwrite the cost of a chair in forest wildlife management.[7]

Nordin endeavoured to cement a deep, and permanent, bond with external agencies, and gain their insight into forestry education, by revitalizing the faculty's moribund advisory committee (renamed the advisory board) that Sisam had inaugurated in the mid-1960s. It met for the first time in the spring of 1974. Although its membership was broader than the previous advisory body, it was still heavily weighted in favour of the dominant players in Canada's forest industry.[8]

One of the advisory board's most important members was Adam Zimmerman, vice-president and comptroller of resource-conglomerate Noranda Mines Limited. Not only did Zimmerman bring a genuine interest in realizing prudent forest stewardship to the board, but he was also a close friend to the newly minted premier, William G. 'Bill' Davis, and the Big Blue Machine that would be ensconced as the 'government party' in Ontario for another decade. Both Nordin and the University of Toronto's senior administrators were acutely aware of Zimmerman's tight ties to the levers of power in Ontario and how they could redound to the Faculty of Forestry's benefit; they no doubt explained his presence on the board.[9]

While the faculty had long been active in international forestry, Nordin pushed it to the forefront of this field in Canada. He negotiated agreements with the Canadian International Development Agency to fund exchanges between the Faculty of Forestry in Toronto and forestry schools in the developing world. In the main, the faculty's role was to act as a mentor and facilitate the establishment of advanced forestry policies and education programs. During his tenure as dean, the faculty benefited from the connections Nordin fostered with nascent forestry programs literally across the globe, including those in Brazil, Peru, and Malaysia.[10]

Nordin's other significant initiative, and masterful stroke, was his successful drive to unite the country's six forestry schools in a campaign to improve their collective lot (the University of Alberta and Lakehead University had added professional forestry programs in the early 1970s). He foresaw this new organization as a formal vehicle for expressing their shared interests and lobbying the public and private sectors for greater support. Just over two years after he launched this effort, the Association of University Forestry Schools of Canada had held its first meeting. It had also published a 'national statement' that echoed the Garratt Report's view that Canada's forestry schools had suffered from grossly inadequate support and required a major cash infusion. Thereafter, the AUFSC's annual meetings became forums for Nordin and his cohorts to publicize this message, and his enthusiastic leadership produced results such as grants in forestry from the Natural Sciences and Engineering Research Council.[11]

Finally, Nordin endeavoured to improve the forester's battered image by wading into the middle of one of the period's most heated environmental conflicts. By the late 1960s Algonquin Park had become the battleground between the local loggers and those who wished to see the end of timber harvesting there. The provincial government attempted to resolve the controversy by taking control over forestry operations in the park out of industry's hands and vesting it in a Crown agency. The Algonquin Forestry Authority (AFA), which officially came into existence on 1 January 1975, was expected to ensure that timber would be cut only in specified areas of the park and according to the strict guidelines spelled out in its *Master Plan*. Nordin was offered and accepted the AFA's chairmanship because he saw it as a golden opportunity to demonstrate that forestry was about far more than simply harvesting trees. Nordin's ability to handle his responsibilities in this capacity for eight years with his typical aplomb played a major part in returning a semblance of peace to the park and raising forestry's profile, at least a touch, in the process.[12]

And so, in the span of only a few short years, Vidar Nordin had improved the Faculty of Forestry at the University of Toronto in sweeping and dramatic fashion. In light of the period's blanket budgetary cutbacks, his achievements were truly remarkable. Nordin sought to complete the faculty's makeover in the spring of 1973 when he informed his staff that 'a bottle of fine Hungarian wine awaits the person who submits the best motto.' It – the motto, that is – was intended to complement the forestry school's new 'official Armoirie' that Nordin was having created and registered at the College of Arms in London, England; J.W. McNutt (3T2) had agreed to cover the cost. Thereafter, forestry's battle cry would be *in requiem tempus arbores hodie* – 'For the Rest of Time, Trees Today' – instead of 'Ring upon Ring.'[13]

Above all else, Nordin invigorated Toronto's Faculty of Forestry and coloured its future in rosy hues. When he argued before the Science Council of Canada in the fall of 1972, for example, that he would need 'an academic staff of at least 50 ... to permit a viable, effective division,' Nordin's self-assuredness – buttressed by his impressive track record of winning major concessions from the University of Toronto – made it seem that his words would indeed become self-fulfilling prophecies. 'I am optimistic that Canadian forestry schools are on the move,' Nordin delighted in telling a colleague from the forestry school in Vancouver in early 1974, 'and will not let up one bit until they have achieved the necessary levels of excellence. The future looks great.' In these exhilarating days of new money, accommodations, and staff for the faculty, the dean had every reason to be upbeat about what lay ahead for his forestry school.[14]

While the rest of Nordin's years as dean (his term was renewed in 1978) were sprinkled with success, they were deluged with challenges. The extraordinary progress he had made had not resolved a few major problems he had inherited and a few new ones arose. In addition, the strengths Nordin brought to the faculty unquestionably redounded to its benefit, but a few came at a high cost. No dean could be all things to all people, and Nordin was no exception.

Nordin's ability to focus on the 'big picture' was a case in point. His aptitude for and interest in orchestrating successful glamorous gatherings was legendary, but it came at the expense of attending to the tedious tasks such as managing the faculty's day-to-day affairs. At a time of dramatic retrenchment in government spending on higher education and general economic decline, it was imperative that someone was closely watching the faculty's budget to ensure that the foresters were making the most of their gains in the early 1970s.

Nordin's strength at being the 'ideas' man was also a double-edged sword. It has already been noted how he came up with the creative titles to myriad courses he anticipated developing for the Faculty of Forestry's graduate program. In no uncertain terms, he was a man of vision. But when the University of Toronto's administration asked for concrete suggestions as to how to build a foundation under some of these castles in the sand, none were forthcoming. Nordin brought in professors to teach courses in 'newer' fields such as wildlife management and recreation, but the heap of new degrees he had projected the faculty awarding, at both the undergraduate and graduate levels, did not materialize.

Likewise, Nordin's ability to take charge of the situation and provide much-needed leadership to the Faculty of Forestry was seen as the missing ingredient it had lacked over at least the previous decade. In the vernacular of the street, it was his way or the highway, and this uncompromising approach was often a good thing for the faculty. Battling successfully against the numerous interests who competed for dollars at Simcoe Hall demanded someone who was unafraid to stand up and be counted. Nordin also had the courage to hand several professors their walking papers shortly after he arrived when he and his colleagues deemed their performance unsatisfactory.[15]

But this purposeful approach also had its drawbacks, and two examples serve to illustrate how it could work to the faculty's disadvantage.

The first involved Nordin's plan to recast the forestry school as a Faculty of Natural Resource Management. The University of Toronto already included an Institute for Environmental Science and Engineering, and numerous other faculties and departments, such as geography and geology, were involved in fields that clearly touched on the broad scope encompassed by 'natural resource management.' Nevertheless, Nordin took unilateral action to rename the faculty with a single-mindedness that clearly ruffled the feathers of these other groups on campus. While all parties were keen to see a revivified forestry school in Toronto, they voiced grave reservations about it proclaiming itself the umbrella under which all related divisions should gather. As a result, Nordin was forced to begin work on, as he put it, '"fence-mending" our suggestion to improve the name of the Faculty' with representatives from these other departments. Moreover, Nordin spoke only in disparaging terms to the university's administration when he discussed the environmental science and engineering institute, and presumptuously informed Simcoe Hall that, to avoid duplication and 'working at cross-purposes,' it would not be 'impossible, following a careful review, that we might consider offering the Institute a "home" in our Faculty environment.' Nordin was apparently blind to how his approach in these instances made his task exponentially more difficult. If he were to succeed in reconstituting the Faculty of Forestry as a highly interdisciplinary entity, it was imperative that he forge tighter bonds with more divisions than ever before. Although he clearly needed these other faculties and departments more than they needed him, one of the first things that Nordin had done as dean was put their collective noses out of joint.[16]

The second example speaks to Nordin's insistence that he was the leader of the Faculty of Forestry, and that meant everyone was under his command. It has already been described how Erik Jorgensen – and his protégé, Willem Morsink – had developed the Shade Tree Research Laboratory in the Borden Building over on Spadina Avenue as an autonomous unit of the faculty. Two of the first moves Nordin made upon his arrival at the University of Toronto were to terminate the laboratory's separate existence and bring in his own urban forester, Martin Hubbes (while Jorgensen was on sabbatical in 1972–3). Jorgensen could see the writing on the wall, and quickly secured alternative employment with the federal government. Morsink also departed soon thereafter.

And with that, the Faculty of Forestry had lost someone who was truly the pioneer in his field. Jorgensen had defined the term 'urban forestry' after all! While his strident individualism made working with him a challenge, Jorgensen was unequivocally *the* expert in this field. If the faculty were going to become an international centre of excellence in this area of study, it was imperative to arrive at a modus vivendi with Jorgensen. Shortly after he and Morsink left Toronto, the Ontario Shade Tree Council discussed how their departure had dealt the Faculty of Forestry a major blow. Dean Nordin vehemently denied this was the case, pointing to the two excellent replacements he had found, but his words had a hollow ring to them. No doubt his approach to dealing with this matter contributed to the University of Toronto's administration recommending in the early 1980s that the faculty consider eliminating urban forestry as a field of study (presumably because of its perceived weakness), an option that would have been practically inconceivable had Jorgensen remained at the faculty.[17]

These aspects to Nordin's approach may not have been as important had the University of Toronto continued to offer the Faculty of Forestry exceptionally strong support and the country merrily floated through a period of economic prosperity that afforded the provincial government the luxury of opening its coffers to all suitors who came knocking. But this was not to be. In fact, by the time Americans were beginning to line up for gasoline in the fall of 1973, it was clear that the Faculty of Forestry's run of good luck had ended. The country, province, and university would be squeezed like never before during the rest of the decade and beyond as a result of seemingly uncontrollable inflation and unemployment.[18]

These were hardly the only factors that caused the Faculty of Forestry to suffer more than its share of setbacks during this period. At the same time as it was grappling with major and continual budget cutbacks, Simcoe Hall was informing Nordin that the faculty's performance was substandard and implying that there was not enough space on the St George campus to operate both the faculty's undergraduate and graduate programs. Furthermore, the faculty's headstrong refusal even to consider seriously logical options that the university presented regarding possible new quarters would prove devastating to its development. In addition, although Nordin repeatedly admitted to the administration that something was fundamentally wrong with the faculty, he chose to operate as if all was well on these fronts. As a result, the forestry school's handling of its situation during this period meant that it played a major role in determining its own fate, a role that was utterly regrettable in hindsight.

The Faculty of Forestry's development between the mid-1970s and mid-

1980s was framed by its struggle with the university's administration over the question of where and how to accommodate it. Other issues like funding entered the mix, but the discussions always returned to the matter of where and how to house the forestry school. The longer the two sides bickered, the clearer it became that the university's view of the faculty was growing less and less favourable.

The autumn of 1973 marked an abrupt end to the Faculty of Forestry's remarkably buoyant times. Thereafter, even though it began to enjoy a significant upsurge in enrolment at both the undergraduate and graduate levels (which translated into more revenue for the University of Toronto based on the funding formula), no longer would the administration increase its financial support for forestry. After a few years of fixed budgets, Nordin was being forced to make, as he put it, 'dramatic cuts' to his program. He remained typically defiant. 'One thing is clear,' he told John Evans, the university's president (1972–78), in the fall of 1974, 'I cannot accept to lose ground in some of the initiatives and innovations which have been begun. We will work on alternatives to keep viable and I am confident at the same time that should some budget relief prove possible, the Faculty will receive appropriate consideration.'[19]

On other fronts, any progress Dean Nordin made was offset by subsequent retreats. In February 1974, for example, he organized a meeting at Queen's Park of the deans of Canada's forestry schools with senior officials from the Ontario government (including three ministers), the Science Council of Canada, and the University of Toronto. The gathering turned into a virtual love-in for the Faculty of Forestry. Officials from the Ontario government agreed that forestry had been 'short-changed over the years,' and all recognized that 'a special infusion of money' was needed to rectify the situation. Just as momentum was building to realize this end, however, the Tories shuffled their cabinet and relegated the project to the back-burner.[20]

It was around this time that Nordin and the Faculty of Forestry became fixed on the goal of finding a new home, and for a while, conditions seemed ripe for such an initiative. Notwithstanding the shock of the 'first' oil crisis, the forest industry was still enjoying reasonably good times. Moreover, Nordin launched, in the fall of 1973, 'a one-man campaign' to solicit the $6 million he believed was needed for the faculty's new quarters. His strategy was to convince Canada's major pulp and paper makers to contribute roughly one-third of this total over the course of a few years. Once news of these commitments spread, Nordin believed that the other 'allied industries' would donate the rest of the funds that he required (the alumni would also be expected to pitch in, to a much lesser extent). When the leading pulp and paper executives, like Abitibi's Tom Bell, conditionally agreed to support Nordin's undertaking, it seemed like everything was falling into place.

But one hurdle stood in his way. In the run-up to the University of Toronto's sesquicentennial in 1977, it was preparing to launch its 'Update' fundraising effort. As a result, it was hesitant to grant Nordin's requests that the donations he solicited be both designated for the Faculty of Forestry's new building and counted as contributions towards the 'Sesqui' campaign. This would set a dangerous precedent for the university, particularly because the faculty's two largest benefactors – its alumni and the forest industry – were exceptionally generous when the university's fundraisers came knocking. In the 'National Fund' campaign in the late 1950s, for instance, Abitibi had contributed 1 per cent of the $15 million the university had raised, and the country's other forest companies had been disproportionately lavish. While the university apparently determined, in mid-1975, that it could acquiesce to Nordin's wish, its refusal to identify the 'product' (i.e., the location and nature of forestry's new quarters) that he could 'sell' to the faculty's benefactors kept his fundraising campaign in a holding pattern. The challenge for the faculty thus shifted to getting the university to release a site on which forestry could erect its new accommodation.[21]

Simcoe Hall again made it clear, as it had in the late 1960s, that it favoured moving the foresters to Erindale College, the University of Toronto's campus in Mississauga. The reasons in favour of pursuing this option were stronger than ever. The provincial government had recently announced an indefinite moratorium on support for new facilities, and Erindale had unused space that the faculty could immediately occupy and convert to its uses. Furthermore, the satellite campus boasted both its own programs in many of the 'support' divisions – such as biology – on which forestry relied and a large tract of undeveloped land (some of which was wooded) that was well suited to serving the faculty's needs for outdoor facilities; this would render its research work at Glendon redundant. While the administration was aware that there were potential disadvantages to moving the Faculty of Forestry from the St George campus, it was squarely behind the idea in principle.[22]

Although Nordin was initially receptive to considering the idea, he soon dismissed it out of hand as an umbrage to his faculty. In his eyes, forestry began on the downtown campus, and there it would remain. He passed along the university's perspective on the matter in mid-1975 to

Dave Love, the faculty's most senior professor and recently appointed associate dean. 'Dr Evans definitely retains the view point that Erindale is an appropriate future location for the Faculty,' Nordin told Love. Because Nordin was leaving for a meeting in Vancouver, he instructed Love to draw up a position paper that compared the propositions of remaining on the St George campus and moving to Erindale. But before Love had a chance to begin drafting his report, Nordin had penned a letter to Evans that bluntly informed the university's president that the Faculty of Forestry was staying put, largely because the Forestry Alumni Association had committed to supporting Nordin's fundraising campaign on the condition that the faculty remain downtown. The St George campus was 'the logical site for our development,' Nordin stated matter of factly, 'and it would be a waste of time to press for a development which you may not be able to support for good reasons involving total University planning. In a country where forestry resources and environmental issues are so significant,' he declared, 'the University of Toronto cannot afford to be without strong representation in forestry resources education.'[23]

Ironically, Nordin's insistence on remaining downtown also reflected the Faculty of Forestry's cognizance that its undergraduate program was in dire straits and may have only a limited life expectancy as a result. At a meeting of the faculty's executive committee on Remembrance Day in 1975, at which the merits of moving to Erindale were discussed, the conversation turned to the faculty's flailing performance in comparison with the country's five other forestry schools, especially its rival in Thunder Bay. 'In view of the location of Lakehead University,' the notes from the meeting read, 'and the physical facilities available to them it is quite possible that Toronto may not be able to easily compete with them on the undergraduate level in the long run ... That we are already in serious trouble on this level,' the document states, 'is indicated by the fact that the 1st year enrolment at Toronto in 1975 was 107 with the minimum Grade 13 mark of 60% while the comparable figures at Lakehead were 134 and 78.3%.' For this reason, the faculty believed that its hope for a bright future at the University of Toronto, if it were indeed going to stay there, lay in fortifying its graduate and research work, both of which it felt required proximity to the other downtown academic divisions. 'Otherwise,' it predicted, 'it is quite possible that in the next ten to twenty years the undergraduate enrolments at Lakehead may be so far above those at Toronto that we may be forced to shut this school down.'[24]

The upshot was a remarkable paradox. Clearly, the Faculty of Forestry recognized that its undergraduate program may fall by the wayside, thereby dramatically reducing the school's need for space. Nevertheless, Nordin continued to reassure the administration that he would be able to raise the money needed for the new forestry building just as soon as the University of Toronto confirmed the downtown site on which it could be erected.

At the same time, Simcoe Hall's shrinking faith in the Faculty of Forestry prompted it to reconsider the faculty's situation in light of the prevailing economic conditions, but again, Nordin was cold to its idea. All concerned were acutely aware that the period's deep cuts to operating budgets were not going away any time soon (a minimum 10% reduction was forecast over the next three years), and there was no hope of receiving significant public money for new university buildings. The administration thus suggested that the funds Nordin was so certain he could raise be spent on the faculty's operating costs instead of a major capital project. George Connell, the university's vice-president of research and planning, met with Nordin in June 1975 and stressed that this might be the only way to strengthen the Faculty of Forestry at the present time. Moreover, Connell noted in a position paper he composed a short while later that 'there has been concern in recent years that the quality of students admitted to Forestry was below the University's normal standards.' In other words, the faculty's rising enrolment was not necessarily a sign of strength and grounds for making forestry a budgetary priority, especially when the University of Toronto was facing the need to fund at least two other projects – the athletic complex and 'improvements in Arts and Science space' – that would benefit a much larger segment of its constituency. In response, the dean agreed that, 'if the University was not going to be able to support a program of a certain size, it would be unwise to plan capital spending which was not appropriate.' At the same time, however, Nordin insisted that it was 'easier to raise money if it is for a specific capital project.' In his mind, the university's suggestion was a non-starter.[25]

From there, the Faculty of Forestry's problems worsened. Repeatedly during the mid- to late 1970s the University of Guelph was suggested as the most logical destination at which to relocate forestry. By this time, the faculty's staff was already teaching two courses there, and Nordin had opened discussions with officials at the former Ontario Agricultural College regarding the possibility of cross-appointing a professor between it and the Faculty of Forestry. Furthermore, several staff members (notably Ken Armson and Bob Fessenden) had raised the matter with Nordin in the early 1970s as a viable option for solving many of the Faculty of Forestry's problems. While this initiative died when senior staff members expressed their aversion to the idea, it resurfaced in the late 1970s when Donald Forster was appointed president of the University of Guelph (Forster had been a strong advocate for the Faculty of Forestry when he had served as a senior administrator in Simcoe Hall). This time the Ontario government pointed out that the Faculty of Forestry's greatest need, regardless of whether it were in Toronto or Guelph, involved capital expenditures, and this money was simply not forthcoming from Queen's Park. In addition, the period's severe economic downturn was now afflicting the forestry industry, and its members were no longer in a position to contribute to a new building for the foresters when the paper and lumber companies were bleeding red ink.[26]

More troublesome for the Faculty of Forestry was the university's ossifying attitude towards it. By the spring of 1977 Nordin had submitted another long-term plan, this one to the university's planning and priorities subcommittee (of governing council's planning and resources committee), and the administration tore it to shreds. Questions were raised about whether the faculty's aim, at both the undergraduate and graduate levels, was breadth or specialization. Simcoe Hall also argued that it was time to consider 'reducing or even discontinuing support' for the faculty's doctoral program because of its low enrolment, dubious quality, and bleak funding prospects. The review of the faculty's plan then levelled a far more stinging criticism. For a long time, it noted, there had been 'a general awareness of the space problems that confront[ed] the Faculty.' After investigating the matter, however, the university now had a 'new understanding' of the situation. 'With the exception of Forest Biology,' the document declared, 'the weaknesses now present in the Faculty's programs are not due to space deficiencies.' The administration was also learning that other senior academics at the university entertained grave doubts about the quality of the Faculty of Forestry's graduate program.[27]

Simcoe Hall's decision to communicate its perspective on the situation to the Faculty of Forestry touched off a heated exchange that clearly demonstrated how the chasm between the two sides was widening. The administration believed that the faculty's present incarnation was below the standard the University of Toronto had set for it, and this deficiency was the stumbling block to the faculty receiving both external and internal support for its new quarters. Harry C. Eastman, the university's new vice-president of research and planning, explained that 'things had deteriorated in the past year so far as Forestry is concerned and that there is presently no priority or plan to satisfy the apparent needs of that faculty.'

President Evans added insult to injury when he asserted that the faculty had set lofty goals and failed to achieve them, and was failing its students by not providing them with a strong background education in the basic sciences. 'There is always a problem that critical or provocative remarks will create the image of the skunk at the garden party,' he told Nordin. 'However, my concern is that the ambitious plans will turn out to be fantasy, not only in the University but in the external sector unless the goals and objectives are extraordinarily progressive, compelling and exciting. If my remarks are taken as an invocation for that type of response, then they have been accurately interpreted.' Nordin was incensed by Evans' comments and by what he saw as the administration's unjustifiably harsh attitude towards the faculty. Moreover, from where he sat the Faculty of Forestry's major hurdle to success was the administration's indifference to the faculty's plight and its refusal to facilitate the construction of new accommodations.[28]

The war of words continued over the next few months, and again, they reflected the University of Toronto's deteriorating opinion of the Faculty of Forestry. In early December, Adam Zimmerman, chairman of the faculty's advisory board and one of its most influential supporters, wrote to Evans about the situation. Zimmerman argued that the enormous role the country's woodlands played in both Ontario's and Canada's economy and society made it incumbent upon the University of Toronto to support the faculty far better than it was doing. While he openly admitted that the Faculty of Forestry's present unsatisfactory situation was at least partly 'self-inflicted,' he added that its grossly inadequate facilities were fettering its success. Evans thanked Zimmerman for his views, but candidly added his concerns about the faculty. 'I do believe,' he explained, 'that for Forestry to play the position of leadership that … there needs to be a very dynamic program in teaching and in research,' which Evans clearly felt was lacking at this time. After learning of this exchange, Nordin rushed to defend his track record at the Faculty of Forestry by listing all the changes he had made since his arrival nearly seven years earlier. These accomplishments, the dean argued, were clear evidence that 'the quality and dynamics of our programmes are well appreciated and constitute no obstacle whatsoever to attracting the support of alumni and the forestry resources community. The real impediment to support is the lack of a clear and appropriate University commitment regarding physical plant facilities which the alumni and resources agencies are willing to support.' Yes, his faculty had weaknesses, but it had produced more than its fair share of successful graduates who had risen to prominence. Evans thanked Nordin for the information but remained unmoved by the appeal. 'It is my assumption that your comments suggest that you do not believe that it would be advantageous to consider alternative directions in your educational programme or a more selective approach to priorities in the scientific areas of the Faculty,' he told Nordin.[29]

With the Faculty of Forestry's relationship with Simcoe Hall descending to new depths, and both sides admitting that the faculty was plagued by problems, Nordin ploughed ahead as if all that was needed to improve the situation was a positive attitude. In his submission to the *Annual Ring* for 1978, the dean pointed out that the Faculty of Forestry was 'again at a crossroads and experiencing a crucial time in its history. The economic climate is unfavourable and severe University budgetary constraints that have been with us for the past three years will continue in the immediate years ahead.' For this reason, he declared that the faculty had to look to external funding partners if it were to strengthen its research and graduate programs and generate the money needed to construct a new home. Increased federal funding would help address the former matters, and the major obstacle as far as the latter was concerned was defining 'the product' to sell to the alumni. With Churchillian doggedness and terminology, Nordin declared that 'we are not going to sit back and flounder!' and he reminded his supporters that 'the pessimistic see only trouble – the optimistic see challenge.' Nordin's views were so out of step with reality that Keith Reynolds, one of the Faculty of Forestry's biggest backers, felt obliged to voice his incredulity at this incongruity. 'I thought you were remarkably restrained' and accented only the positive, Reynolds told Nordin. 'On the other hand,' he asserted, 'I felt you understated what must surely be your concern at the apparent low level of priority accorded to your Faculty in the University of Toronto.'[30]

In the meantime, several new options for accommodating the faculty came to light, but predictably, the university rejected the location the faculty desired and the faculty rejected the one the administration desired. The Faculty of Forestry was keen to embrace 1 Spadina Crescent as both its new home and a clear 'product' it could sell to its benefactors. The University of Toronto thwarted this plan, however, on the grounds that it was both too expensive to renovate and too large for forestry's needs. As an alternative, the university suggested that the faculty occupy part of the Sandford Fleming Building after it was renovated following a fire that had occurred in early 1977. Although this would leave forestry spread between two locations – it would still retain its original quarters at 45 St George Street, the sites were relatively close and would allow for a significant concentration of its activities. Nordin was adamant, however, that such a move would sound a virtual death knell for the faculty by tightly limiting the total space available to it. He also warned Evans that, 'if, for severe budgetary and other planning constraints, this is the best that can be done for Forestry at the University of Toronto as a permanent arrangement, then I suggest we consider finding another home for the Faculty at an institution that might be in a better position to provide the support that would more properly reflect the significance of forestry education in relation to the importance of Canadian forestry resources and manpower needs.'[31]

The Faculty of Forestry favoured pursuing another avenue that the administration, try as it might, was simply unable to block. In the mid-1970s the university was considering how best to redevelop a large tract in the southwest corner of the campus. The existing divisions in this area were in need of improved accommodation, and the university was keen to utilize this precious space far more efficiently. It thus planned a new 'multi-purpose complex' for the 'Southwest Campus.' Although a large number of supplicants vied for space in the facility, the task force that investigated the matter recommended in October 1977 that the best proposal involved divisions that included botany, geography, and forestry; they wished to turn the site into the 'Earth Sciences Complex.' No sooner had the report been rendered than the university unilaterally removed forestry from the plan. Vehement protests from other members of the Earth Sciences group, however, compelled the university to acquiesce.[32]

The problem for Nordin was that the era of government funding cutbacks to Ontario's universities was worsening, making his challenge much more daunting. James Ham, the former dean of the Faculty of Engineering and of the School of Graduate Studies, succeeded Evans as the university's president in 1978 (and held that office until 1983). His approach to coping with what was now chronic underfunding was to transform the university into what Friedland describes as 'an intellectually leaner and tougher place.' This included reducing undergraduate enrolment even though it meant less money from the government, and focusing increasingly on graduate work. If the forestry school were going to survive on the St George campus in this environment, it would have to follow this lead. While Nordin and his staff had already identified this as the Faculty of Forestry's likely future path, how they handled events during his second term as dean ensured that its mandate included little room for educating undergraduate students.

The situation for Nordin and the faculty turned from bad to worse over the course of late 1978. The grim projections for funding turned out to be accurate. In describing the situation to Michael D. Sandoe (7T6), who was doing graduate work at Oregon State University, Nordin let down his habitual sanguine guard and admitted that 'it does not look good and most people are worried, including myself.' Around the same time, Nordin learned that Ottawa was about to slash its funding to the country's forestry schools.[33]

Far more serious, and telling as far as the problems within the Faculty of Forestry were concerned, was the dramatic attrition rate that afflicted its staff and the university's refusal to rectify the situation. In early September 1978 Donald A. Chant, vice-president and provost of the university, felt obliged to alert Ham of an 'alarming situation that has developed in the Faculty of Forestry. Over the last 18 months,' Chant explained, the faculty had 'lost no fewer than 9½ full-time academic staff (out of a total of about 23) through resignations and deaths.' While the latter were unavoidable, the former reflected a fissure that divided the members of the faculty. By this time, a solid clique – led by Ken Armson and Bob Fessenden – believed that the Faculty of Forestry ought to focus on the more traditional field of 'boreal forest management.' With Nordin favouring a fundamental shift towards urban and international forestry, the Armsons and Fessendens found gainful employment elsewhere. To fill this gaping hole, Nordin initially asked for the university to approve six tenure-stream appointments. In an internal memorandum on the subject, Chant declared that there was 'absolutely no way that I can approve replacement recruitment at this level in Forestry. However, I will have to give them something.' This turned out to be one new position.[34]

Simcoe Hall also doled out an even larger dose of tough love to the faculty at this time. Major space had become available at Scarborough College, the University of Toronto's eastern satellite campus, as a result of lower than expected enrolment. Coincidentally, the size of the space roughly matched that which the forestry school required. Ham thus asked Nordin to consider moving to the suburban location. Although the University of Toronto's administration was acutely aware that the Faculty of Forestry would refuse this option kicking and screaming, implicit in the university suggesting this option was a clear message. If the faculty wished to sustain its undergraduate program, it would have to do so somewhere other than the downtown campus.

The administration was convinced that the advantages of moving the faculty to Scarborough far outweighed the disadvantages. It would finally silence the Faculty of Forestry's griping about inadequate quarters and consolidate the foresters in one place. Furthermore, Scarborough needed a new library, and the Faculty of Forestry housed its own sizable collection of materials in its building and would require space in the suburban campus for its books and journals if it moved there; this gave the university a compelling reason to invest its limited capital in a new library in Scarborough. The relocation would also provide the Faculty of Forestry with a golden opportunity to develop 'unimpeded by present spatial problems,' give its nursery and research work (which was still tenuously operating at Glendon) a permanent home, greatly reduce the university's costs in supporting the faculty (the space it rented at 203 College Street was costing the University of Toronto roughly $600,000 per year), and presumably strengthen Scarborough. Moreover, the university was considering relocating all its 'non-Arts and Science professional divisions,' and saw moving the forestry school as a first step towards developing 'multi-faculty operations' at the suburban campuses. Finally, the university increasingly felt that 'Forestry does not have an intrinsic role in the activities to be carried out in the South-West Campus as presently conceived.' While the administration committed to investigating the feasibility of moving the Faculty of Forestry to Scarborough, it was apparent that the only location where the faculty would continue in its present incarnation – with vibrant graduate, research, *and* undergraduate programs – was in Toronto's east end.[35]

Also at this time, the University of Toronto informed the Ontario government that it was contemplating jettisoning the Faculty of Forestry's undergraduate program unless it received greater support from Queen's Park. In mid-1979 Ham wrote to Keith Reynolds, the Faculty of Forestry's long-time supporter and Ontario's deputy minister of natural resources, to investigate whether 'there [was] sufficient requirement for Professional Foresters to justify the continued participation of the University of Toronto.' While senior ministry officials were downright indignant about the inquiry and suggested that the faculty ought to find a more appreciative host university, they could not produce concrete evidence to dismiss the pointed question posed by the president of the university. Ham was even more forthright when he met with Bette Stephenson, Ontario's minister of education. He argued that Lakehead's forestry school was catering to the demands of the more junior levels of the profession, and it was essential that the 'research-based professional school [in Toronto] become more effective. Its development can take place in the University of Toronto or possibly at another university such as Guelph.'

Ham reminded Stephenson how the present financial crunch was forcing the university to review all of its programs and reduce its staff at a rate of 2 per cent per year despite constant enrolment. In this milieu, he declared, it was 'very difficult to make new staffing commitments for any purpose. Nevertheless, if a significant long-term public need for improved forestry education is confirmed, the University is prepared to consider the issue of staffing priority provided the Faculty can be adequately housed and is enabled to deliver an appropriate range of research services to the Ministry of Natural Resources and of course, to private industry.' In Ham's eyes, the nub of the matter was clear. 'The overall objective of achieving more effective professional educational support for a sustained provincial policy thrust in forest research and development,' he contended, 'cannot be attained in any location without a measure of designated government support.'[36]

In responding to Ham's concerns, the Ontario government made it clear that forestry and forestry education, especially in Toronto, were simply not policy priorities. Stephenson and James Auld, the minister of natural resources, told Ham that, 'for reasons of political balance,' there was 'an underlying desire to further the development of forestry education at Lakehead.' At the same time, all recognized that it would be some time before Lakehead University developed 'the ambiance for a first class professional school capable of serving the future needs of forest management.' Furthermore, Auld predicted that, with the inauguration of the forest management agreements in Ontario signalling the dawning of a new day in silviculture, 'forests as a renewable resource are to be clearly recognized in public policy as having major long-term importance.' Unfortunately, he despaired, 'I am not wholly sanguine that political priorities will permit such a policy to be clearly ennunciated [sic] and sustained.' The only way for 'an initiative in forestry education and research at the University of Toronto,' the two ministers reasoned, was if it were set in 'the context of a government policy initiative originating from Natural Resources.' Stephenson echoed this view, arguing that there was no prospect of capital grants to the Faculty of Forestry in the short term 'unless there were a Government initiative which assigned a public priority to forestry,' a development that seemed highly unlikely.[37]

Under the circumstances, moving to Scarborough held out the only real hope for the Faculty of Forestry to gain its new quarters in the near future and retain a substantial undergraduate program, but the faculty – and particularly Nordin – refused to give the idea a fighting chance. This was especially important in light of the faculty's most recent five-year

plan, submitted in 1978–9, that stressed how its undergraduate program
had 'the highest priority in the Faculty.' But instead of spearheading
a rational and thoughtful assessment of the Scarborough proposition,
Nordin both actively and passively did all he could to kill it.[38]

In fact, the Faculty of Forestry had set the tone for its reaction to this
suggestion long before the University of Toronto had even begun investigating the matter. In February 1979, just as the university was striking
its Task Force on the Feasibility of Relocating the Forestry Faculty to
Scarborough College, the faculty's graduate students despatched a stern
protest against the idea to John Evans, the university's president. They
also expressed their view that the real issue in this instance was not moving out to the suburbs but the university's refusal to support the faculty
adequately.[39]

Nevertheless, the task force went ahead. It reported in August that,
indeed, it was feasible for the forestry school to move to Scarborough.
Because it did not comment on the desirability of pursuing this option,
Evans wrote to both Joan Foley, the principal of Scarborough College,
and Dean Nordin asking them to assess the desirability of such a relocation. Before Foley and Nordin had officially responded to Evans' request, the task force was invited to present its findings to a meeting of
the Faculty of Forestry's council. The atmosphere was electric as Harry
Eastman, the university's vice-president of research and planning, traced
the history of the task force's activities and how two major needs – Scarborough College's for a library and the Faculty of Forestry's for 'consolidated facilities' – would be met by moving the faculty to the suburbs. At
the same time, he stressed that the formation of the task force did not
indicate that the University of Toronto felt that forestry 'rated as a first
priority for improvement by the University; other divisions are equally
poorly off or worse.'[40]

The events at that meeting in October 1979 exemplified the faculty's
emotional response to this issue at a time that demanded a reasoned
and detached one. For her part, Principal Foley presented a cautious
yet sensible approach to the matter. 'Generally,' she declared, 'Scarborough College welcomed the Task Force's recommendations and felt that
there would be advantages academically in having other disciplines on
campus.' At the same time, however, she wished to learn the university's
long-term plans for the Faculty of Forestry if it moved to Scarborough. In
contrast, Nordin and some of his colleagues did all they could to quash
the idea by raising specious objections to it. Veteran staff member Dave
Love asked, for example, if there were another smaller division on the

St George campus 'who are in a more difficult situation than Forestry space-wise. Would it not be feasible for one of those other divisions to move to Scarborough?' Some of his colleagues followed by trying to turn some of Scarborough's strengths into potential weaknesses. They spoke derisively of the college's facilities and the quality of its professors, criticisms against which Foley easily defended.

The university's senior administrators showed no patience for the Faculty of Forestry's petulance, and outwitted the faculty's carping at every turn. When a few Faculty of Forestry professors argued that, despite the better space Scarborough offered in terms of both quality and quantity, 'facilities don't necessarily make for good research,' Ronald W. Missen, vice-provost of the university and chairman of the task force, shot back that it 'has been the Faculty of Forestry itself that has been saying that poor quality space and facilities are affecting its research programmes.' Likewise, when the faculty endeavoured to argue that relocating in Scarborough would be a major obstacle to attracting undergraduate students, Eastman rightly pointed out that the data simply did not support this interpretation. 'Forestry on the St George Campus is not that attractive to undergraduate applicants now,' he declared contemptuously. 'Twice as many applicants with 80% and over indicate Lakehead Forestry as first choice as do Toronto Forestry.' Consequently, he sternly warned Nordin and his colleagues that they 'must face the President's questions [about the move to Scarborough] with seriousness and not conjecture.'

But this the Faculty of Forestry refused to do, and it instead launched a major public relations campaign to remain downtown. The Foresters' Club led one arm of the offensive by striking an ad hoc committee that produced a survey that, it declared, demonstrated that there was 'overwhelming' resistance to moving to Scarborough. The committee's evidence, however, was patchy at best. As far as data for travel time went, for instance, it admitted that it had none to show that it would take more time for students to get to school or how many students would actually enjoy a shorter commute because they lived in Scarborough. Nevertheless, the committee concluded that 'students have commented that they fear an increase in both travel time and travel expense if the Faculty were to relocate.' Likewise, many of its other conclusions were couched in such ambiguous language that they were virtually meaningless. The Foresters' Club survey argued, for example, that relocation would have an 'unknown' impact on housing and transportation, and 'change' the quantity and quality of undergraduate studies. In the same vein, the club convened a meeting in late November to discuss the matter, and invited

Foley to speak to the issue. The Foresters' Club proved to be a poor host, as its repeated insistence that Scarborough was literally in another time zone left Foley little choice but to declare repeatedly that 'it isn't really far away at all.' This line of attack reflected the view of most of the Faculty of Forestry's staff as well. On one occasion when they were discussing the possible move, one professor snidely suggested that, if the faculty migrated to Scarborough, it 'may as well be in Siberia.'[41]

Just at the height of the faculty's intense and heart-felt discussions about the matter in the fall of 1979, a sagacious 'voice in the wilderness' endeavoured to recast its perspective on the issue. Dave Love had been appointed chairman of the 'Forestry Relocation Response Committee,' and he had invited the faculty's staff and students to comment on the subject. At least one professor, Victor G. 'Vic' Smith, used the opportunity to take a step back from the brink and proffer a sober second thought.

Smith presented sound and rational reasons for viewing the move in a different light. He took issue with the faculty's contention that it was imperative to remain on St George Street because of its tight relations with other downtown divisions and officials in government and industry. 'I believe that the strength of our Faculty's programmes rests primarily with the individuals within the Faculty and the interactions that take place among them,' Smith's memorandum began. 'Of secondary importance is the strength we draw from our contacts outside the Faculty. If the outside contacts were uppermost, it implies an inherent weakness among ourselves as individuals in that we cannot survive without the outside support.' As far as enrolment was concerned, Smith argued that 'a serious attempt should be made to estimate the enrolments we can expect if we were to move to Scarborough – recognizing that it is still part of the U. of T. but in a suburban setting that may attract students we are now losing to Guelph, our greatest competitor, or to Lakehead who competes with us for forestry students.' Likewise, he pointed out that there was value in pushing the university's administration to foster growth at Scarborough College in areas 'that better match Forestry's graduate and research needs,' an aim that could be accomplished with cross-appointments between Scarborough and the Faculty of Forestry. Moreover, he noted that the faculty had 'scheduled with great ease' the engineering courses for students in the wood science program. There was thus every reason to believe that the same could be done to ensure that, if the faculty moved, these students 'could spend entire days on the St George Campus rather than attempting to shuttle between the two campus's [sic] for courses on the same day.'

Finally, Smith hit on the crux of the matter. Evans had urged the Faculty of Forestry to assess the potential advantages of moving to Scarborough, something it had refused to do. Smith delivered the same message. 'The committee *must* acknowledge in its report,' he insisted to Love, 'that Scarborough does merit some consideration and that it does have a number of good points besides the space benefit ... The proximity of our staff to others in the College under the same roof should lead to some exciting possibilities; these should be identified.'[42]

Unlike the Faculty of Forestry, Joan Foley and others at Scarborough College carefully evaluated the proposal and concluded that the advantages outweighed the disadvantages. On the one hand, the arrival of the forestry school would create a monumental practical headache for her and her colleagues. Furthermore, she informed the university's president that 'some concern has been expressed in the College that, in the past, the research programs of the Faculty of Forestry appear to have lacked the vigour characteristic of many divisions of the University of Toronto.' This was an especially serious concern, she underscored, because the recent report of governing council's planning and priorities subcommittee 'recognizes Scarborough College as one of the divisions of the University which is highly active in research and which accords research development a high priority.' On the other hand, those at Scarborough boasted strengths in areas related to forestry (such as biology and environmental geography), and that, 'in our opinion, Scarborough College has much to offer academically to undergraduates of a professional division.' Foley was thus ready to reach out to Nordin and his team. As she put it, the 'College perceives Forestry research and development as an area of crucial national concern deserving of the University's full support and hence welcomes the opportunity to become more closely involved if that is to occur.'[43]

The Toronto media picked up the story of the warm embrace Foley was offering the Faculty of Forestry and portrayed it as Romeo courting Juliet. A headline in the *Toronto Star* in early December 1979 described how Scarborough 'College beats drum for forestry,' and how the satellite campus was on 'pins and needles over the proposed relocation' of the faculty to its campus. The article also explained that the University of Toronto's president had revealed that Scarborough would 'welcome the Faculty of Forestry' because it saw the foresters' arrival as delivering a string of attendant benefits to the campus. It also pointed out that the move eastward would help the faculty address its accommodation issues. The piece closed by noting, however, that the Faculty of Forestry had not yet submitted its formal response to the task force's recommendations.[44]

When it did so, it predictably rejected outright the suggestion that it move to Scarborough. The faculty argued that it was imperative that it remain on the St George campus because its graduate and research programs required the downtown campus's 'mature academic environment' and the undergraduate program depended on the 'interdisciplinary resources' there. Moreover, the faculty presented a 'better alternative' to moving to Scarborough, namely, expanding its original quarters at 45 St George Street in the short term and moving into the Earth Sciences Complex when it was built. 'In other words,' the faculty's official response concluded, 'there is nothing in the Scarborough proposal which is superior to that in the St George proposal; conversely, the St George Campus offers an academic environment which is clearly superior to the Scarborough Campus.'[45]

Foley was disgusted by the Faculty of Forestry's response, but the resonance between her criticisms of the faculty and those levelled by Simcoe Hall indicate that her views were not simply those of a jilted lover. She admitted that she was disappointed but not surprised by the forestry school's response, as she understood why it would wish to remain downtown. But, she stressed, all the reasons in favour of doing so were known before the task force had undertaken its study. More importantly, the forestry faculty's 'response does not discuss at all the benefits of a move to Scarborough compared with the present situation of the faculty, but focuses only on the disadvantages.' Its response had also ignored Scarborough's major strengths in fields related to forestry. In addition, Foley remarked that many of the letters that the faculty had submitted to support its rejection of the move to Scarborough had come from deans and administrators who had a vested interest in seeing the Faculty of Forestry remain downtown because its removal from 'the S-W campus plan would be a severe blow' to their aspirations to realize this development.

For Foley, the faculty's approach to dealing with this particular issue reflected the general malaise that afflicted forestry. 'I cannot help but note,' she confided to Ham, 'that the character of the response and the supporting documentation which ... [the Faculty of Forestry presented to make its case] is revealing with respect to questions which have been raised about the quality of the Faculty. A particularly dismal piece of information is that only one full-time graduate student out of twenty-three chose to enrol here because of the reputation of the Faculty. If Scarborough College were to enter a close partnership with the Faculty, it would be a challenge, to say the least. I do not pretend to know what the proper resolution of the Forestry question is, but this field is of such obvious national concern that I must say that, if this University is to keep a Faculty of Forestry, it *must* find some way of upgrading the operation.'[46]

Incidentally, at this time the Faculty of Forestry could afford to disregard Foley's views. It had unknowingly snubbed someone who would, however, come to play a fundamental role in the school's development in the years to come.

Simcoe Hall, and especially President Ham, were profoundly frustrated with the Faculty of Forestry's behaviour in this instance and communicated that it would have serious repercussions. Even before Nordin had officially informed Simcoe Hall of the faculty's response to the Scarborough proposition, Ham had urged the dean to recognize that the forestry school 'might be missing an opportunity to establish a much improved unit on Scarborough Campus,' specifically in terms of consolidating its activities under one roof and ensuring that its undergraduate program remained vibrant. Conversely, staying downtown would mean both an increasing focus on graduate and research work and probably the end to undergraduate enrolment. Ultimately, when Nordin and his colleagues proved recalcitrant on this occasion, it was they – and not the university – who determined the Faculty of Forestry's fate. In mid-1980 Ham confirmed that the University of Toronto would accept the faculty's decision regarding the move, but he euphemistically informed the faculty that the university would be looking to a 'steady state' as far as the Faculty of Forestry's undergraduate student body was concerned, and that 'where opportunities exist, emphasis will be placed on upgrading standards.'[47]

Simcoe Hall did not stop there. It was so vexed by the faculty – and especially by Nordin – that it launched a series of verbal assaults against the foresters in general and their dean in particular. After having attended a number of Faculty of Forestry meetings in the fall of 1980, the university's vice-president and provost, Ronald Missen, complained to Nordin that the agenda for each rendezvous had been dominated by one topic, namely, the 'physical plant facilities for the faculty.' He thus asked Nordin whether this obsession had allowed the Faculty of Forestry to neglect 'more fundamentally important matters.' The university's senior administration also compiled data that indicated that, compared with Canada's other universities that hosted English-language forestry schools, the University of Toronto ranked favourably in several crucial categories, including the ratio of full-time students to professors. President Ham wasted little time in using this information both to defend the university's position vis-à-vis the forestry faculty and attack what he saw as its languishing directorship. He sent the information to E.F. 'Ted' Boswell, vice-president of woodlands for E.B. Eddy and a critical member of the forestry faculty's advisory board. 'Since it is common practise [sic] these days for darts to be thrown our way with regard to the question of our commitment of Forestry,' Ham explained, 'I thought you might be interested in this comparison.' Ham then went on the offensive. 'What thoroughly concerns me is the quality of leadership and development in Forestry,' he opined. 'These fundamental issues seem to have been buried under the Faculty's single minded concern for space.'[48]

Simcoe Hall then censured the faculty when it reviewed its latest 'Plans and Priorities.' It shot down the Faculty of Forestry's hope for diversifying its activities, suggested discontinuing urban forestry and transferring wood science to the Faculty of Engineering, and insisted on the faculty improving the quality of its program. This would result in a much leaner faculty, but the administration offered it a carrot stick. If the Faculty of Forestry shaped up, the administration would raise addressing its inadequate facilities to a 'high priority.' This would involve vacating its rented space at 203 College Street and moving into the Sandford Fleming Building in the short term, and taking up residence in the anticipated southwest campus Earth Sciences Centre in the long term.[49]

No sooner had the administration riddled the forestry faculty's latest long-term plan full of holes than Nordin and his colleagues began receiving some much-needed good news. In early 1981 the university's governing council approved developing the southwest campus as a 'centre for Resource and Environmental Studies' that included forestry. Roughly one year later, persistent lobbying by James Ham and Adam Zimmerman, whose friendship with Premier Bill Davis proved invaluable in the effort, bore fruit when the Ontario government announced its commitment to fund roughly two-thirds of the project's $42 million cost. Industry and the university would provide the rest of the money. The problem for all concerned, however, was that the economy was in another downward spiral and the major fundraising campaign among the 'natural resources' sectors was, in Zimmerman's words, 'a non-starter at this time.' The faculty then learned that it would be receiving exponentially more in the way of financial support from the federal government. This propelled its total research funding from roughly $200,000 in late 1978 to more than five times this total by 1982, the same year in which the Faculty of Forestry celebrated its seventy-fifth anniversary. The university agreed to mark this occasion by awarding a pair of iconic foresters the degree of doctor of laws, honoris causa, and the faculty could take immense pride in the fact that one went to one of its own, Des Crossley (3T5).[50]

While the Faculty of Forestry was enjoying the festivities that marked its diamond jubilee, the University of Toronto was crystallizing the direction in which it saw itself and the faculty moving; this path would have severe consequences for Dean Nordin and his colleagues. In March 1982 Ham described how the funding crisis left the University of Toronto little choice but to 'limit the range of its endeavours by focussing its strengths in research and teaching in key areas.' For the Faculty of Forestry this meant concentrating its resources on forest biology and management, regardless of what Nordin believed the faculty's focus should be. Around the same time, Frank Iacobucci, the university's provost, tabled his *White Paper Task Force*. It recommended that the University of Toronto make better use of its scarce resources by trimming or terminating weak programs and striving to make the university, if it was not already, the 'Harvard of the North.' As the report put it, 'it is widely accepted that the University of Toronto is now and should continue to regard itself as a research-based university.' To ensure that the Faculty of Forestry understood the policy's implications for its activities, Roger N. Wolff, Iacobucci's successor as the university's provost, set them out as George Connell, the university's new president, prepared to meet with the faculty in the fall of 1984. 'State that the University, through Jim Ham's support two years ago,' Wolff carefully instructed Connell, 'recognized the important role Forestry can play in the University of Toronto. The role is one of acting as an interface between the disciplines of Botany, Geology and Geography, etc., and the forestry [sic] industry. The Faculty should continue,' Wolff emphasized, 'to develop its strength in research and graduate study and become the leading research oriented Faculty of Forestry in the country.' In this vision, there was clearly little room for teaching undergraduates.[51]

Long before this point in Nordin's tenure had been reached, the Faculty of Forestry had graduated hundreds of students whose accomplishments continued a series of traditions that had begun long ago. In terms of international connections, the faculty was stronger than ever, as its students and staff reflected Canada's increasingly multicultural society. This trend was particularly pronounced in terms of Asian representation in the student body, although natives of the Caribbean and Africa were also more prevalent at the faculty during these years. This period also saw the

Faculty of Forestry receiving a steady stream of graduate students from India who were drawn to Toronto by their interest in studying under their fellow countrymen – namely, Nautiyal and Roy – who were members of the faculty's staff.

Nordin's years as dean also saw the Faculty of Forestry deepen its involvement in international forestry. Its graduates increasingly plied their skills in the developing world, including William Aspinall (7T7) in Costa Rica and Peru, Robert K. Harris (7T7) in St Lucia, Paul J. Martins (8T3) in Brazil and Honduras, and Walter Sarafyn (7T5) and Jeremy St John Williams (7T9) in Malaysia. Others chose to work in the Western world, including Guillaume Gignac (8T5) in Sweden and Richard E. Cooper (7T3) in Scotland. And a handful of graduates worked in both realms. As a student, Donald W. 'Don' Gilmore (7T2) served an internship in Finland, and after graduation he worked in China. Chong L. Tan (8T4) returned to his native Malaysia after graduation to work in the woodlands end of the furniture business. Following a short stint in China, Tan began working as a senior manager for Ikea in Sweden. Martin O. Kemmsies (8T5) set the standard as far as being a citizen of the international forestry scene goes. A native of Washington, DC, and graduate of the Faculty of Forestry's wood science stream, he is a polyglot (he is competent in nine languages) whose career has included stops in Norway, Brazil, New Zealand, Sweden, and Germany. In 2002 he was given responsibility for establishing a branch operation in Brazil for the Swedish firm, Casco Adhesives.[52]

Lineage was still an important factor in attracting students to the Faculty of Forestry. Douglas C. 'Doug' Drysdale (8T1) and John C. Stewart (8T4) graduated one generation after their senior namesakes, and Garth I. Jameson (7T9) and Andrew G. Chewpa (8T5) followed in their fathers' footsteps by earning their bacheor of science in forestry degrees in Toronto. This period graduated its share of brother teams, including Harold M. (7T7) and Winifried R. Armleders (8T0), Raymond Richard (7T3) and Kenneth V. 'Vic' Robinson (7T1), John A. (7T8) and Eric E. Salo (8T1), and Tomir J. 'Tom' (7T1) and Andrzei B. 'Andy' Tworzyanski (8T0). It also witnessed the Faculty of Forestry's only sister team, Hedy A. (7T9) and Susan A. Wiecek (8T5), and sister-brother combination, William E. (7T4) and Elizabeth M. 'Liz' (7T8) Raitanen.[53]

The Faculty of Forestry at the University of Toronto also continued to produce graduates who distinguished themselves in their professions, both outside and inside the forestry realm. Laird M. Miller (8T4) blazed a remarkable trail. He earned a master of business administration degree from the University of Toronto in 1986, and became a chartered accountant specializing in forest companies. Roughly a dozen years later, he became the chief financial officer of London Drugs, a Vancouver-based, private firm. Likewise, John M. Duncanson (7T4) has spent a career in the financial end of the forest industry, and he is one of the world's top experts in this field. In the academic realm, Sandy M. Smith has enjoyed remarkable success after earning her doctorate from the Faculty of Forestry in 1985 and joining its staff three years later. She quickly established herself as one of Canada's leading forest entomologists (her specialization is biological control of forest insect pests), and the Entomological Society of Canada recognized her extraordinary abilities when it awarded her the coveted C. Gordon Hewitt Award in 1993. Eddie Bevilacqua (8T4) also entered the Ivory Tower, teaching forestry at Syracuse University's College of Environmental Science and Forestry. Likewise, Gary R. Bortolotti (7T7) and Robert A. 'Bob' Rosebrugh (8T5) joined the professorial ranks (the former at the University of Saskatoon and the latter at Mount Allison University), but their expertise lay outside forestry: Bortolotti specializes in avian biology and Rosebrugh in mathematics and computer science. Others, such as Stephen J. 'Steve' Colombo (7T7), Donald A. McGorman (7T9), and Alexander J. Mosseler (7T7), have enjoyed successful careers in the civil service. Similarly, Richard M.U. Ubbens (8T5) has earned international distinction for his work as Toronto's city forester, particularly for his work in managing and restoring forest ecosystems in the heart of the country's most densely populated region. Finally, Murray G. Ferguson (7T4), David G. Milton (7T4), Brian D. Nicks (7T7), and Rob G. Tomchik (7T8) have worked their way up the ranks of the 'traditional' forest industry's managerial ladder. Ferguson is the third generation in his family to work for the pulp and paper mill in Dryden, Ontario, and has long sat atop its forest stewardship team. Milton has occupied a similar position within the Ontario Lumber Manufacturers' Association, Nicks with E.B. Eddy/Domtar in Espanola, Ontario, and Tomchik with Donohue/Abitibi in northeastern Ontario.[54]

And just like previous eras, this one witnessed the Faculty of Forestry producing graduates who made landmark contributions to what has become known as 'environmentalism.' James G. Robb (8T4), for instance, is a native of Scarborough, Ontario, and acted as chairman of the Save the Rouge Valley System. It succeeded in protecting from development a major swath of this watershed just outside Toronto's eastern boundary. Unrivalled in this field of endeavour is Solon L. 'Monte' Hummel. A committed environmentalist since his youth, he came to the Faculty of Forestry as a mature student in the late 1970s, with an academic background in philosophy (both a B.A. and M.A.). On the vanguard of the Canadian environmental movement – Hummel co-founded Pollution Probe in 1969 – he realized he was working in a field without the requisite technical knowledge. To fill this lacuna, Hummel enrolled to do a master of science in forestry degree at the Faculty of Forestry, which he earned in 1979 (he had been acting as coordinator of Innis College's Environmental Studies Program since 1975). By this time he was already president of the World Wildlife Fund Canada, an organization he helped build. In recognition of his lifelong dedication to protecting the country's environment, Hummel was named an officer of the Order of Canada. Tellingly, although Hummel had often confronted foresters in his work, he openly admits that, 'arguably North America's greatest conservationists, people such as Aldo Leopold and Douglas Pimlott, were trained as foresters.'[55]

By the mid-1980s Nordin's reign as dean had run out, and the difficulties that the university encountered in replacing him spoke to the major challenges that the Faculty of Forestry faced. Over the course of late 1982 and into the middle of the next year, a decanal search committee was struck but it failed to find a suitable candidate. Apparently, the word had gotten out that the Faculty of Forestry was beset by problems, and acting as its dean was a most undesirable assignment. The committee thus convinced Nordin to stay for another year.

When it reconvened over the course of 1983–4, again it could not find a qualified – and willing – successor to Nordin, even though it had been granted permission to pursue international candidates. With its back against the wall, the commitee asked Dave Love, the most senior faculty member, to serve as acting dean for 1984–5. This would buy the Faculty of Forestry another year to find a new leader.[56]

Nordin's parting comments reflected his trademark sunny disposition in the face of trying conditions. In one of his last addresses to the faculty advisory board in November 1983, Nordin listed a string of triumphs that his time as dean had witnessed. These included a ten-fold increase in research funding (it now topped $2.5 million), two new tenure-stream appointments, and the promise that the Faculty of Forestry would soon occupy part of the yet-to-be-built new Earth Sciences Complex. While it was difficult to argue with his numerous accomplishments, Nordin's peroration had typically ignored mentioning the portion of the glass that was half-empty. Unfortunately for the Faculty of Forestry, Nordin's panache for the job of leading it had not been enough to fill that goblet.[57]

Chapter 9

'Forestry at U. of T. Is Not Dead Yet,' 1985–2005

Several major forces shaped the Faculty of Forestry's development during the 1985–2005 period, and most were unfavourable. These were years of unprecedented social awareness of the environment and the ever more urgent need to care for it wisely. The United Nations convened landmark conferences and produced a number of reports on the issue, and these helped prod governments in Canada to buttress their environmental policies. In 1992, for example, a national forest strategy (i.e., *Sustainable Forests: A Canadian Commitment*) was published and a few provincial governments followed up with their own practical initiatives; Ontario enacted its Crown Forest Sustainability Act in 1994. In this milieu, the 'traditional' approach to practising forestry was outdated. Because decisions regarding how to manage woodlands could no longer be made in a vacuum within which only silvicultural factors were considered, the skill set with which the Faculty of Forestry was providing its undergraduates was less and less valuable and marketable.

At the same time, although the Canadian economy enjoyed intermittent growth spurts during this period, it generally stumbled. Not only did this make less public funding available for universities, but the difficulties hit the country's forest industry especially hard. At home it faced rising costs, trade conflicts and ever more devastating attacks by radical environmentalists, and increasingly stiff competition from new players in the field abroad. And although Ontario – and the other provinces, for that matter – dramatically improved their forest management practices during this period, this did not translate into major employment opportunities for foresters. The upshot was a paucity of job openings, especially permanent ones in the public sector, which was historically where many of the faculty's graduates had spent their careers. In a position paper on the depressing situation, faculty member Fred J. Keenan noted that it had deteriorated so much by the mid-1980s that Canada's six forestry 'schools will have to deal with the questions of whether too many foresters are being educated in Canada, and whether voluntary and coordinated reductions in undergraduate enrollments [sic] should not be contemplated.'[1]

The situation at the University of Toronto was little better for the Faculty of Forestry, as challenges arose from both new and old quarters. The era of steeply declining funding for post-secondary education in Ontario not only continued, but it worsened. In response, the University of Toronto stepped up its drive to become an internationally renowned centre of graduate studies and research. Achieving this goal entailed making some tough decisions in terms of chopping programs, however, and this made for tumultuous times on campus. Not only did the affected parties protest vehemently, but many, including influential donors, questioned the University of Toronto's ability to govern itself in light of the mayhem it was creating. While the university endeavoured to address the problem by democratizing its governance, Simcoe Hall could not help but appear heavy-handed each and every time it threatened to drop the axe on another facet of its operations.[2]

It was within this broader context that two very different deans guided the Faculty of Forestry. The first, Justin R. 'Rod' Carrow (6T1), steered it through the choppiest waters it ever encountered (1985–94). The major disturbance was its undergraduate enrolment, which had slipped to unhealthy low levels. When Carrow took the helm in 1985, it hovered at about half its annual target of sixty new students. Five years later, the dean could count the incoming class on his fingers and toes *and* have a few digits left over. Conversely, the faculty's graduate program was thriving in terms of both students and research funding. This dichotomous situation, combined with the continued and severe provincial budget cuts and the university's over-riding agenda of becoming an institution centred on research and graduate work, gave Simcoe Hall a prime window of opportunity to achieve a long sought-after goal vis-à-vis the Faculty of Forestry: terminate its undergraduate program.[3]

Although Carrow and the faculty's supporters fought the valiant fight against the effort, Joan Foley ensured that the university not only realized this aim but did so in a most callous manner. As the University of Toronto's provost, she was now in a position to affect fundamentally the faculty that had, in 1979, haughtily rejected the proposal to move to Scarborough College, where she was principal. As provost, Foley certainly demonstrated her unequivocal commitment to pursuing the university's agenda to cut costs and concentrate on graduate studies and research by rendering a number of tough decisions and demonstrating that due process would not be a hurdle to her success. Nevertheless, the evidence suggests that her actions in dealing with the Faculty of Forestry were fuelled by a single-mindedness to enjoy the last laugh at the hands of an academic division that had rebuffed her more than a decade earlier.[4]

While Carrow was instrumental in facilitating the Faculty of Forestry's recovery when he chose not to abandon ship immediately after the University of Toronto cut the Faculty of Forestry's undergraduate program, his successor, Rorke Bryan (1994–2005), was largely responsible for fathering the faculty's rebirth as a fundamentally transformed unit of the university. Bryan arrived with a clear idea of the direction in which he believed the Faculty of Forestry ought to move, and he unilaterally and zealously pushed his agenda during his two terms as dean. In the main, he was overwhelmingly successful in this effort, although it came at a cost. Nevertheless, the upshot was a viable faculty with a future, a description that would have seemed most inappropriate a dozen years earlier.

In many ways, Rod Carrow was a most unlikely candidate for the deanship of the Faculty of Forestry. After an undistinguished record as a student at the faculty, he earned graduate degrees from the University of British Columbia and Cornell University while specializing in forest entomology. Thereafter, he began a highly successful career in the civil service at both the provincial and federal levels. When the Faculty of Forestry – specifically Dave Love and Vidar Nordin – asked him to serve as dean in 1985, he was thoroughly enjoying his job as assistant deputy minister in New Brunswick's Department of Natural Resources. Despite his relative lack of experience in academia, Carrow's initial disinterest in assuming the post (it was common knowledge that the faculty was in trouble), and his satisfaction with his present employment, Carrow agreed to take up the challenge.[5]

Like most of his predecessors, Carrow initially enjoyed a honeymoon period, albeit an abbreviated one. The Davis government (1971–85) had committed to fund the lion's share of the new Earth Sciences Centre, in which the University of Toronto had promised to locate the Faculty of Forestry. When the Liberals dethroned the Conservatives – for the first time since the Second World War – in the 1985 provincial election, rumours began to swirl that the new premier, David Peterson, would withdraw the government's backing for the project. Adam Zimmerman, the faculty's long-time and superbly connected adviser and advocate, would hear none of it and arranged to see the premier, together with George Connell, president of the University of Toronto, and Carrow. 'David,' Zimmerman forthrightly opened the rendezvous with Peterson, 'I can't believe we are having this meeting.' Within short order, the funding was restored, albeit for an attenuated facility (Zimmerman saw to it that his firm, Noranda, paid the lion's share of industry's contribution). This unleashed a flurry of planning related to the faculty's move into the new home, an effort that generated tremendous optimism about its future. At the same time, although the declining enrolment in the undergraduate program was troubling, the Faculty of Forestry's graduate and research programs were booming. Consistently during Carrow's tenure, more than fifty master's and doctoral students were studying annually at the faculty, and the funding to support them remained at a high level, reaching roughly $2.5 million by the end of the 1980s. At the same time the faculty forged ever-stronger international connections. It scored a major coup during the mid-1980s when it entered into agreements with several forestry schools in China, the world's up and coming – yet still relatively closed – superpower.[6]

Word that the Faculty of Forestry was definitely moving into the Earth Sciences Centre coincided with it undertaking yet another review of its curriculum. There were at least two driving forces behind this. The impending move to the Earth Sciences Centre prompted Carrow to deem it the appropriate time both to reflect on the faculty's present orientation and, far more importantly, update and improve it in a meaningful way. At the same time, the administration let the Faculty of Forestry know that its status in its new quarters would be determined by its willingness to effect substantial alterations to its pedagogy. 'There are resource implications for the Earth Science Building,' Carrow's notes from a meeting with the provost in 1987 explain, 'if the Faculty does not change the structure of its programmes.'[7]

While this process was under way, the Faculty of Forestry was forced to wrestle with some old nemeses. Most notable were the renewed calls for the faculty to terminate its undergraduate program, especially in light of its shrinking student body, and the University of Toronto's obvious support for this proposal. While the university was debating closing its Faculty of Architecture and Landscape Architecture in early 1986, Robert G. Rosehart, president of Lakehead University, spoke out in that faculty's defence. He wrote President Connell and argued that the university should support the Faculty of Architecture and Landscape Architecture but that 'I would not make the same statement with respect to Forestry. It would seem realistic to have one major Forestry effort at the Lakehead.' When Rosehart's views became known, they aroused the ire of those who supported the faculty in Toronto, especially Carrow. 'A prompt response from you to Dr Rosehart, reaffirming the University's commitment to Forestry on this campus,' the dean demanded of Connell, 'is necessary to allay the concern generated by Dr Rosehart's comment.' Connell clearly only did so, however, to placate the faculty. He informed Rosehart that the latter's comments regarding the Faculty of Forestry were 'unacceptable,' but not because Connell disputed them. Connell instead stressed that, in principle, he felt it was inappropriate for university presidents to meddle in each other's business. Moreover, he explained that the university had decided a few years earlier that the Faculty of Forestry's role at the University of Toronto would be to concentrate on research and graduate work. In a follow-up exchange, the contents of which Connell did not relay to Carrow, he openly admitted that the forestry school 'remains a very serious concern in my mind.'[8]

The Faculty of Forestry also continued to face deep budget cuts and an attendant attrition in its complement of professors. Just before Carrow had officially taken over as dean in mid-1985, the faculty learned that it would be forced to incur a more than 2 per cent reduction in its budget the following year. 'There is absolutely no fat in the system as far as we are concerned,' faculty veteran Dave Love informed Simcoe Hall on learning the news. Nonetheless, the faculty's budget continued to shrink over the next few years, albeit at a slightly lower rate. In early 1988 Carrow decried how the situation was preventing the faculty from both renewing its teaching staff (since 1979 it had suffered a 32% loss in complement and was now the smallest in Canada) and hiring professors to deliver new courses. At the same time he sent a clipping from the *Globe and Mail* that announced that the University of British Columbia was appointing six new tenure-stream forestry professors. 'This is our competition!' Carrow exclaimed to Simcoe Hall.[9]

Despite these ominous signs, the Faculty of Forestry toiled away on its long-term strategy, and by mid-1988 it had produced a document – *Towards 2000: A Development Plan for the Faculty of Forestry* – that, for the first time since the Second World War, laid out a plan for reorienting it in a meaningful manner. Recognizing that forestry had, over at least the previous decade if not much longer, undergone dramatic changes, the document stressed that the faculty's location at the University of Toronto provided it with a unique opportunity to develop a 'well-defined education and research profile that is distinctive, and takes advantage of both the academic strengths of several departments at the University of Toronto, and the corporate presence of the natural resource sector in Toronto.' In practical terms, this meant that its general goals over the next ten years would be to focus on broadening existing knowledge of sustainable forest management, forest biology, and utilization of wood fibre. At the undergraduate level, although the two streams – forest science and wood products science – would be retained, the predominant aim was clearly to improve the program's quality. This would entail gradually raising the admission standard, making the curriculum much more rigorous (especially in terms of basic sciences), and introducing new courses that would better equip graduates to deal with modern ethical and social issues related to natural resource management. The graduate program, which was already successful, was to continue striving for excellence but also introduce a non-thesis master of science in forestry degree, and gradually increase the size of its student body. As far as research was concerned, the plan identified four areas in which the faculty would specialize in order to maximize both funding opportunities and its relevance: tree growth and survival, environmental stress, forest management systems, and pest management systems.

The Faculty of Forestry's strategic plan also outlined the resources that would be needed to achieve these goals, and tangible means by which its success would be measured. Most notably, it sought three tenure-stream professors, one to replace a vacant position in timber design and two new ones in forest management science and wood science, and $20,000 to help promote its undergraduate program. *Towards 2000* also laid out specific milestones that the Faculty of Forestry would be expected to hit in terms of enrolment, admission standard, and new pedagogy. Finally, and most importantly in light of future developments, the plan called for the new program to undergo a comprehensive review in 1993–4 to determine if it were hitting these targets.

The release of the plan in June 1988 fuelled a wave of excitement that the Faculty of Forestry rode for over a year, as many saw it as the elixir that had eluded the foresters for so long. In August 1988 the faculty hosted a special evening to mark the beginning of 'Forestry Renewal,' which was described as 'a new chapter in the history of Forestry at the University of Toronto.' Then in January 1989, the faculty co-sponsored – in concert with the School of Continuing Studies – a symposium on Old Growth Forests, and the turnout of over five hundred greatly surpassed expectations. Six months later, the faculty officially marked the 'inauguration of [its] new, consolidated facilities' (i.e., the Earth Sciences Centre) with a week-long celebration. In no uncertain terms, Carrow and the Faculty of Forestry's supporters were gripped by a vivacious spirit that they felt signalled a pronounced upturn in the faculty's fortunes.

This cheery outlook was buttressed by a resounding endorsement from Adam Zimmerman, the chairman of the faculty's advisory committee and steely yet supportive critic of the University of Toronto's forestry school. In conveying his views on *Towards 2000* to Connell, Zimmerman recognized that 'perhaps a half dozen years ago we had a faculty that had an uninspiring staff, hopeless quarters and an inappropriate curriculum and thus a declining enrolment.' Nevertheless, as he stressed to the president of the university, the faculty had improved. For this reason he, and the committee, fully supported the goals outlined in *Towards 2000*, and urged Connell to help the Faculty of Forestry replenish its depleted teaching ranks. 'We accept as a given,' Zimmerman asserted presumptuously, 'that the University is committed to the Faculty of Forestry and we wish to offer the best advice we can as to improvements.'[10]

But Zimmerman, and the Faculty of Forestry's supporters, were dead wrong in their assumption. Ironically, it was at this very time, with the faculty having finally formulated a strategy for effecting a bona fide renewal, that the University of Toronto began making it clear that it was committed neither to the realization of the goals outlined in *Towards 2000* nor even to sustaining the faculty in its present incarnation.

The first sign that, yet again, Simcoe Hall and the faculty were on different wave lengths came in the fall of 1988. In an effort to buttress his staff's shrinking ranks, Carrow hoped to solicit financial support from forest companies to help cover the cost of 'NSERC' Industrial Research Chairs in Forestry. The administration reviewed his plan and rejected it on the grounds that it would interfere with the university's 'Breakthrough Campaign.' Carrow shot back that the two initiatives were hardly in conflict, as his effort would be appealing to the firms' research divisions while the university sought support from corporate donation offices. After another heated exchange, Carrow dejectedly told Joan Foley, the provost, that the administration's attitude towards the matter 'left me more bewildered than ever before.' President Connell sent a private note to Foley that patronizingly informed her that 'Rod [Carrow] clearly needs help in seeing the bigger picture.' A few days later, Foley attempted to reassure Carrow that the administration did 'not wish to stifle fund-raising efforts by the Faculty,' but central control of the process was essential to its effectiveness.[11]

Then Simcoe Hall lambasted *Towards 2000*. Dan Lang, assistant vice-president (planning) and university registrar, was responsible for vetting the plan, and he found it wanting on nearly every front. In the first place, Lang pointed out that the forest industry was in disarray, there was a declining demand for 'graduates of the B.Sc.F,' and that the persons whom industry was hiring were often graduates of non-forestry programs. So why, he wondered, was the Faculty of Forestry planning on increasing undergraduate enrolment? Moreover, Lang was highly sceptical about the faculty's ability to support an undergraduate program with a broad curriculum, a graduate diploma, a master's degree with two options, a doctoral program, and a research agenda with four streams. Furthermore, the document called for more resources for the faculty at a time when neither the University of Toronto nor the Ontario government were in any position to provide them. In conclusion, Lang was dubious about the usefulness of *Towards 2000*. 'It is in some respects less realistic than the previous plan,' he argued, 'particularly in regard to enrolment and capital funding. To the extent that it rests on questionable assumptions, it is likely to be ephemeral.'[12]

The faculty responded to the administration's criticisms and, after making some concessions, emerged with the essence of its plan intact. By October 1989 Foley and governing council had reviewed and approved the document. While the faculty agreed that permanent appointments would be delayed because of funding constraints, it immediately began implementing the new curriculum and working towards the milestones outlined in its plan. The ultimate litmus test of the strategy's success would be the program review, which was still set for 1993–4.[13]

Just as it appeared the Faculty of Forestry had laid the foundation for its renaissance, Foley sent Carrow a Trojan horse. In an innocuous letter, she conveyed to the dean the disappointing news that her office projected budget cuts totalling 10 per cent over the next six years. In reality, she *assigned* the Faculty of Forestry a 10 per cent reduction, even though the University of Toronto's long-term funding guidelines for the 1990–1 to 1995–6 period indicated an across-the-board cut of 4.8 per cent was needed. Nevertheless, she asked Carrow what impact the 10 per cent cut would have on his faculty. Recognizing the severity of this measure, but believing that Foley would never act on his prediction, Carrow grimly forecast that it would spell the end to the undergraduate program.[14]

That was all Simcoe Hall needed. On 13 March 1989 A.H. Melcher, the university's vice-provost, delivered the bleak news to Carrow. By way of background, Melcher stressed the downward trend in the Faculty of

Forestry's undergraduate enrolment, the existence of other forestry programs in Canada and Ontario, the limited employment opportunities
for graduates with a bachelor of science in forestry degree, and the need
to concentrate resources during a time of cutbacks on 'what we do well.'
Despite the optimism that was generated while preparing *Towards 2000*,
'it is now difficult to see how the Undergraduate Program can achieve
distinction ... All of this supports the conclusion,' Melcher declared candidly, 'that the University can no longer afford to support the program
given our budget projections and the present arrangements for funding
of universities by the Provincial Government. Your statement concerning the effect on your professional program of a 10% reduction in your
budget over six years reinforces this pessimistic judgement.' Recognizing that this was 'a most sensitive, important and complicated question,'
Melcher asked the dean to prepare – 'as expeditiously as possible' – a
proposal that outlined how the Faculty of Forestry could both phase out
its undergraduate program and concentrate its efforts on developing
'high-calibre' graduate students and research programs.[15]

And so, the battle lines had been redrawn. No longer was the Faculty
of Forestry lobbying merely to defend its budgetary allotment. Now it
was engaged in the fight of its life to save its undergraduate program.
Although Carrow would do all he could to win the day, the evidence
indicates that there was nothing he could have done to succeed in this
endeavour. Although his relative inexperience in dealing with the Byzan-
tine world of university politics did not help his cause, nor did his refusal
to assume a bulldog stance when attacked, the University of Toronto –
particularly Foley – were determined to chop the faculty's professional
forestry stream.

The enrolment situation at this time undermined the strength of Carrow's case. Although the Faculty of Forestry's undergraduate student
body had been shrinking for some time, by 1990 it had plummeted to
depths not seen in nearly four decades. The first-year class consisted of
only thirteen, and total full-time enrolment was just over seventy.

These distressing numbers drove Carrow to come up with novel solutions that would save the undergraduate program and deliver significant
cost-savings, but the administration met each suggestion with a dismis-
sive counterproposal. The provost had struck a Working Group on the
Environment to investigate means by which the University of Toronto
could better utilize resources in environmental education and research,
for example, and the Faculty of Forestry was one of the most active con-

tributors to the effort. The group ultimately recommended creating a new Faculty of the Environment, with forestry as a department, but Simcoe Hall rejected the proposal. Then came an initiative from the Faculty of Arts and Science to establish a Division of the Environment, and the Faculty of Forestry began developing a plan – *New Directions* – that would closely link it to the new unit. Carrow, in consultation with Caroline Tuohy, the vice-provost, also came up with the idea of turning forestry into a 'second entry' program, akin to the faculties of law and medicine. While these ideas were still being fleshed out, Foley asked the Faculty of Forestry to consider becoming an institute within the School of Graduate Studies.[16]

By the end of 1991 Carrow had had enough of what he saw as Foley's games; given the degree to which the administration had toyed with the Faculty of Forestry, the torrent of frustration he sent her way was understandable. He began by declaring Foley persona non grata in the faculty. As he told her, 'I believe it would be counterproductive, and perhaps very destructive, for you to meet with members of the Faculty to discuss future options for the Faculty of Forestry at this time.' He then provided a plethora of reasons to support his perspective. Upon becoming dean back in 1985, he explained, Roger Wolff, the university's vice-provost, had encouraged him to develop a plan for the Faculty of Forestry, and the result had been *Towards 2000*. While the University of Toronto had been reviewing the document, Simcoe Hall had assured Carrow that, 'in a university the size of Toronto, the number of students entering forestry, whether it be 25 or 50, was insignificant ... instead, that maintenance of quality in the student body was of primary importance.' Carrow then listed the series of events that, since the time that governing council had approved the faculty's long-term strategy, had undermined his efforts to achieve its aims. These included the assignment of a 10 per cent budget reduction only months after the University of Toronto had approved *Towards 2000*, the university's request that the faculty consider a new model to deliver its programs within the new constraint, and the string of new program initiatives over the course of 1990–1, all of which Simcoe Hall had shot down. 'Thus, in the two years since *Towards 2000* was approved,' Carrow decried to Foley, 'there have been five different proposals from your office relating to the design and delivery of our programmes.' Not only had this disrupted the Faculty of Forestry's internal planning, Carrow explained, but it had seriously damaged morale. 'You have expressed some concern about my reluctance to act as Dean, but I can assure you that there has been no proposal from your office in recent years that has remained viable long enough for me to see it through to completion. I should add,' Carrow continued, 'that the events of the last two years represent only the latest chapter in a long story in which the university has attempted to expel forestry from the St George campus.'

The dean then made it clear that he was at wit's end. Although he recognized that recent events were 'indeed unfortunate,' he nevertheless argued that they had

> largely destroyed the trust and good faith that I thought had been established between your office and the Faculty. It has forced me to re-examine the existence of forestry as an area of study at the University of Toronto. History shows that, for several decades now, there has been a lingering discomfort with forestry as a distinct area of study, both within and outside Simcoe Hall. This discomfort manifests itself not as outright animosity, but more like periodic indigestion. Over the past 25 years, in an attempt to relieve this indigestion, there has been a variety of attempts to get forestry off the St George campus; these attempts have been argued on the basis of some convenient and visible criterion, such as enrolment, which on critical examination, could not be defended. I think it is time for the University of Toronto to settle the question about the role of forestry at University of Toronto, so that we can all get on with our lives, and do something useful to society.[17]

While Carrow closed by outlining a number of ways in which the matter could be decided, Foley had her own plan in mind. The Faculty of Forestry's strategic plan called for its record to be reviewed in 1993–4, but Foley was prepared neither to wait that long nor allow any office but her own to carry out the investigation. She thus seized the initiative in February 1992 by appointing the 'Working Group on the Future of the Faculty of Forestry,' chaired by James Edgar Till, a professor in the Department of Medical Biophysics. Its mandate was not to recommend the future path down which the Faculty of Forestry should travel. Instead, the working group, in which the faculty was well represented and whose work the faculty initially supported, was charged with assessing a range of possible options for the faculty and analysing the implications of each for the University of Toronto, the Faculty of Forestry, and other directly affected divisions.[18]

The group submitted its report in mid-April 1992, and its assessment of the situation underscored the dramatic improvements that Carrow had made in the Faculty of Forestry's operations. The document stressed that these were troubled times for the forestry profession in general, but that Carrow had done a wonderful job of changing with the times and strengthening his school even with the diminishing resources at his disposal. It also recognized that most of the problems, especially declining undergraduate enrolment, were not of the faculty's making (Lakehead was struggling with the same issue). Most importantly, it declared that, not only was the undergraduate program sound, but the graduate program had received an 'A' rating in its last review by the Ontario College of Graduate Schools. Simply put, the issue was no longer the quality of the faculty's work, but rather the underutilization of its resources, specifically in the area of undergraduate teaching.[19]

To address this issue, the group presented five possible models for the Faculty of Forestry's future development, four that included providing it with far fewer resources. Option 1 entailed eliminating the undergraduate program and creating a strictly graduate faculty. This would allow for savings on academic staff but would create myriad problems, including upsetting the alumni and the 'external community,' and possibly not providing enough staff to sustain the present high level of research funding. Option 2 foresaw offering a 'Major Program' in the Faculty of Arts and Science instead of a bachelor of science in forestry, and retaining the graduate program. The group felt that this option would provide more flexibility than Option 1. Option 3 included establishing a single, four-year program in forest conservation and environmental science leading to the bachelor of science in forestry, continuing the graduate program, and collaborating with Scarborough College in offering a program in community and international forestry. Although pursuing this plan would be more costly than the present set-up, it would substantially increase the faculty's teaching 'productivity.' Option 4, which would see the Faculty of Forestry become a department in either the Engineering or the Arts and Science Faculty, was stillborn because the host faculties objected to it. Finally, Option 5 suggested combining Options 2 and 3. It would involve establishing a 'Major in Forestry' in Arts and Science, and continuing the Faculty of Forestry's graduate program. The attractiveness of pursuing this avenue was that it came with 'no apparent disadvantages' and would be the most cost-effective solution and the one that utilized the Faculty of Forestry's teaching resources most efficiently.

Although the report was generally complimentary of the Faculty of Forestry, it drew heated criticism from both the faculty's academic and non-academic staff. Their attack centred on the document's cursory nature (the report had been submitted after barely two months of study), but they also took issue with its call for a significant reduction in their numbers.[20]

Foley endeavoured to assuage their concerns by delivering a solemn pledge. In August 1992, she wrote to the Faculty of Forestry to explain that she, too, could see the weakness in the group's report. She agreed that its analyses were 'incomplete and form a less than adequate basis for any decision. I intend,' Foley assured the faculty's staff, 'that any decisions will follow full discussion by all interested parties who will have access to the relevant information.'[21]

Concomitantly, the faculty mounted a campaign to make the best of a bad situation. Realizing that all five options that the group had presented would cause the Faculty of Forestry to shrink in size, it decided that its best hope for the future lay in 'Option 5,' which would at least keep its undergraduate program alive. Carrow thus orchestrated a campaign to pressure the administration into choosing this path. The Faculty of Forestry prepared a form letter outlining its case, which its members were encouraged to send to their local MPPs. In addition, the faculty's other supporters – particularly its alumni and other forestry advocates – wrote Simcoe Hall in defence of the faculty's undergraduate program. Ironically, one of the letters came from the head of Lakehead University's forestry school, a scenario that would have been unimaginable a few decades earlier![22]

As usual, one of the most insightful appeals came from Adam Zimmerman, whose comments hit on the dichotomy that has traditionally marked the relationship between Canadians and their forests: espousing a profound affection for them but refusing to back it up with meaningful action. In response to a request from both Foley and Carrow for his views on the matter, Zimmerman argued that the University of Toronto should not only retain the Faculty of Forestry's undergraduate program, but lengthen it by one year to allow for more practical experience and increased training in business and communications. He explained that forest companies are generally run by people 'who have had to learn forestry on the run,' a process he felt indicated 'that pure professional forestry must be embellished with engineering, accounting, law and communications to make it attractive. Thus it is hard to oppose the folding in of forestry with a much broader training and degree.' At the same time Zimmerman castigated the country's citizenry for their hypocritical attitude towards their woodlands. 'I do remain mystified by the vast and strong public interest in the forest industry in contrast to the intellectual disinterest in becoming a forester. I can only draw the conclusion,' Zimmerman declared disappointedly, 'that a strong, almost visceral interest in our forest remains, but somehow the forestry faculties have been unable to trade on that.'[23]

Unbeknownst to the Faculty of Forestry and its supporters, while it was lobbying its cause, the University of Toronto was deciding its future. In early December 1992, Foley and the university's new president, Bob Pritchard, met with Dr Richard Allen, Ontario's minister of colleges and universities, and Dr B.J. Shapiro, the deputy minister, to present the university's plan to terminate the Faculty of Forestry's undergraduate program. Although the details are sketchy, the government supported the initiative. Shapiro granted that it was hardly ideal to split the undergraduate from the graduate program, but he deeply appreciated the University of Toronto's distinct willingness to make the difficult choices that the times demanded and its continued move towards concentrating its resources on graduate studies, an area in which it excelled. The next day, on behalf of the university Prichard thanked Allen for the meeting, which he found 'very helpful. It will certainly shape our thinking in the days and weeks ahead,' he told the minister.[24]

The University of Toronto's senior administrators were not kidding, but the problem was as much their means as their end. They, and more specifically Foley, were not only bound and determined to terminate the Faculty of Forestry's undergraduate program, regardless of the reasons in favour or against doing so, but to do so in a dictatorial manner that generated at least as much anger and frustration as the decision itself. If ever there was an instance in which there was good reason to shoot the messenger, this was it.

It would have been difficult to imagine a more heartless way of delivering the news. At noon on 6 January 1993, Joan Breckenridge, a reporter with the *Globe and Mail*, contacted the Faculty of Forestry because of a tip that had come her way. Having heard that Foley was recommending that forestry become a strictly graduate and research division by terminating its undergraduate program, Breckenridge wished to learn of the faculty's reaction to this information. The Faculty of Forestry's personnel gathered later that day to learn more of the story, and met with Foley at 1 p.m. the next day to ask her questions about it. Three hours later, Foley read her statement to the academic board of the university's governing council.[25]

Foley rested her case on dollars and cents. Stressing the urgent need to reduce the university's costs, she explained that cutting the program would save $500,000 – largely by trimming academic and non-academic staff – from the Faculty of Forestry's total budget of $2.5 million. Although she commended the faculty for its initiatives in responding to 'the changing environment' over the previous half-decade, its enrolment was 'very low,' with an average of merely twenty first-year students annually. Finally, she argued that this attenuation in the size of the Faculty of Forestry would free up space in the Earth Sciences Centre, into which the University of Toronto would be able to move other related divisions, such as the Institute for Environmental Studies. For these reasons, she believed it was best for the Faculty of Forestry to cease admission to the bachelor of science in forestry program, continue its existing graduate programs and develop a new professional master's degree in forestry, and find new means by which the faculty's professors could 'participate in undergraduate education.'[26]

Foley's haste in making this recommendation spoke to her resolve to terminate the Faculty of Forestry's undergraduate program come what may. Her statement mentioned her wish that governing council's planning and priorities committee hear her proposal at its 19 January meeting, after which it would require the approval of the academic board of the governing council to become effective. So after she had established a 'Provostial Working Group' to investigate options for the Faculty of Forestry back in February 1992, and allowing it merely two months to complete its work, Foley had rashly chosen the option that would decide the faculty's fate practically unilaterally and rushed to implement it. More importantly, when the faculty's members had collectively protested against the working group's report back in mid-1992, Foley had not only concurred with their views but promised them that no decisions would be made until all stakeholders had their say. Her haste to pull the trigger on this execution suggests that the working group's efforts were simply part of her preordained plan to cut the Faculty of Forestry's undergraduate program.

Not surprisingly, Foley's decision to fire this salvo touched off a war of words that intensified with each passing day. In the immediate wake of her announcement, Carrow endeavoured to rally the troops. To provide his Faculty of Forestry colleagues and supporters with the data to refute Foley's case, the beleaguered dean explained that the faculty's operating budget for the present year was roughly $2.6 million, or less than 0.5 per cent of the university's total budget. Consequently, terminating the faculty's undergraduate program, which the administration estimated would save $500,000, would represent a cut of less than 0.1 per cent in the University of Toronto's annual operating budget. Moreover, he pointed out that, under the university's long-term budget plan for 1990–1 to 1995–6, 'Forestry absorbed the largest budget cut (over 10%) of any academic division at University of Toronto. Much of this has already been implemented.' In other words, his faculty had already shouldered far more than its fair share of the budgetary burden, and now it was being asked to endure an additional 20 per cent cut. Furthermore, Carrow pointed out that the university could still save over $250,000 if it agreed to implement the option the faculty favoured, namely, creating a new master's in forestry and 'second entry' undergraduate program. To demonstrate the utility of the bachelor of science in forestry degree, Carrow listed many of the faculty's more accomplished graduates and their current positions.[27]

Protests of various sorts also followed on the heels of Foley's announcement. The University of Toronto's student newspapers excoriated the administration for its decision. At a time of unprecedented concern about the environment, they wondered how the university could terminate forestry's undergraduate program. In a captivating piece in the *Varsity*, Naomi Klein, its editor, argued that the University of Toronto's approach to this matter was symptomatic of the neo-conservative agenda to 'streamline education' according to the mantra of 'bigger is better.' Entitling her piece, 'If a Tree Falls ...,' she urged her readers to protest against Foley's recommendation at the 19 January meeting of the planning and priorities committee. She warned them that, if they did not speak up now, sooner or later, they too would be touched by this maelstrom. The Faculty of Forestry also organized several events to publicize its cause prior to the meeting. A news conference at the Earth Sciences Centre, headlined by leading forest industry official Mike Innes (6T5), and a protest rally that marched on Simcoe Hall, drew widespread attention to the Faculty of Forestry's plight.[28]

Because the issue proved too contentious for the planning and priorities committee to decide during only one meeting on 19 January, it met again one week later, and ultimately, it ended up supporting Foley. Over the course of these get-togethers, dozens of individuals made personal appeals on behalf of the Faculty of Forestry, while roughly seventy written submissions were presented. Foley stood her ground, reiterating her case that 'the program has been unable to raise its significantly low enrolment levels over the past few years and that the University must reduce its expenditures due to severe funding cuts from the province.' At the same time, she rejected the faculty's proposal for pursuing one of the Till's working group's options on the grounds that it would entail increasing the Faculty of Forestry's budget and staff complement, something she felt was impossible under the circumstances. While the faculty's supporters naturally attacked her decision, many of them felt that the real issue was the 'process and rationale underlying the Provost's recommendation.' Foley also had her supporters, such as Desmond Morton. As principal of Erindale College, he had been forced to cut numerous programs because of low enrolment, and felt that the Faculty of Forestry ought to follow suit. In the end, a motion to refer the matter back to the provost to develop additional options was defeated by a vote of 15 to 4, and her recommendation passed 14 to 5.[29]

This outcome unleashed another round of reflections from the faculty's supporters, and a few of their appeals were noteworthy but for diametrically opposing reasons. Adam Zimmerman wrote R.J. McGavin, chairman of the governing council, and offered his trademark forthright and cogent analysis. As a veteran member of the faculty's advisory board, he recognized the numerous and varied impediments that the Faculty of Forestry had traditionally faced in attracting students. These included terrible facilities, 'inability to fully staff with the right people,' and most importantly, the recent paucity of job opportunities. Although he understood the difficult decisions the University of Toronto was forced to make at this time, he suggested it find a 'better way to solve the dilemma of the Faculty at the moment.' Other letters defended the faculty but on weak grounds. Dave Love, who had retired after having taught at the faculty for four decades (1946–85), pleaded with the academic board to find another solution to the Faculty of Forestry's challenges. He explained that his forty years on campus had taught him that 'Simcoe Hall was "hard nosed" but fair, and always ready to listen to a reasoned argument.' To illustrate his case he recounted how, when the proposal had been presented roughly a dozen years earlier to move the faculty to Scarborough College, it had been subjected to a 'detailed examination' that had indicated that it would be best for the faculty to remain downtown. He thus urged the administration to be equally scrupulous in examining this matter, a process he felt certain would result in the faculty's undergraduate program surviving. Ironically, had the Faculty of Forestry truly given this option a 'detailed examination,' it probably would have avoided the predicament in which it now found itself.[30]

Two of the most moving entreaties to the administration came from students. As a native of France, James F. Dat (9T3) explained that he had been keen to attend a Canadian forestry school. After receiving his letters of acceptance from the University of New Brunswick, the University of British Columbia, Laval University, and the University of Toronto, he had chosen to attend the latter. The reason, he stressed, was that 'the B.Sc.F. at the University of Toronto has an *international* reputation.' He added pointedly that it had taken '86 years, 7 Deans, 101 professors, and 1,850 graduates to built [sic] this international reputation,' and urged the university not to throw it all away now. Cindy Hutchinson (9T6) took a fundamentally different approach. In a very simple letter, she presented a unique case for rejecting Foley's recommendation. She listed all the institutions that considered her solely as a number, everything from the federal government to the Ontario Health Insurance Plan, and from the University of Toronto to the Ontario Student Assistance Program. Then, in bold letters, she wrote a one-line sentence: 'I am Cindy to the faculty and staff of the University of Toronto Faculty of Forestry.' She then asked a series of rhetorical questions, one that included: 'doesn't everyone deserve a personal and humanizing education?'[31]

During the run-up to the meeting of governing council's academic board on 11 February, at which Foley was to present her recommendation, the University of Toronto enhanced the perception that it was acting in a dictatorial and harsh manner. With the faculty's precarious situation splashed across the *Globe and Mail*'s front-page headline on 9 February, the administration decided that it would not allow the Faculty of Forestry to present more than the standard three submissions to the board's meeting. This prompted many of the faculty's supporters who formed part of the crowd that day to wear gags. Others chose to hand out seedlings and flyers that set out their case to those entering council chambers. Moreover, a few days before the meeting the university's Faculty Association voted unanimously to support the Faculty of Forestry in its fight with Simcoe Hall. The Faculty Association argued that the university's academic integrity was at stake because Foley had ostensibly based her decision on financial – not academic – considerations.[32]

Predictably, the meeting was a raucous affair, but Foley heightened the crisis considerably when she began waffling as she explained the basis on which she had rendered her recommendation. The two-hour debate was electric, and the *Varsity* described how 'supporters and opponents of the cut battled it out with rhetoric, theatre, motions and counter-motions.' The pith of the matter became 'the question of how the university should make decisions on eliminating academic programs,' with the administration being vilified for doing so on a strictly financial basis. In the face of this barrage, Foley fought back. She denied that this was the sole reason for chopping the Faculty of Forestry's undergraduate program, although she contended that the $500,000 savings she projected was 'an important matter.' But when Peter Rosenthal, a member of the academic board, asked her about the other criteria that had influenced her decision, Foley stonewalled. As the *Varsity* put it, 'the provost refused to answer, telling him [i.e., Rosenthal] to read her report.' With the board unable to render a decision, it scheduled an emergency meeting for 25 February to decide the matter.

Foley's equivocation only exacerbated an already tense situation, and it elicited an appeal that captured the repulsion that supporters of the Faculty of Forestry felt towards Simcoe Hall. As president of the World Wildlife Fund Canada, Monte Hummel was arguably the faculty's most renowned environmentalist, having earned a master of science in forestry there in 1979. After attending the academic board's 11 February meeting, he was so crestfallen – 'bordering on disgusted,' was how he put it – by what he had witnessed that he conveyed his views to R.J. McGavin, chairman of the governing council. His principal complaint was that no clear reason was being provided for terminating the undergraduate program. 'The Provost has put financial concerns on the table,' Hummel asserted, 'but then confounds the issue by saying categorically the decision is not being made on financial grounds. Well, is it or isn't it? Why provide all this information if finance is not a factor? On the other hand, if it's not a financial decision, then it must be an academic one. Yet, not a single academic reason has been expressed for terminating this program.' If this were the problem, he demanded that the academic board identify it. 'In short,' he despaired, 'those of us interested in preserving the undergraduate program are reduced to wrestling with phantoms. I think this process reflects badly on the University as a whole. Frankly, I sat there in disbelief that an apparently mature intellectual community could conduct its affairs with such little focus and competence. If the undergraduate program in Forestry at the University of Toronto is going to be terminated, Canadians inside and outside the University deserve to know why.'[33]

Foley successfully concealed from public critique her true animus, namely, her belief that the Faculty of Forestry's undergraduate program was weak and had to be wound down, but she conveyed it surreptitiously to the key decision-makers during this critical time. Back in mid-January 1993, on the eve of the meeting at which the planning and priorities committee had approved her recommendation, Foley had written the committee's chairman, James F. Burke, in an effort to denigrate the Faculty of Forestry's undergraduate program. The data she provided indicated that, over the past three years, fewer than 20 per cent of those entering its graduate program had earned their undergraduate degrees there. Her implication was that Carrow's undergraduates were generally incapable of succeeding at the next academic echelon, even though this was hardly a fair standard by which to judge the effectiveness of the faculty's undergraduate program. Then, only two days before the academic board's 25 February meeting, she again endeavoured to disparage the faculty's operation. This time, she resorted to chicanery to achieve her end by forwarding to the board carefully manipulated data that made it seem as though the Faculty of Forestry's undergraduate program was astronomically expensive to operate. According to her figures, faculties such as physical and health education and pharmacy, respectively cost $4,019 and $5,320 per student, and these were at the low end of the scale. Music was at the high end, at over $14,000 per student. Forestry was in an opulent orbit all its own, Foley contended, at a cost of over $32,000 per student. Foley's calculations were specious, however. She had divided the faculty's total budget (i.e., for graduate and undergraduate education, and administration and support for all of the Faculty of Forestry's activities) by its undergraduate student body instead of computing the additional cost of delivering the undergraduate program over and above the cost of operating the graduate department. After all, Foley's recommendation to terminate the undergraduate program included her full support for the faculty's graduate and research work, at a total cost of $2.2 million. This made the actual cost of delivering the current undergraduate program only $5,900 per student. In addition, Foley employed data from the faculty's worst years in four decades, 1989 and 1990, to support her contention that its undergraduate program was suffering from declining enrolment (years in which it had 15 and 13 first-year students respectively). She thus intentionally avoided citing the figures from 1991 and 1992 (25 and 29 respectively) that laid plain that the faculty's student body was growing again. If the empirical evidence had been strong enough per se to support the decision to terminate the Faculty of Forestry's undergraduate program, Foley would not have been driven to knead it in this manner.[34]

While the deck was stacked against the faculty at the academic board's emergency meeting on 25 February, the discussion revealed that Robert Pritchard, the university's president, fully supported Foley's proposal (she may, in fact, have simply been his henchman, so to speak). Again, the debate was impassioned, with Carrow endeavouring to defend his program in the face of unyielding senior administrators. He pointed out that his faculty, because of its relatively high graduate enrolment and research funding, actually generated a surplus for the University of Toronto of $300,000 each year. Carrow also expressed his frustration at having been given in 1989 four years to reach the targets set out in his strategic plan, but then 'the energy of the Faculty had been deflected by 5 different initiatives and reviews.' He added that the university had not considered the option of increasing the faculty's enrolment by opening its undergraduate courses to students in the Faculty of Arts and Science. But the administration – specifically Foley and Pritchard – dismissed his analyses and arguments, with the latter stressing that program cuts were a normal part of contemporary university life. When the votes were counted, the tally was 54 to 30 in favour of Foley's recommendation. This left the Faculty of Forestry with one last gasp. Governing council would vote on 11 March to confirm the board's ruling.[35]

With less than a fortnight remaining to earn a stay of execution, the faculty and its supporters circled their wagons one last time. Some from its pantheon of distinguished graduates endeavoured to leverage their influence to convince McGavin, chairman of the university's governing council, to reject Foley's recommendation. In so many ways, their professional positions reflected the value of the undergraduate experience that the faculty had given them. Tom Buell (5T6), the second generation in his family to graduate from the Faculty of Forestry and an eighteen-year CEO of Weldwood of Canada Limited, urged McGavin to reconsider, particularly in light of Buell's demonstrated loyalty to the faculty and to the university. Over the previous few decades, he had not only been very active in the Faculty of Forestry's Alumni Association, but he – and his father – had sponsored several student awards. John Duncanson (7T4), one of the country's leading forest industry analysts and head of his own investment research firm, tried to reach McGavin speaking *mano-a-mano*, one financial expert to another (McGavin was a senior executive with the Toronto Dominion Bank). Having observed business and market trends over the past two decades, Duncanson argued that the timing of the University of Toronto's decision could not be worse because he presciently predicted that the forest sector was in the early stages of 'a very explosive cycle' (he was correct). The University of Toronto's Faculty Association expressed its profound concerns with the precedent that it felt that the decision was establishing. Arguing that the academic board was supposedly the university's instrument for expressing its academic priorities and determining its economic mandate, the Faculty Association maintained that it was 'the key component in the University's governance system. But recent events – including the forestry debacle – show that it is in danger of becoming captive to "the Simcoe Hall agenda."'[36]

Finally, Carrow launched one last appeal to the administration, one that bore his typical unemotional and straightforward assessment of the situation. He emphasized that Foley had skewed the data to buttress her case in terms of the per student cost of the Faculty of Forestry's undergraduate program and undergraduate enrolment. In the same breath he reminded Simcoe Hall that, although it had set a university-wide budget reduction target of under 5 per cent for the 1990–6 period, Foley had imposed a 10.7 per cent cut on his faculty, which was the highest of any division. 'It is clear that the justification for the recommendation to eliminate the undergraduate forestry degree program at the University of Toronto cannot withstand careful analysis. Whether it be on the basis of cost saving, cost of the programme, undergraduate enrolment trends, or the alternative of a graduate faculty,' Carrow maintained, 'the arguments for program elimination are not convincing. As an alternative, the Faculty has developed a proposal which would allow it to assume a 16% budget reduction, and still deliver a broadly endorsed undergraduate program which responds to future environmental, social and economic priorities. The Faculty asks for approval to introduce this new program, on the condition that it be reviewed in five years to determine whether the program has been successful enough to warrant continuation of undergraduate forestry education at the University of Toronto.'[37]

But it was not to be. At the governing council's jam-packed meeting on 11 March, the administration made its case more forcefully than ever. It reminded everyone that it had cut fifty programs over the previous five years, and denied that it had rendered its decision regarding the Faculty of Forestry's future in a manner that was either hasty or clandestine. It also reiterated that the period's financial constraints made it impossible for the university to sustain an area of study that had fewer than one-quarter of the students enrolled in the next smallest 'first-entry' undergraduate professional program. In addition, the decision to turn the Faculty of Forestry's full attention to graduate and research work was congruent with the university's mission. For these reasons, Pritchard argued that there was no reason to reopen the debate on the issue, and it was time for the University of Toronto to cross the Rubicon. By a count of 27 to 12, the university's governing council voted in favour of adopting Foley's recommendation.[38]

Foley wrote Carrow a simple letter that officially conveyed the news. It explained that the governing council had resolved at its meeting on 11

March that 'the University of Toronto focus its efforts in the field of forestry on research and graduate studies, that the admission of students to the BScF program be suspended indefinitely as of the current admission cycle, and that its delivery be phased out over the next three years.'[39]

This decision expectedly produced some toxic fall-out. The student press was outraged, as it viewed the move as another step in the inexorable decline of university education. A trenchant editorial in the *Varsity* vilified Simcoe Hall – 'the Pritchard Monarchy' – for having strong-armed members of the governing council into voting for Foley's recommendation. The piece also reported that, prior to the meeting, at least one professor (John Mayall) had overheard 'a member of the Provost's office urging a dean to "Make sure your academic board members are on side for this motion,"' a charge to which neither Foley nor Prichard would respond during the meeting. Fourth-year student Thomas S. Werner (9T3) echoed these views in a letter entitled, 'The Day After.' Explaining that 11 March 1993 'seemed like the beginning of the end for me and my ideas of what purpose the university serves,' he asserted that the university's administrators did not have the courage to oppose Pritchard's 'Kingdom' for fear that they might 'be the next to go.' Bill Graham, president of the Canadian Association of University Teachers, rightfully emphasized that Foley had reneged on her promise to hold a full and fair review of the Faculty of Forestry in 1993–4 in favour of carrying out her own highly flawed review, and then topped it off by choosing her own option.[40]

The alumni directed a stream of barbed letters to Simcoe Hall that expressed their authors' decisions to turn their backs on the university, particularly in terms of financial support. Ian B. Mackenzie (8T1) described to the university's development office how he had heretofore been proud to have been a University of Toronto graduate. He had thus gladly donated to his alma mater, particularly because the most recent 'Breakthrough' campaign had supported the Faculty of Forestry's move to better facilities (i.e., the Earth Sciences Centre). 'The University, in its wisdom,' Mackenzie declared disgustingly, 'is not supporting the Faculty of Forestry, so I hereafter will not support the University ... I simply cannot support such a short-sighted organization.' Michael R. 'Mike' Rosen (8T4), one of the faculty's most exuberant advocates, echoed these sentiments and provided insight into the depth of emotion he and his peers felt at this time. In writing to the university's Office of Alumni and Development, he expressed how the university's decision to cancel the undergraduate program 'has left me (and many others) with bitter feelings about my affiliation with the university.' Although he would continue to support the Faculty of Forestry in its new incarnation as a strictly graduate faculty, he explained that he was convinced that 'the future of the Faculty, the university's commitment to sustainable development/professionalism and the Applied Sciences, remain greatly prejudiced by the removal of the undergraduate program. For these reasons,' he declared simply, 'I do not wish to continue my ties with your department.' Consequently, he asked that the University of Toronto remove him from its alumni contact list so that his 'memories of the University's betrayal of 86 years of superior, undergraduate forestry education will not be rekindled every time you send out a mailing.'[41]

When a flare of hope raced across the sky in April, Pritchard ensured its flame was quickly extinguished. The idea of having the Faculty of Forestry operate as part of the Ontario Agricultural College in Guelph predated the faculty itself. With Carrow suggesting in April 1993 that the faculty could relocate to another university, and officials from the University of Guelph expressing support for the move in principle, it appeared that the Faculty of Forestry might migrate slightly westward. At its meetings in the spring of 1993, the Ontario Council of University Affairs had heard 'the rumours about a move to Guelph' by the faculty although no formal proposal was on the table. Nevertheless, the council made its view known that, 'if a program is transferred from one institution to another, the funded BIU's [basic income units] travel with it.' Dan Lang, one of Simcoe Hall's most senior administrators, forwarded a summary of these developments to Pritchard on which the latter had jotted down a simple marginal note that made it clear that the faculty would not be moving anywhere under his watch. Pritchard declared definitively that 'I've told Guelph we're not withdrawing from MA + PhD.'[42]

The student newspaper, the *Varsity*, provided a laconic bookend on this affair that captured the essence of the gaping wound that the University of Toronto had inflicted on one of its own. In the newspaper's annual review of the 'Best and Worst' of the 1992–3 school year that was drawing to a close, it noted sarcastically that the 'most deafening silence' award went to 'cutting the Faculty of Forestry. If a tree falls does anyone care?' it asked rhetorically. 'Not at the University of Toronto.'

The University of Toronto's decision to terminate the Faculty of Forestry's undergraduate program meant that the class of '96 would be the faculty's last. With the number of undergrads shrivelling each year, a pall was cast over the Faculty of Forestry as its staff stood on a death watch, which, incidentally, ended not in 1996 but eight years later. John E. Glatt, who had enrolled in the faculty in 1980, had departed the program a few credits short of earning his bachelor of science in forestry. After taking equivalent courses (at the University of Waterloo) to those he lacked at the University of Toronto, he earned the distinction of being awarded the faculty's final undergraduate degree in 2004.[43]

Although Carrow was hardly the type to wallow in self-pity, several forces ensured he would not do so. For one, Pritchard made it clear that he would not countenance such behaviour. Not long after the university had terminated the undergraduate program, Carrow, William Harris (5T5), and Adam Zimmerman met with Pritchard to discuss the situation. With regard to the University of Toronto's decision to cut the faculty's undergraduate program, Pritchard forthrightly told Carrow to 'get over it' and focus on strengthening the Faculty of Forestry's graduate program. Then a far friendlier face – Nancy A. Daigle (nee Kliewer) (9T3), one of the faculty's last graduates – exhorted the beleaguered dean to keep his chin up. In mid-1993 she wrote him to express her dismay at not having seen anything about the faculty in the 'University News' section of the recent issues of the *Forestry Chronicle*, the journal of the Canadian Institute of Forestry. 'Don't you think it is crucial that the University of Toronto still have strong involvement with the C.I.F.?,' she inquired. 'I think it is very important that we keep the public aware that forestry at U. of T. is not dead yet … I think there is a lot going on in the faculty which we can be proud of and *The Forestry Chronicle* is a great way to publicize it.'[44]

Carrow characteristically put his nose to the grindstone and soldiered on, even though it must have been remarkably difficult to do so. With many of the faculty's long-time and faithful benefactors turning their backs on it, he endeavoured to pick up the pieces. From mid-1993 to mid-1994, he and his colleagues worked with the provost's office and the university's School of Graduate Studies to formulate yet another plan for the Faculty. Entitled *Towards 2001*, this strategy called for it to retain its focus on its existing 'research-based' graduate programs, namely, the master of science in forestry and the doctorate, and to introduce a new master of forestry degree. At the same time the faculty would endeavour to offer undergraduate courses in other faculties, principally arts and science and applied science and engineering. *Towards 2001* set out milestones for the faculty to meet with regard to both enrolment in its graduate programs and a review of the master of forestry program three years after its scheduled introduction in 1996.[45]

For the first time in living memory, the University of Toronto agreed to respect the Faculty of Forestry's strategic plan as a binding contract between the two sides. In the spring of 1994 the provost's office, now occupied by Adel S. Sedra, committed to hiring four new tenure-stream professors to replace those who were retiring from the Faculty of Forestry before the new millennium. As the faculty's part of this quid pro quo, it was obliged to meet the goals set out in *Towards 2001*.[46]

While Carrow's work and the university's ostensible commitment ensured the Faculty of Forestry's survival and facilitated its growth in a new direction, he would not be around to oversee its development. His term as dean had been renewed in 1991, but only for three years. By mid-1994 no one could have faulted him for believing that it was time for him to bid adieu to the faculty.

Rorke Bryan came to the Faculty of Forestry with a unique professional background. A native of Ireland and still bearing a sharp homeland accent, he had been involved in developing the university's Scarborough College during its formative years. He was a geographer who specialized in soil erosion and degradation, specifically in dry lands, and had been instrumental in starting Scarborough's renowned international development program; he also acted as chair of social sciences there from 1981 to 1985. During those years, he had become associated with the Faculty of Forestry in a number of ways (ironically, he had been part of Scarborough's welcoming committee for the faculty's proposed move to Scarborough in 1979). He had befriended Vidar Nordin, insisted that Scarborough's international development program include forestry courses (for which its students had to venture downtown), and led a major land rehabilitation project in Kenya in the mid-1980s that soon involved the Faculty of Forestry's graduate students. In the run-up to Carrow's departure, Bryan had been invited by several of the faculty's senior members – Vic Timmer and Terry Blake – to apply to be its new dean, and he succeeded in this quest.

Bryan was, on so many levels, precisely what the faculty needed. He was its first dean who was neither a forester nor a part of the tight-knit Canadian forestry community, and was hardly bothered by it. In fact, he seemed to relish being an outsider. In his eyes, this left him unburdened by the baggage that had heretofore fettered the attempts of both the profession and the faculty to change with the times. Bryan also brought his distinctive boundless energy to his new job. He was a bull in a china shop if there ever was one, a one-man show who was prepared to do everything he felt was needed to rebuild the faculty, and to do it all himself. In addition, unlike Carrow, he was a veteran university administrator who had been integral to the creation and development of innovative and unique academic programs at Scarborough. And he adeptly tapped this well of experience to foster a strong and enduring tie with Sedra, the university's provost. Bryan undeniably benefited immensely from Sedra's commitment to support the deans he appointed, of whom Bryan was the first.

But more than anything, Bryan's strength lay in his willingness to take the Faculty of Forestry into terra incognita, a departure that recent events now made possible. Whereas many in the forestry community were still lamenting the demise of the undergraduate program, Bryan saw it as a golden opportunity. Previously, he, and others in the faculty, felt that it had been unable to make a complete break with the 'tradition-al' approach to forestry, which focused on raising tree crops, because the school was constrained by the need to teach its students the curriculum that would qualify them for accreditation by the professional organiza-tions like the Ontario Professional Foresters Association. In contrast, the death of the bachelor of science in forestry program had, for the first time in the Faculty of Forestry's history, freed it to head off in whichever direction it pleased.

Bryan had a very clear idea of where the faculty needed to go. He was convinced that, in general terms, the entire country had to move in a new direction in terms of managing woodlands. In his eyes, forests were just too darn important to accept the status quo. His many years work-ing on international projects had shown him how foresters had been on the vanguard of the environmental movement in places like Africa and Europe, and he felt that his mandate was to recreate this paradigm in Canada. It was time to marry forestry and 'environmentalism' by break-ing Canadian forestry out of the slumber, specifically by severing its tight ties to 'industrial' forestry. Instead, he drove towards transforming for-estry into a domain whose guiding principles were ecological sensitivity and social awareness. In terms of rebuilding the Faculty of Forestry, these ideas drove nearly his every move.

Naturally, implementing this revolutionary approach ruffled more than a few feathers, and Bryan's manner of proceeding only exacerbated the situation. The university's decision to terminate Faculty of Forestry's undergraduate program in 1993 had driven a fissure, in many cases an irreparable one, between the faculty and the lion's share of its alumni. In the wake of the decision, droves of previously devoutly loyal and sup-portive graduates cut their ties to both the faculty and the university.

After Bryan became dean and as he began recasting the faculty, he concomitantly could have reached out to the 'Old Guard' – the faculty's profoundly loyal cadre of traditional professional foresters – in an effort to draw them back into the fold. Granted, this would have been immensely challenging under the circumstances, but he could have made the effort. Instead, he both widened and cemented the gulf the university had already created between itself and its Faculty of Forestry graduates. First and foremost, he blatantly ignored the need to rekindle the alumni's allegiance; for him, this was not a priority. In addition, his penchant for formulating faculty policy unilaterally or only in consultation with those who shared his vision left little room for graduates to offer counsel, as they had traditionally done. Moreover, his black-and-white views concerning the faculty's past (its traditional approach bad, his new one good) cast aspersions upon much of the faculty's proud heritage, which its alumni had long celebrated. The upshot saw a reconstituted Faculty of Forestry heading off in a bold new direction with few progeny there either to witness or support the undertaking.

Nevertheless, Bryan channelled all his vim and vigour into realizing his agenda. His overarching goal was to transform forestry education into a broader and more international field of study, one in which he felt the faculty had a particular role to play. In terms of making a practical difference, in the fall of 1994 he articulated how he envisioned the Faculty of Forestry at the University of Toronto acting as a catalyst to facilitating effective sustainable forest management in Canada, playing a stronger role in international forestry and development, and working closely with the country's Aboriginals in order to increase their involvement in forestry work. To assist in achieving a few of these aims, he organized several conferences in 1994 and 1995. The one in 1995, entitled 'Conservation of Northern Forests: Canada's Role in Responsible Stewardship,' spoke specifically to his determination to transform forestry into a broader field than ever before. The presenters spanned the gamut from industry leaders such as Adam Zimmerman to radical environmentalists such as Elizabeth May.[47]

Within the University of Toronto community, Bryan also charted a bold new course for the faculty, and he saw its mandate there as encompassing three main functions. It was obviously responsible for promoting first-class research and teaching in selected areas of specialization. Second, the faculty ought to provide 'objective' general education on forest issues, and work towards augmenting the base of informed citizenry. Third, he felt it ought to develop 'innovative, practically oriented "professional" teaching programmes' that built on the university's existing competitive advantages and, in the process, made unique contributions to forest conservation and sustainable management.[48]

Work was already well under way towards realizing the third goal, but Bryan reformulated the effort very much in his own image. The university had already agreed to enhance the Faculty of Forestry's graduate program by creating a master of forestry. Under Carrow, the faculty had decided that it would be course-based, include a significant field trip or practicum component, and be marketed to non-forestry and forestry graduates and international students. Bryan did not feel, however, that this plan was radical enough to break the faculty out of its doldrums. He thus spearheaded the creation of a program that was unlike any other in the world, one that would attract strong students from a variety of academic and cultural backgrounds. Consequently, he redesigned the program using Scarborough's international development program as a template, thereby shifting its focus to international activity and internships. He even rechristened the undertaking – the master of forest *conservation* – to distinguish it from existing like efforts.

Bryan also endeavoured to reorient the faculty's existing graduate programs, namely, its research-based master of science in forestry, and its doctoral program, to reflect his 'broadening out' agenda. Prior to his arrival, they had been grouped into three general fields that had parallelled the undergraduate program: forest management, forest biology, and wood science. But Bryan felt that most of the world's most important forestry-related research topics did not fit easily into this framework. 'In fact,' he explained in 1998, 'this structure served to artificially constrain students' choice of courses and their interaction with those of different interests and, in effect, became an obstacle to the high quality research currently needed.' Bryan thus merged the three fields into one whose emphasis was, predictably, on forest *conservation*. At the same time he oversaw the creation of two other master's degree programs. The first one – a master of science in wood engineering – came into being in the fall of 2000 in collaboration with the Faculty of Engineering. Bryan oversaw the introduction of another, more avant-garde endeavour roughly five years later. The master of international trade in forest products is an interdisciplinary degree program, offered in conjunction with the University of Toronto's Rotman Business School, Faculty of Law, and Institute for Policy Analysis, and modelled after the university's highly successful executive master of business administration program. Designed for senior industry and government officials, it is financed entirely by fees and structured in six highly intensive modules taught over sixteen months.[49]

Bryan effected a much more radical transformation at the undergraduate level. On his arrival in 1994 Bryan had bluntly informed his colleagues that he would not work towards restoring the old bachelor of science in forestry program, but he set as a goal creating the possibility of reintroducing an accredited undergraduate program down the road. In his mind, the Faculty of Forestry could only reach this destination by drastically expanding forestry's appeal. The first step was continuing to teach forestry courses in other faculties, principally in the Faculty of Arts and Science. Then, just at the turn of the new millennium, Bryan jumped at the chance to make a major breakthrough on this front. At this time the Ontario government faced a daunting task in terms of accommodating a one-time, 'double-cohort' enrolling in universities and colleges. The decision to eliminate Grade 13 in the late 1990s meant that, in 2002–3, students would graduate simultaneously from both the last class of the traditional five-year and the first class of the new four-year streams. In light of this unprecedented need for spaces at post-graduate institutions, the government was open to suggestions regarding new courses. Bryan seized the moment and won approval for new undergraduate programs – based squarely on forest *conservation* and not forestry – in the Faculty of Arts and Science that were introduced in September 2000. These included streams that were intensive (a specialist in forest conservation bachelor of arts or forest conservation science bachelor of science), moderate (a major in forest conservation science), and generalist (a minor in forest conservation or forest conservation science). Tellingly, the word 'forestry' does not appear in any of the degree titles.[50]

Bryan also endeavoured to foster more external ties, particularly international ones, and benefited enormously from some fortunate circumstances. The first involved Haliburton Forest and Wildlife Incorporated, the owner and manager of roughly 25,000 hectares of mixed-wood forest lands not far from the Faculty of Forestry's former University Forest just south of Algonquin Park. German-born forester Dr Peter Schleifenbaum had been managing the area as a multiple-use, sustainable woodland since the early 1960s, and his was the first forest in Canada to receive certification from the Forest Stewardship Council. In the early 1990s, when Carrow was still dean, it began actively cooperating with the Faculty of Forestry on a range of research projects. It soon became the faculty's de facto 'university forest,' with the added bonus of burdening the forestry school with none of the attendant costs of managing such a large tract.

Another propitious development involved Murray Koffler, the founder of Shoppers Drug Mart. In the mid-1990s Koffler donated his 400-hectare 'Jokers Hill Estate' on the Oak Ridges Moraine just north of Toronto to the University of Toronto. This gave the faculty, and other divisions, access to another living research laboratory, and a particularly valuable one at that. It encompassed one of the few remaining residual patches of Carolinian forest near Toronto, and quickly became a major centre for faculty research. Finally, Bryan signed formal agreements with a number of international forestry, and related, schools literally across the globe. By the time Bryan left the dean's office in mid-2005, the Faculty of Forestry had formal arrangements with institutions in Austria, China, Germany, Greece, Mexico, India, and Sweden.[51]

Finally, Bryan was adamant that the Faculty of Forestry sine qua non be world-class and relevant research, and much of it international in scope. The termination of the undergraduate program freed the faculty's academic staff from worrying about its traditional research 'clients,' namely, industry and government, and created a situation in which they could be creative academics. This environment was thus highly conducive to excellent work, as was Bryan's success in recruiting dynamic new professors and expanding the focus of the faculty's research activities. As a result, the Faculty of Forestry continues to boast expertise in the ecology of the Great Lakes–St Lawrence Forest Region but now this renown also includes tropical forest conservation.

In pushing his agenda Bryan enjoyed éclat in nearly every realm. Enrolment in the master of forest conservation program exceeded targets practically from the time it was introduced, and it has stabilized at the prescribed optimal level of fifteen students per year. Although it experienced the predictable birth pains, the program's most important attribute is how readily its graduates have found meaningful and rewarding forestry-related employment. The traditional research streams – the master of science in forestry and the doctorate – have seen more changes than growth since 1994. While the number of students pursuing these degrees has hovered steadily around seventy, the ratio of master of science in forestry to doctoral students has shifted dramatically. Whereas in 1994–5 the former outnumbered the latter by a ratio of roughly two-to-one, one decade later the situation had nearly reversed itself, with just over 1.5 doctoral students to each master of science in forestry student. Likewise, the ratio of females to males has risen from roughly 0.4 to 1 to one-to-one during this period. At the undergraduate level, enrolment in courses offered through other faculties has increased slowly but steadily, from ten in 2000–1 to thirty-one six years later. Bryan's efforts have thus laid the foundation for the reintroduction of an 'accredited degree' program in the future, if this is deemed a desirable path to follow. In addition, the Faculty of Forestry's connections to both the Haliburton Forest and the international forestry community have produced prodigious benefits. The former, for example, has resulted in the faculty receiving significant financial support from several benefactors – most notably Americans Mr and Mrs Carl Brown – who have made major donations to support a collaborative research program between the two partners. Finally, the Faculty of Forestry is recognized internationally as a first-rate research institution, even though it has undergone, and continues to undergo, a major renewal process. Notwithstanding its continued move towards replacing senior professors with junior ones whose research portfolios are presumably still in their nascent stages, the faculty's staff is, on a per professor basis, winning larger than ever research grants and publishing at a prodigious rate that ranks it among the continent's most productive forestry schools.[52]

To be sure, everything Bryan touched did not turn to gold. The efforts to increase the Aboriginal community's interest in the Faculty of Forestry's activities, for instance, have borne little fruit. As for the master of forest conservation, it has not attracted many international students and is unable to offer students *any* financial support at a time when most post-graduate degrees include such benefits as a matter of course. Likewise, the proposed master of science in wood products engineering has also encountered major difficulties in getting off the ground, specifically in terms of arranging for industry sponsorship.[53]

Far more distressing, however, is the Faculty of Forestry's financial situation, which, in a stroke of ultimate irony, Bryan conceived of in a manner that flew in the face of the core message that foresters have traditionally preached. Faced with continued budget cuts, and the faculty having sustained itself during the early 2000s only because of the School of Graduate Studies' largesse, Bryan projected in his 2004 *Academic Plan* that the faculty's total debt would be nearly $1.5 million by 2010. One means of addressing it, he pointed out, was tapping the 'Faculty's trust funds. While this is certainly not desirable, precluding use of these funds for other purposes such as scholarships,' he reasoned, 'we do not consider it to be irresponsible, or a "band-aid" solution. It does provide the possibility to "buy time" until the next tranche of retirements in 2010 and 2013.' In expressing such willingness to spend the principal instead of only the interest accruing from the Faculty of Forestry's bank of funds,

Bryan transgressed the cardinal rule of sustained yield management that all of the Faculty of Forestry's previous deans had inculcated in their students. A more fitting example of the break Bryan had effected at the faculty would be difficult to find.[54]

As much as Bryan had completely reoriented the Faculty of Forestry, the graduates from his and Carrow's years as dean continued to tread down the same paths that their predecessors had long been travelling. In terms of international connections, its students still came from and worked around the world. Dinesh Misra, for example, earned both a master of science in forestry and a doctorate in Toronto before returning to India to become the chief conservator of forests in Gujurat, while Moses O. Imo took his doctorate back to Kenya, where he joined the staff of Moi University in Eldoret. Others, like Jose S. Laureano and Dinesh Mohta, used their graduate degrees from the Faculty of Forestry respectively to take up posts in Mexico and Singapore. Likewise, the faculty's staff was as multicultural as ever, with Shashi Kant and Ning Yan hailing respectively from India and China.[55]

Faculty of Forestry graduates continue to distinguish themselves in a vast array of fields. Many, such as Brad Farquhar (8T6), Martin Streit (8T6), Parthena Fotiadis (9T1), and Robert Schuetz (8T7), entered the 'traditional' forest industry, while Marshall Buchanan (8T9) became a prominent player in urban forestry in Ontario. A string of Faculty of Forestry alumni from these years joined the forestry and non-forestry professorial ranks, including Marie R. Coyea (8T6) at Laval University, Daniel Kneeshaw (8T9) at the University of Quebec at Montreal, and Adam M. Taylor (9T6) at the University of Tennessee. Ken Dewar (8T7) has carved his own niche in this area. He earned his master of science in forestry from the faculty and doctorate in forestry from Laval University, but his area of expertise is the sequencing of human genes at McGill University and the Genome Quebec Innovation Centre. Yet again, Faculty of Forestry graduates were drawn to research in both the public and private sectors, especially the former. Taylor Scarr (8T6) became Ontario's provincial forest entomologist, James P. Brandt (8T7) worked in the same field with the Canadian Forest Service, and Ken Farr (8T9) and Suzanne Wetzel (8T6) were also employed as scientists with the CFS. Gordon D. Nigh (8T9) works for the British Columbia Forestry Service. His field is growth and yield modelling, specifically with respect to the impact of climate change and the recent pine beetle infestation. Despite the recent attrition in the ranks of the civil service, many graduates

earned jobs with the government. Doris C. Krahn (8T6) has worked for
Ontario's Ministry of Natural Resources since graduation, as does her
classmate, Silvia Strobl (8T6); the latter's projects have included helping
to formulate a policy for conserving the Oak Ridges Moraine in southern
Ontario. A number of graduates from this period are also civil servants
but not in forestry, including Steve Spezzaferro (8T6), with Ontario's
Ministry of Municipal Affairs and Housing, and Stephan Zuberec (8T6),
with Industry Canada. Finally, in addition to those noted above and who
went on to apply their forestry knowledge in the name of environmental
stewardship, David Pearce (MFC 1999) worked for the Wildlands League
and Colin Anderson (MFC 1999) for the World Wildlife Fund Canada.[56]

Finally, a few Faculty of Forestry graduates did what their colleagues
had always done in terms of taking their degrees in new directions. Ruth
J. Carvey (nee Pond) (8T5) earned degrees in women's studies and law
after leaving the faculty, and worked for one decade as the executive
director and legal counsel for the HIV/Aids Legal Clinic in Toronto.
Lastly, Marianne Karsh (8T6) truly cut her own swath. After earning her
master of science in forestry from the faculty, she worked as a research
scientist in Canada and Iceland and then became deeply involved in
melding forestry and spirituality. In 2003 she founded Arborvitae, a company that 'creates deeply reflective opportunities for people to nurture
their spirituality through nature and realize the benefit of connecting
with earth, air, light and water for their physical, mental, emotional and
spiritual selves.'[57]

Ultimately, Bryan was undeniably the tonic for what had long ailed the
Faculty of Forestry. His bold and decisive leadership style certainly made
few friends, but he had unquestionably solidified the faculty's standing
at the University of Toronto and beyond. 'Ten years ago,' Bryan reflected
in 2004, 'the Faculty of Forestry passed through a very difficult period
... The role of forestry education in southern Ontario was unclear, and
many within the University and the external forest sector were doubtful
that the Faculty could survive. Progress since this nadir has been substantial,' he contended: 'excellent new, young faculty from different parts
of the world have been recruited; new, highly innovative graduate and
undergraduate programs have been established – these programmes attract an excellent and diverse student group from a wide disciplinary
background; numerous new research initiatives have been launched, by
individual faculty members, and as interdisciplinary or inter-institutional
initiatives; the Faculty has attracted many new international partners.'

While he admitted that there was still much to be done, and despite the severe budgetary difficulties, he predicted that the faculty would 'continue to occupy a leading place in global forestry education in the future, as it has done for nearly one hundred years.' If the Faculty of Forestry at the University of Toronto has indeed achieved this goal, it owes no small debt to Bryan's two terms as dean.

Conclusion

All That Is Old Is New Again

If James H. White (0T9), the Faculty of Forestry's first graduate and forty-year staff member, were to have attended its 'Holiday Season' potluck dinner in December 2006, he would have felt out of place, at least initially. Back in the early 1900s his professors, and later his professorial colleagues had all been white men of European background who had taught subjects specifically related to preparing their charges to manage Canada's forests on a sustained yield basis. In contrast, the professors who were enjoying the feast in 2006 were both men and women. They hailed from places as diverse as India and China, and they specialized in training students to carry out a wide range of activities, spanning the gamut from developing natural fibre-plastic composites for use in making cars to studying small mammal communities in littoral areas of the Amazon River watershed. Tellingly, the Faculty of Forestry's silviculturist no longer focuses on regenerating spruce crops in places like Kapuskasing, Ontario. Instead, he – namely, Sean Thomas – deals with forest management issues in Malaysia, Africa, *and* North America. And although there are but a few accredited professional foresters on staff, the Faculty of Forestry's vast array of expertise reflects the myriad fields that now fall under forestry's dramatically expanded rubric.

The students White would have seen also seemingly bore little resemblance to those who had both sat beside and studied under him. Not only did they come from all over the world and study what appeared to be every subject under the sun, most were working towards graduate degrees. More importantly, there were nearly as many women as men, and Devlin Fernandes personifies this 'new age' forestry student. Born of Indian and Irish-Canadian parents, she completed an honours bachelor of arts degree at the University of Toronto, majoring in criminology, before earning her master of forest conservation degree in late 2006. She was drawn to the Faculty of Forestry by the half-dozen years she had spent working in the tree-planting industry across Canada, trying but rewarding days that inspired her to become involved in improving forest stewardship in this country. Her master's thesis examines the degree to which recent legislative changes in Ontario have succeeded in bringing the province's Aboriginals into the forestry fold. Within weeks of graduating, she had secured a job in Prince Rupert, British Columbia, with the Ministry of Forests as a member of the North Coast Forest District's Compliance and Enforcement team.

As much as today's version of the Faculty of Forestry may have appeared prima facie to White as something very different from what it had been during his time there, it should be clear that several main themes have defined Canada's oldest forestry school since its doors first opened. Yes, its focus is clearly on international forestry today in terms of staff, research, and students, but it has always been connected to woodlands issues around the world, such as they were during the era in question. Furthermore, the present faculty's emphasis is 'forest conservation,' yet it has always preached this message; the problem has been that few of its students were given the opportunity to put this idea into practice in the field. Most importantly, the faculty continues to produce graduates who excel in their chosen fields. Last but not least, blood lines still tie those associated with the Faculty; Fernandes's uncle, Brian John Myers (7T0), graduated from the same institution nearly four decades before her.

Furthermore, while the Faculty of Forestry exists in a milieu that appears fundamentally different from the one in which it was founded, the major forces that shaped its development have shown remarkable resilience. Twenty-first-century technology has undoubtedly made the world a much smaller and quicker place, one in which natural resources are growing exponentially in value. By the same token, public funding for both the enforcement of environmental regulations and sustainable projects remains relatively small compared with other priorities such as health care and infrastructure. Although the media continues to harp on how 'the environment' is now prominent on Canadians' radar, the political parties that have a bona fide shot at shaping our country's future steer clear of taking meaningful action for fear of losing their traditional constituencies. In an article on this subject, in the *Globe and Mail* in late June 2007, Jeffrey Simpson poignantly asked, 'Are Canadians great stewards of their land?' to which he provided a monosyllabic response, 'Hah!' In an eerie bit of irony, Bernhard Fernow's description of the factors he believed were needed to bring about genuine guardianship of the woodlands in the early 1900s is as applicable today if one were to substitute the word 'energy' for 'forest.' 'Only a radical change in attitude,' the first dean of the new Faculty of Forestry postulated, shortly after it opened, 'a realization that forest conservation is a present necessity, and that existing methods are destructive of the future, will bring forward the needed reform.' Likewise, although the University of Toronto has recently been supportive of it, the Faculty of Forestry still faces many of the same challenges it did in the early 1900s. Not surprisingly, the most important of these is the lack of financial support the university doles out to the faculty, specifically for its undergraduate teaching.[1]

Nevertheless, the Faculty of Forestry has seemingly bucked the odds, and survived almost in spite of the circumstances and itself. In fact, although the faculty reoriented in the early and mid-1990s under duress, its ability to do so with such success thrust it to the head of the pack in terms of forestry trends in Canada. Even before Dean Rorke Bryan had left office, the country's professional forestry associations had recognized that they, just like the Faculty of Forestry at the University of Toronto had some time earlier, were on the verge of becoming anachronistic. Forestry was being practised so differently in the early twenty-first century than it had been even twenty years earlier that they were in danger of having the profession pass them by. They had to move in a bold new direction that embraced the broad range of talented and well-educated persons – everyone from GIS (geographic information system) specialists to ecologists – who were working in forestry but were generally ineligible for entrance into the profession. To remedy this situation, the Canadian Forestry Accreditation Board, established in 1989 to set a national standard for entry into the profession, launched the aptly titled 'Inclusivity Project' in 2003. Although it was a difficult process, by late 2006 the CFAB had hammered out broader certification standards and academic requirements for anyone wishing to be a certified 'professional' forester (in doing so, the CFAB used as a model the engineering profession, which had come to the same realization years earlier and reacted accordingly). Work remains to be done, but the profession will soon more accurately reflect the range of jobs done by those working in 'the woods.' In many respects, the forestry profession is now catching up to the Faculty of Forestry, which could potentially ask to have its undergraduate programs evaluated to see if they meet the new standard. Succeeding in this effort would return the faculty to the status of an undergraduate-degree granting institution.[2]

   Likewise, the Faculty of Forestry's ability to orchestrate a Phoenixlike recovery after the 'Crisis of '93' gave it a major head start on the country's other forestry schools in another respect. Nearly all have been suffering from declining undergraduate enrolment since just before the turn of the millennium. Canada's second-oldest forestry school, at the University of New Brunswick, for instance, has seen its undergraduate enrolment plummet from 549 in 1997–8 to 181 in 2006–7. Declining enrolment has left the affected schools at the proverbial fork in the road that the Faculty of Forestry at the University of Toronto passed years ago. The declining interest, the *Toronto Star* reported in May 2007, has been attributed to the 'perception problem,' whereby forestry at best is equated with the 'plaid-shirted Paul Bunyan type' and at worst a cutover the size of Nebraska. 'Forestry programs have had a harder go of trying to seem relevant again,' the article argues, especially because they have been overtaken by the rise of environmental studies in the 1980s. The piece applauds the faculty for having replaced 'its undergraduate forester program' with one centred on forest conservation in which the emphasis is 'on broader issues of climate change, plant ecology, and aboriginal studies.' Tellingly, Faisa Moola, director of science for the David Suzuki Foundation, commended the Faculty of Forestry for having recognized relatively 'early on that forests have values that far exceed the commodities – timber or pulp – that we extract.' The same article noted that most agricultural programs were suffering from the same malaise, but the few exceptions to this general trend reinforced the wisdom of the course that Toronto's forestry faculty had charted in the mid-1990s. 'McGill University's Agricultural and Environmental Sciences faculty halted the declining enrolment of a decade ago after broadening the program,' the *Toronto Star* declared.[3]

   It has fallen to C. Tattersall 'Tat' Smith, who in mid-2005 was named the Faculty of Forestry's eighth dean, to continue strengthening the faculty. A native of Pennsylvania, with a background in forest ecology and soil science, Smith brought to the faculty a very different mindset from that of his predecessor. Although cognizant that the faculty could not slip back into its old ways if it were going to remain relevant – let alone influential – in the early twenty-first century, Smith came with an appreciation for the forestry tradition. And whereas Bryan was a one-man machine, Smith is a consensus-builder who endeavours to engage the entire team in the decision-making process.[4]

   Not surprisingly, one of his first priorities was to rekindle the ties with the Faculty of Forestry's traditional constituencies; now that the forestry profession has changed, there is a general feeling that it is time to 're-connect.' The most important group to win back is, as Smith puts it, 'the alum.' He accepts that some of forestry's fervent traditionalists will never be able to swallow the University of Toronto's decision to termi-nate the undergraduate program because it was the ultimate slap in the face to their profession. Nevertheless, his efforts to reconcile with the faculty's long-time supporters are convincing an increasing number to let bygones be bygones. The fact that Rod Carrow (6T2) – and a handful of other members of the 'Old Guard'– returned to the faculty to help facilitate the healing and celebrate its centennial anniversary speaks to Smith's success in this regard. The hope is that the Faculty of Forestry can also rebuild bridges to industry in a way that does not compromise its integrity.

With this process already well under way, 'Tat' Smith and his faculty colleagues were reminded of the enormous tangible benefits accruing from the profound devotion its graduates have traditionally felt towards 'the Faculty family.' Some cohorts, such as the class of '68, still gather once a year to renew old acquaintances. Thomas C.E.H. Buckley (4T7), who earned a master of science in forestry from the faculty three years after earning his bachelor of science in forestry at the University of To-ronto, passed away just as the faculty was planning its hundredth an-niversary. Not long after his death, the faculty learned that Buckley had bequeathed a sizeable fortune – over $2 million – to it. This news was a revealing testament to the Faculty of Forestry's trademark esprit de corps.

Despite these propitious signs, the Faculty of Forestry still faces major problems. Although its redesigned undergraduate programs are growing in popularity, the faculty does not receive funding to support its teaching in the Faculty of Arts and Science. This remains a major issue that the Faculty of Forestry would like to see remedied, and it plays no small part in keeping the faculty in a deficit situation that will only worsen. Further-more, the faculty's focus on graduate studies and research – individualis-tic pursuits if there ever were any – may undermine its ability to foster its legendary familial spirit. This does not bode well for the faculty in terms of its contemporary alumni feeling the same inclination to give back to it to the same degree as the 'Old Guard' has; bequests like the one Buckley recently made are probably doomed to being a thing of the past.

And so, the Faculty of Forestry at the University of Toronto looks to-wards the future with many of the same sentiments that marked its birth a century ago. Questions about the suitability of its location in an urban,

liberal arts university, and its public, political, and intra-university support still exist. At the same time, the faculty has clearly been on the vanguard of the transformation that forestry has recently undergone, and its
staff includes a disproportionately large number of the world's leading
'neo-forestry' researchers. For those interested in what lies ahead for the
Faculty of Forestry, just like for those one hundred years earlier, there is
cause for equal amounts of optimism and anxiety.

Notes

Introduction

1 A.B. McKillop, *Matters of Mind: The University in Ontario, 1791–1951* (Toronto: University of Toronto Press, 1994), parts 1–3. M. Friedland, *The University of Toronto: A History* (Toronto: University of Toronto Press, 2002), ch. 13. University of Toronto Archives (UTA), Faculty of Forestry (FF), A2004-0017/031. (G.R. Morrison), 24 Jan. 1958, J.D. Coats to Morrison.
2 H.V. Nelles, *The Politics of Development: Forests, Mines and Hydro-Electric Power in Ontario, 1849–1941* (Hamden, Conn.: Archon Books, 1974), 212.
3 UTA, Private Fonds (PF), B1983-0022/008 (J.H. White), 23 Feb. 1926, J. Walton to White.
4 UTA FF, A2004-0017/026 (D.D. Lockhart), 30 May 1988, Lockhart to J.R. Carrow.
5 Ibid., /039 (A.L.K. Switzer) and /032 (P.J. Johnson), all documents. Henceforth, where no specific or individual document is mentioned for the file or box, the source is all documents in that file or box.
6 UTA, Office of the President/Executive Assistant to the President (OTP), A1968-0007/089, 29 Nov. 1951, J.W.B. Sisam to S.E. Smith.
7 UTA FF, A2004-0017/007 (R.D. Carman), 15 Nov. 1984, Carman to D.V. Love.
8 Ibid., /025 (L.M. Lein), 10 Aug. 1938, Lein to C.D. Howe.

1: 'There Is Nothing in It Practically for the Government,' 1894–1907

1 A.D. Rodgers, *Bernhard Eduard Fernow: A Story of North American Forestry* (Princeton: Princeton University Press, 1951), chs. 1–6. K. Johnstone, *Timber and Trauma: 75 Years with the Federal Forestry Service, 1899–1974* (Ottawa: Supply and Services Canada, 1991), chs. 1–2.

2 R.P. Gillis and T.R. Roach, *Lost Initiatives: Canada's Forest Industries, Forest Policy and Forest Conservation* (New York: Greenwood Press, 1986), ch.4.

3 Queen's University Archives (QUA), W.L. Goodwin Fonds, 15 Jan. 1903, 'Forestry – School of Mining and Agriculture, Kingston, Ontario,' William Harty.

4 Ibid. *Toronto Globe*, 1 May 1902. *Statutes of Ontario*, 1901 Ch. 44 Edw. VII. Archives of Ontario (AO), RG2-42, 1902/296 (Guelph and Forestry 1902), 15 Oct. 1901, R. Harcourt to B.E. Fernow, and 21 Oct. 1902, Fernow to Harcourt.

5 *Dictionary of Canadian Biography*, vol. 14 (Toronto: University of Toronto Press, 1998), 'James Loudon.' J.G. Greenlee, *Sir Robert Falconer: A Biography* (Toronto: University of Toronto Press, 1988), passim. C.W. Humphries, *'Honest Enough to Be Bold': The Life and Times of Sir James Pliny Whitney* (Toronto: University of Toronto Press, 1985), 108. Loudon alleges in his memoirs that he had begun discussing the subject of the forestry school with provincial authorities as early as 1900, but his claim is dubious in light of the evidence.

6 UTA PF, B1972-0031/016 (11), 'The Memoirs of James Loudon, President of the University of Toronto, 1892–1906' (hereafter Loudon Memoirs), 29 Oct. 1902, Loudon to Harcourt.

7 AO, RG4-32, 1902/1372, 17 Sept. 1902, J. Mills to J.M. Gibson. Loudon Memoirs, passim. QUA, Goodwin Fonds, 1, 9, and 30 June 1903, W. Lockhead to G.Y. Chown, and 4 July 1903, H.L. Hutt to Chown.

8 UTA PF, B1972-0031/007 (7), ca. 1905 notes by J.H. Faull. Loudon Memoirs, passim. QUA, Goodwin Fonds, 1, 9, and 30 June 1903, Lockhead to Chown.

9 *Educational Monthly*, Jan. 1903. *Toronto World*, 9 Dec. 1903. UTA PF, B1972-0031/005 (35), 12 Jan. 1903, W.H. Fraser to The College Editor, *New York Post*, and encl.; ibid., (12) 6 and 28 Jan. 1903, Mills to Loudon.

10 Loudon Memoirs, 3–5.

11 Ross is cited in Friedland, 193. This subject is treated at length in D.J. Ayre, 'Universities and the Legislature: Political Aspects of the Ontario University Question, 1868–1906' (Doctoral dissertation, University of Toronto, 1981); McKillop, ch.7; H. Neatby, *'And Not to Yield': Queen's University*, vol. 1, *1841–1917* (Kingston and Montreal: McGill-Queen's University Press, 1978), ch.14; D.D. Calvin, *Queen's University at Kingston: The First Century of a Scottish-Canadian Foundation, 1841–1941* (Kingston: Trustees of the University, 1941), chs. 3 and 4, esp. 139–47.

12 Loudon Memoirs, passim. UTA PF, B1972-0031/005 (12), 3 Feb. 1903, Loudon to Mills, and 6 Feb. 1903, Mills to Loudon. Friedland, chs. 3 and 17.

C.B. Sissons, *A History of Victoria University* (Toronto: University of Toronto Press, 1952), 219.

13 *St Thomas Evening Journal,* 7 Mar. 1903. *Globe,* 27 and 28 May, 3 and 22 June 1903. Friedland, 186–9. UTA PF, B1972-0031/005 (12), 13 June 1903, Mills to Loudon.

14 Loudon Memoirs, 6 June 1903, Loudon to Harcourt, and 10 June 1903, Harcourt to Loudon, 15–16.

15 QUA, Goodwin Fonds, 1, 8, and 14 Aug. 1903, E.J.B. Pense to Goodwin.

16 Calvin, 142–5. QUA, Goodwin Fonds, 1 and 20 July 1903, G.M. Macdonnell to Goodwin.

17 Loudon Memoirs, 17–18 [24–5]. Calvin, 154.

18 *Globe,* 24 Mar. 1904. Loudon Memoirs, 15–16 [22–3]. Friedland, 193–6. AO, RG2-29, MS2630/68, 14 Mar. 1904, E.A. James to Harcourt.

19 AO, RG2-42, 1903/2510, 20 Feb. 1903, G.W. Ross to Pense. *B.C. Lumberman,* Apr. 1904.

20 UTA PF, B1972-0031/005 (12), 28 Jan. 1903, Mills to Loudon; /014 (9), 24 Jan. 1906, J. Mills to J.P. Whitney, encl. 18 Jan. 1906, H.S. Graves to J.F. Clark.

21 Gillis and Roach, ch.4.

22 R.S. Lambert and P. Pross, *Renewing Nature's Wealth: A Centennial History of the Public Management of Lands, Forests and Wildlife in Ontario, 1763–1967* (Toronto: Ontario Department of Lands and Forests, 1967), 184–92. E.J. Zavitz, *Recollections, 1875–1964* (Toronto: Department of Lands and Forests, 1964), 3–5.

23 For an example of Clark's work during these years, see AO, RG3-2-0-77, 3 Feb. 1906, Memorandum for the Minister, J.F. Clark.

24 Humphries, 113–16 and 126–8. Friedland, ch.18.

25 UTA PF, B1972-0031/008 (17), 10 Apr. 1905, B.S. Walker to J.P. Whitney. *Canadian Forestry Journal* 2(3) Aug. 1906, 104 and 150. *Report of the Canadian Forestry Convention Held at Ottawa, January 10, 11 and 12* (Ottawa: Government Printing Bureau, 1906), 172.

26 *Report of the Royal Commission on the University of Toronto* (Toronto: King's Printer, 1906), 12.

27 UTA PF, B1972-0031/005 (12), 28 Jan. 1903, Mills to Loudon; /014 (9), 24 Jan. 1906, Mills to Whitney, encl. 18 Jan. 1906, Graves to Clark; /008 (16), 24 Jan. 1906, Whitney to Loudon.

28 AO, F5 (J.P. Whitney), MU3118, 16 Feb. 1906, Clark to Whitney.

29 Gillis and Roach, 95–6, argue that one of Clark's indiscretions that had made him persona non grata with the Ontario government was his decision to deliver this speech, but this seems unworthy of the significance the

authors accord it. Thomas Southworth, Ontario's Director of Forestry, was
the author of the paper Clark read in his address, and the Convention's
program clearly indicated this. Clark had merely stepped up to the micro-
phone because Southworth was allegedly suffering from an illness that had
afflicted his throat. *Globe*, 11 Jan. 1906.

30 *Globe*, 9 Mar. 1906.
31 AO, F5, MU3119, 27 Apr. 1906, Memo Favouring University Forest, J.F.
    Clark, and 30 Apr. 1906, Clark to Whitney; ibid., MU3118, 16 Feb. 1906,
    Clark to Whitney, and RG8-5/7, ca. April 1906, J.F. Clark to W.J. Hanna. Bilt-
    more Company Archives (BCA), Forestry Department Manager's Records
    (FDMR), Series C, 7.0045, 9 and 22 June 1906, Clark to C.A. Schenck. UTA
    OTP, A1967-0007/001 (J.H. Faull), 1 Aug. 1906, Faull to M. Hutton.
32 *Forestry Quarterly* 4(2), 1906, 'The Measurement of Sawlogs.' H.S. Graves and
    E.A. Ziegler, *The Woodsman's Handbook* (Washington, DC: USDA, Forest Ser-
    vice, 1910), 13–14 and 25. F. Freese, *A Collection of Log Rules* (USDA Forest
    Service General Technical Report, FPL 1: ca. – General Technical Report 1,
    1973), 17–18 and 22–3.
33 AO, F5, MU3119, 30 Apr. 1906, Clark to Whitney. UTA FF, A1972-0025/023,
    20 Mar. 1924, Notes.
34 *Globe*, 29 Oct. 1906. BCA FDMR, Series C, 7.0045, 5 Nov. 1906, Clark to
    Schenck and encl. Rodgers, 398.
35 *Globe*, 16 Nov. 1906, in which the *American Lumberman* is cited. *Canadian For-
    estry Journal* 2(4), 1906, 175.
36 Fernow's view pervades his writings on the subject, esp. 'Lecture IX: Princi-
    ples and Methods of Forest Policy,' 69, in B.E. Fernow, 'Lectures on Forest-
    ry: Delivered at the School of Mining, Kingston, Ontario, January 26th-30th,
    1903,' which was reprinted in 'Report of the Bureau of Forestry for 1905,'
    Ontario Sessional Papers, 1905. Rodgers, 439.
37 UTA OTP, A1967-0007/001 (B.E. Fernow), 16 Feb. 1907, Fernow to Hut-
    ton. UTA, Board of Governors (BOG), A1970-0024/013, 10 Jan. and 14 Feb.
    1907, Min. of BOG Mtg. *Globe*, 18 Mar. 1907.
38 UTA OTP, A1967-0007/001, 23 Mar. 1907, Fernow to Hutton, original em-
    phasis. The following paragraphs are based on this document.
39 Ibid., 25 Mar. 1907, Fernow to Hutton. UTA FF, A1972-0025/186, 23 Feb.
    1927, C.D. Howe to W.S. Wallace.
40 UTA OTP, A1967-0007/001, 25 Mar. 1907, Fernow to Hutton.
41 UTA BOG, A1970-0024/013, 10 Jan. and 28 Mar. 1907, Min. of BOG Mtgs.
42 UTA OTP, A1967-0007/001, 30 Mar. 1907, Fernow to Hutton, and 2 Apr.
    1907, F.A. Moure to Fernow.

2: 'The Child of My Creation,' 1907–1919

1 UTA FF, A1972-0025/187, 13 Apr. 1907, 'Do We Need a Forestry College?,' Thomas Southworth; /190 (Letterbook), responses to Southworth's letters to Ontario lumbermen.

2 In 1899 the Dominion government established the forerunner to the Canadian Forest Service as a tiny branch within the Department of the Interior. Numerous times over more than the next century Ottawa would change its name and position within the national bureaucracy. For simplicity, it will be referred to here as the 'Dominion Forest Service.' Details can be found in Johnstone.

3 UTA FF, A1972-0025/190, 1908 List of Forest Engineers in Canada; /210, 'Forest Resources and Problems of Canada, B.E. Fernow, Delivered to the Society of American Foresters, 28 Dec. 1911,' 144; /187, 'Report of Senate Cte on FF, 1906–1907.'

4 Ibid., /193, ca. July 1909, Fernow to W.E. Wilson.

5 UTA BOG, A1970-0024/013, 11 Apr. 1907, Min.of BOG Mtg, 144. UTA OTP, A1967-0007/001 (Fernow): 29 Apr. 1907, Fernow to Hutton; *President's Report of the University of Toronto for the Year Ending June 30, 1909* (hereafter *President's Report, year*), 23.

6 UTA FF, A1972-0025/189, 16 May 1910, Fernow to W.L. Bray.

7 UTA PF, B1972-0031/003 (9), ca. 1903, notes regarding forestry schools in North America and the proposed forestry curriculum for the University of Toronto. UTA FF A2004-0017/036 (F.B. Robertson), 19 June 1909, Fernow to Robertson; A1972-0025/189, 16 May 1910, Fernow to Bray. *President's Report: 1910*, 13; *1911*, 17; *1914*, 20; *1916*, 24.

8 UTA FF, A2004-0017/007 (J.R. Chamberlin), 27 Dec. 1910 and 21 Nov. 1911, Fernow to Chamberlin; /038 (G.S. Smith), 30 Nov. 1910 and 16 Aug. 1911, Fernow to Smith.

9 *Torontonensis*, 1913, 235. The Class of 1913 had eleven graduates, but one of these was a senior student who had entered the program after 1913 and was given advanced standing for his previous university work.

10 UTA OTP, A1967-0007/001, 24 Apr. 1907, A.H.D. Ross to Fernow; /003, 15 Jan. 1908, Fernow to R.A. Falconer.

11 Ibid., /028, 9 Sept. 1913, Fernow to Ross.

12 AO, F1095, MU2556, autobiographical notes regarding Ross's career. UTA OTP, A1967-0007/028, 9 Sept. 1913, Fernow to Falconer; UTA BOG, A1970-0024/016, 24 Mar. 1914, Min.of BOG Mtg, 347. UTA FF, A1972-0025/190, 25 Nov. 1914, Fernow to H.H. Chapman.

13 UTA FF, A2004-0017/012 (T.W. Dwight), 11 Dec. 1956, G. Tunstell to D. Fayle; /038 (R.D.L. Snow), 16 Apr. 1918, Fernow to R.H. Campbell.

14 Ibid., /038, 16 Aug. 1911, Fernow to Smith; /030, 22 Nov. 1911, Fernow to A.L. Mills, and 23 Nov. 1911, Mills to Fernow. Due to an oversight, the Faculty officially conferred Mills' degree 34 years after he had first enrolled.

15 Ibid., /014 (E.H. Finlayson), 14 May 191[3], Finlayson to Fernow.

16 Ibid., /022 (J. Kay), 1 Mar. 1923, Kay to C.D. Howe.

17 *Torontonensis, 1913*, 176, and *1910*, 255. *President's Report, 1909*, 23.

18 UTA OTP, A1967-0007/001, 30 Apr. 1907, Fernow to Hutton, and 24 Jan. 1908, Falconer to Fernow; /003, 15 Jan. 1908, Fernow to Falconer. UTA BOG, A1970-0024/013, 13 Feb. 1907, Min.of BOG Mtg, 273, and 9 May 1907, Min.of BOG Mtg, 152–3. UTA FF, A1972-0025/190, 4 June 1907, G. Creelman to Fernow.

19 UTA OTP, A1967-0007/003, 14 Feb. 190[8], Fernow to Falconer.

20 UTA BOG, A1970-0024/013, 28 May 1908, Min.of BOG Mtg, 304. D. MacKay, *Empire of Wood: The MacMillan-Bloedel Story* (Vancouver: Douglas and McIntyre, 1982), 30–1.

21 *President's Report: 1909*, 23; *1910*, 13; *1912*, 17; *1913*, 13 and 21; *1914*, 20; *1915*, 20.

22 Friedland, 234 and ch. 21. R. White, *The Skule Story: The University of Toronto Faculty of Applied Science and Engineering, 1873–2000* (Toronto: Faculty of Applied Science and Engineering, 2000), 64–7.

23 UTA FF, A1972-0025/187, 21 Sept. 1907, W.L. Goodwin to Fernow, and 19 Oct. 1907, Fernow to Goodwin. Fernow delivered this same message to his students: A2004-0017/036 (F.B. Robertson), 19 June 1909, Fernow to Robertson; /041 (W.J. Vandusen), 22 Mar. 1909, Fernow to Vandusen.

24 Ibid., A1972-0025/190: 27 Jan. 1908, Fernow to F. Cochrane; 23 Mar. 1908, Fernow to R.H. Campbell; and 6 Feb. 1908, Cochrane to Fernow. /189, 8 June 1912, Fernow to H.R. MacMillan. *Toronto News*, 8 Apr. 1908. *University Magazine* 6 (Dec. 1907), 497–506.

25 UTA FF, A1972-0025/190, 4 June 1907, Fernow to J.Q. Gulnac. Rodgers, 435–9. Gillis and Roach, 100.

26 UTA FF, A1972-0025/187, (copy of) Resolution of the Senate of the University of Toronto, Moved by Dean Howe, Dr Fernow; /189, 19 Feb. 1909, Fernow to W. Saunders.

27 M. Girard, *L'écologisme retrouvé: Essor et déclin de la Commission de la Conservation du Canada* (Ottawa: Les Presses de l'Université d'Ottawa, 1994).

28 UTA FF, A1972-0025/190, 28 Dec. 1912, Fernow to C. Sifton. AO, F6, MU1309, Envelope 4, 'Conditions in the Clay Belt of New Ontario,' 1913, Fernow. *Globe*, 23 and 24 Jan. 1913.

29 *Globe*, 27 Jan. 1913. UTA OTP, A1967-0007/023 (W.H. Hearst), 27 Jan. 1913, Falconer to Hearst, and 31 Jan. 1913, Hearst to Falconer. UTA FF, A1972-0025/189, 27 Jan. 1913, Fernow to Hearst, and 31 Jan. 1913, Hearst to Fernow.

30 *Globe*, 8 Feb. 1913. AO, RG1-68, Memoranda 176–82. UTA FF, A1972-0025/195, 8 Feb. 1913, Fernow to J.F. MacKay.

31 C.D. Howe and J.H. White, *Trent Watershed Survey* (Ottawa: Commission of Conservation, 1913).

32 *Globe*, 1 and 2 Dec. 1913. More than two decades later, the degree to which Fernow's comments had stirred up a hornet's nest was still legendary. In the late 1930s, Gordon Cosens (2T3), a junior Faculty member who was about to become its dean, was discussing with Ed Bonner (3T4) the advisability of establishing 'forest communities' in cooperation with the Ontario government near Kimberly-Clark's newsprint mill in Kapuskasing. Cosens was directing Bonner as to how to draw up a report to support this proposal, and strongly advised Bonner 'to avoid any reference to the social, moral or mental condition of the inhabitants. Do you remember hearing about the difficulty Dr Fernow got into over the Trent Watershed Survey?' UTA FF, A2004-0017/004 (E. Bonner), 14 Oct. 1939, G.G. Cosens to Bonner.

33 *Globe*, 12 Dec. 1913.

34 UTA FF, A1972-0025/194, 28 Nov. 1914, Fernow to S. Coulter. UTA OTP, A1967-0007/028, 30 and 31 Mar. 1914, Fernow to Editor, *Globe*.

35 University of British Columbia Archives (UBCA), H.R. MacMillan Personal Papers (hereafter MacMillan Papers), Box 36, File 11, 14 Jan. 1918, J.H. White to MacMillan. J.W.B. Sisam, *Forestry and Forestry Education in a Developing Country: A Canadian Dilemma* (Toronto: University of Toronto Press, 1982), 30–1. M. Kuhlberg, '"By Just What Procedure Am I to Be Guillotined?": Academic Freedom in the Toronto Forestry Faculty between the Wars,' *History of Education* 31(4), 2002 (hereafter 2002b), 354.

36 Fernow strenuously resisted the campaign to release White to work for the Ontario government, and his persistence resulted in White finding the time to deliver 'special lectures' at the forestry faculty. UTA OTP, A1967-0007/042, 21 Dec. 1916, Fernow to Falconer; /042 (J.H. Faull), 26 Feb. 1917, Faull to Falconer, and 28 Feb. 1917, Falconer to Faull; /045a (J.H. White), 28 Feb. 1917, White to Fernow.

37 UTA FF, A2004-0017/043 (J.H. White). Lambert and Pross, 193–4. P. Oliver, *G. Howard Ferguson: Ontario Tory* (Toronto: University of Toronto Press, 1977), passim, and esp. ch. 6. Nelles, ch. 9.

38 UTA FF, A2004-0017/020 (H.M. Hughson), 23 Mar. 1917, O. Fernow to Hughson.

39 Friedland, 259–61. Immediately after the war's outbreak, an unsuccessful movement arose at the OAC to have it dismiss E.J. Zavitz, but he continued to teach there after his appointment as Ontario's second Provincial Forester.

40 *Globe*, 15 and 16 Jan. 1915. *Toronto News*, 15 Jan. 1915. UTA OTP, A1967-0007/034, 30 Jan. and 8 Feb. 1915, Falconer to Fernow; /038, 13 Nov. 1915, Falconer to Fernow.

41 UTA OTP, A1967-0007/047a, 22 June 1918, Falconer to Fernow.

42 UTA FF, A1972-0025/186, 8 July 1957, R.S. Hosmer to Sisam.

43 *Torontonensis, 1918*, 134.

44 *President's Report, 1916*, 24. *Torontonensis, 1918*, 134, and *1920*, 1200. Friedland, ch. 22.

45 UTA FF, A2004-0017/011 (W.A. Delahey).

46 Ibid., /038 (J.L. Simmons), 19 Apr. 1915, Fernow to Simmons; /015 (A.V. Gilbert), 19 Apr. 1916, Fernow to Gilbert. *President's Report, 1916*, 24.

47 For details, see D. Avery, *Reluctant Host: Canada's Response to Immigrant Workers, 1890–1930* (Toronto: McClelland and Stewart, 1979).

48 UTA FF, A2004-0017/032 (O. Nieuwejaar).

49 Ibid., /005 (W.J. Boyd), 10 Apr. 1919, Boyd to Fernow, and 15 Apr. 1919, Fernow to Boyd. UTA PF, B1983-0022/010, 25 May 1915, L. Coleman to White; 7 June and 5 Oct., White to R.G. Robinson; 26 Oct., White to H.A. Greene. UTA FF, A1972-0025/189, 4 Mar. 1911, Fernow to R.S. Woodward.

50 UTA FF, A1985-0036/001, A. Meighen to S. Smith, encl. is Cameron's CV.

51 *Torontonensis, 1911*, 293. The information dealing with Ellis has been gleaned from correspondence with Michael Roche (Massey University, Palmerston North, New Zealand) and 'Ellis, Leon Macintosh 1887–1941,' *Dictionary of New Zealand Biography*, vol. 4; *Australian Timber Journal*, Dec. 1941, 'Vale – Leon Ellis Macintosh, Forester'; UTA FF, A2004-0017/013 (L.M. Ellis).

52 UTA PF, B1983-0022/005 (208), 11 Apr. 1930, Ellis to White.

53 Friedland, ch. 20. UTA FF, A1972-0025/142, 8 June 1918, Fernow to MacRoberts.

54 Library and Archives Canada (LAC), RG32, C-2, 608, E.H. Finlayson. UTA FF, A2004-0017/014 (E.H. Finlayson); /036 (W.M. Robertson). Johnstone, passim.

55 UTA FF, A2004-0017/027 (E.C. Manning), esp. *Globe and Mail*, 8 Feb. 1941.

56 Ibid., /029 (P. McEwen); /032 (F.S. Newman); /022 (R.A.N. Johnston). Lambert and Pross, 198–9 and 235.

57 UTA FF, A2004-0017/016 (J.D. Gilmour), esp. 21 Mar. 1934, Howe to F. Reed. *Forestry Chronicle*, June 1926.

58 UTA FF, A2004-0017/026 (R.W. Lyons).

59 MacKay, passim. UTA FF, A2004-0017/041 (W. Vandusen).

60 UTA FF, A1972-0025/187, 15 May 1907, 'FF UofT, Preliminary Notice Sent to Principals.'

61 Ibid., A2004-0017/024 (W. Kynoch), esp. 9 Feb. 1929, Kynoch to Howe, and 19 Feb. 1929, Howe to Kynoch. *Globe and Mail*, 18 July 1966.

62 *President's Report: 1916*, 24–5; *1917*, 22; *1919*, 20.

63 UTA BOG, A1970-0024/016, 13 Jan. 1916, Min.of BOG Mtg, 206, and 27 Apr. 1916, Min.of BOG Mtg, 246.

64 MacKay, ch. 2. Gillis and Roach, ch. 6.

65 UTA OTP, A1967-0007/058a, 27 Sept. 1917, Fernow to Falconer. MacMillan Papers, Box 36, File 4, 27 Sept. 1917, Fernow to MacMillan.

66 MacMillan Papers, Box 36, File 11, 29 Sept., 2 and 20 Nov. 1917, and 14 Jan. 1918, White to MacMillan.

67 Ibid., File 4, 2 Mar. and 18 Apr. 1918, Fernow to MacMillan, and 9 May 1918, G.C. Creelman to MacMillan.

68 Ibid., 22 Dec. 1918, Fernow to MacMillan; File 6, 24 Feb., 11 Mar., and 7 Apr. 1919, MacMillan to Fernow; 1 and 14 Mar. 1919, Fernow to MacMillan.

69 MacKay, 42–3 and 93. UTA FF, A1972-0025/189, 2 May 1916, MacMillan to Fernow.

70 UTA OTP, A1967-0007/052 (Fernow), 8 and 12 May 1919, Fernow to Falconer. UTA BOG, A1970-0024/017, 8 and 29 May 1919, Min. of BOG Mtgs.

3: 'We Cannot Progress in Forestry Very Much Ahead of Public Opinion,' 1919–1929

1 K. Norrie and D. Owram, *A History of the Canadian Economy* (Toronto: Harcourt, Brace Jovanovich, 1991), ch. 17. M. Bliss, *Northern Enterprise: Five Centuries of Canadian Business* (Toronto: McClelland and Stewart, 1987), ch. 14.

2 UTA FF, A1972-0025/151, 11 June 1924, C.D. Howe to M.A. Grainger; A2004-0017/009 (A.B. Connell), 24 Feb. 1922, Howe to Connell.

3 Ibid., A1972-0025/186, 1926 acceptance speech by Howe; /141, CFA 1926–33, 28 Oct. 1932, Howe to G.M. Dallyn.

4 See, e.g., B.E. Fernow, *Forest Conditions of Nova Scotia* (Ottawa: Commission of Conservation, 1912); Howe and White; Canada, Committee on Forests, *Forests of British Columbia* (Ottawa: Commission of Conservation, 1918); and Howe, *Forest Regeneration on Certain Cut-Over Pulpwood Lands in Quebec* (Ottawa: Commission of Conservation, 1918). UTA FF, A1972-0025/055 (Dean C.D. Howe), 'The Chairman of the Ontario Forestry Board,' A.R.R. Jones; /146 (C.D. Howe), obituaries. LAC, RG39, 351, 47600 (Ontario Research Silvics, Spanish River Pulp and Paper Mills), 15 Jan. 1926, W.M. Robertson to D.R. Cameron.

5 UTA FF, A1972-0025/146, 7 June 1921, Howe to W.N. Millar (2); /139, 17
May 1923, Howe to H.P. Baker; /143, 7 May 1923, Howe to W. ab-Yberg.
6 Ibid., /147, 9 Dec. 1924, Howe to R.M. Watt; /148, 3 May 1921, Howe to
Watt.
7 Ibid., /146, 11 Aug. 1914, Howe to Fernow.
8 Ibid., /148, 8 May 1923, Howe to W.J. Dunlop; /147, 29 Nov. 1922, Memo
for the President on the Matter of a New Building for the FF, C.D. Howe.
Ibid., A2004-0025/014, *Faculty of Forestry Calendar, 1921–1922* (hereafter *FF
Calendar, year*); /001 (H.W. Allen), 24 Sept. 1930, Howe to Allen. Kuhlberg
2002b, 345–55.
9 UTA FF, A2004-0017/025 (L.M. Lein), 10 Aug. 1938, Lein to Howe, and
31 July 1941, Howe to Lein; /010 (A. Crealock), 9 June 1928, Howe to Crealock; /013 (H.P. Eisler), ca. 17 May 1931, Eisler to Howe.
10 Kuhlberg 2002b, 355. UTA FF, A1972-0025/150 (Foresters' Club); /149, 4
Oct. 1922, Howe to E. Wilson; /145 (Save the Forest Week, Apr. 1926); /139
(AAAS, Toronto Mtg, Dec. 27 + 28, 1921).
11 Ibid. /127, 23 Feb. 1923, W.N. Millar to Howe.
12 *FF Calendar, 1920s.*
13 Ibid., A2004-0017/003, Student Records and correspondence 1909–1911.
14 Ibid., esp. 6 Feb. 1924, Howe to Bayly, and 14 Jan. 1927, Howe to J. Brebner
from which the citations are taken.
15 C.M. Johnston, *E.C. Drury: Agrarian Idealist* (Toronto: University of Toronto
Press, 1986), 54–60 and ch. 12. P. Oliver, *Public and Private Persons: The Ontario Political Culture, 1914–1934* (Toronto: Clarke, Irwin, 1975), ch. 3. UTA
FF, A2004-0017/016 (J.D. Gilmour), 31 Dec. 1920, Howe to Gilmour.
16 AO, RG6-2/8 (Drury 1919–23), 7 Jan. 1921, 'The Basis of a Forestry Policy
for the Province,' E.H. Finlayson, encl. in 2 Feb. 1921, Drury to B. Bow-
man; RG3-4 (Reorganization of Lands and Forests 1922), 12 Aug. 1922,
J.F. Clark to Drury (2). UTA FF, A1972-0025/139, 14 Nov. 1921, Howe to
E. Wilson; /149, 26 Sept. 1921, Howe to Wilson; A2004-0017/016, corre-
spondence1920–1, esp. 24 Sept. 1921, Howe to Gilmour. Drury's ability to
implement the changes he intended was undermined by the Phoenix-like
renaissance of Howard Ferguson (the leader of the Conservative Opposition and former minister of lands and forests) during the Timber Commission's hearings over the course of 1920–2.
17 *FF Calendar, 1922–3* and *1923–4*; UTA FF, A1972-0025/029 (P. McEwen),
material submitted for Faculty's sesquicentennial book. M. Kuhlberg, 'Ontario's Nascent Environmentalists: Seeing the Foresters for the Trees in
Southern Ontario, 1919–1929,' *Ontario History* 88(2), 1996, 119–43.
18 UTA FF, A2004-0017/009 (A.B. Connell), 24 Feb. 1922, Howe to Connell.
19 Kuhlberg 1996, passim; Kuhlberg 2002b, passim.

20 M. Kuhlberg, '"We Are the Pioneers in This Business": Spanish River's
    Forestry Initiatives after the First World War,' *Ontario History* 93(2), 2001,
    150–78; M. Kulberg, '"We Have 'Sold' Forestry to the Management of the
    Company": Abitibi Power and Paper Company's Forestry Initiatives in On-
    tario, 1919–1929,' *Journal of Canadian Studies* 34(3), 1999, 187–210.
21 UTA FF, A1972-0025/144, 10 Feb. and 11 Mar. 1925, Howe to B.F. Avery;
    /144, 17 Nov. 1926, Howe to C.A. Schenck; /142 (OFB, Legislation), 31
    Dec. 1927, W.H. Finlayson to Howe.
22 Ibid., /149 (DFS), 31 Mar. 1928, E.H. Finlayson to Howe; /139, 23 Mar.
    1928, Howe to P.Z. Caverhill; /139, 17 May 1928, Howe to H.R. Christie.
    UBCA, MacMillan Papers, Box 1, File 4, 23 Mar. 1927, Howe to MacMillan.
23 M. Kuhlberg, 'A Failed Attempt to Circumvent the Limits on Academic
    Freedom: C.D. Howe, the Forestry Board, and "Window Dressing" Forestry
    in Ontario in the late 1920s,' *History of Intellectual Culture* 2, 2002a, 1–23; M.
    Kuhlberg, '"In the power of the Government": The Rise and Fall of News-
    print in Ontario, 1894–1932' (Doctoral dissertation, York University, 2002),
    436.
24 Germane documents can be found in UTA FF, A1972-0025/021–022; /025,
    Dec. 1933, 'Summary – Report on Forest Research – Ontario Forestry
    Branch, E.H. Finlayson,' on which is written, 'For the use of the Ontario
    Forestry Branch only.' While the Ontario government may have shelved the
    surveys, Brodie put them to good use. They served as the empirical founda-
    tion for his master's in forestry ('Vegetation and Soil as a Basis for Classifica-
    tion of Forest Types in Ontario'), which the forestry faculty awarded him
    in 1945: see A1972-0025 /005 (J.A. Brodie), esp. 18 Nov. 1938, 'The Forest
    Types of Ontario' (draft).
25 AO, RG1-BB1/7 (Interview, J.A. Brodie), 11 June 1965, transcript of Inter-
    view with J.A. Brodie, H.V. Nelles. UTA FF, A1972-0025/151, 21 Oct. 1927,
    Howe to H.S. Graves; /147, 2 Apr. 1929, Howe to F.D. Mulholland, and 28
    Jan. 1929, Howe to W. Metcalf; /148, 20 Feb. 1929, Howe to J.W. Toumey;
    /147, 15 Apr. 1919, Howe to Muskingum Debating Association.
26 UTA FF, A1972-0025/143, 23 May 1921, Howe to Falconer, and 8 June 1921,
    Falconer to Howe. UTA OTP, A1967-0007/065, 10 June 1921, Howe to Fal-
    coner. UTA FF, A2004-0017/012 (A.B. Doran), 16 Nov. 1921, Howe to L.M.
    Ellis.
27 UTA FF, A2004-0017/016 (J.D. Gilmour), correspondence 1920–21.
28 Ibid., /012 (T.W. Dwight). UTA OTP, A1967-0007/077, 10 Oct. 1922, Howe
    to Falconer.
29 UTA FF, A1972-0025/142, 8 June 1921, Falconer to Howe, and 7 Mar. 1922,
    Howe and J.H. Faull to Falconer.
30 *Globe*, 17 Dec. 1923. UTA FF, A1972-0025/147, 29 Nov. 1922, 'Memo …

New Building for the FF,' Howe. UTA OTP, A1967-0007/083a, 16 Jan. 1924, Howe to Falconer, and 15 Jan. 1924, Howe to G.H. Ferguson.

31 *Canada Lumberman*, 1 Jan. 1924, 57, and 1 Feb. 1924, 58.

32 UTA OTP, A1967-0007/083a, 18 and 20 Mar. 1924, Howe to Falconer, and 19 and 24 Mar. 1924, Falconer to Howe.

33 UTA FF, A1972-0025/142 (all from 1924): 11 Feb., Howe to T. Marshall; 11 June, Howe to H.J. Cody: 9 June, Howe to Falconer; 11 June, Falconer to Howe. /148, all correspondence between Howe and the Toronto Board of Trade. *Globe*, 16 Jan. 1924. *Canada Lumberman*, 1 Feb. 1924, 66. AO, RG3-4 (Forest Fire Protection 1920), 22 Dec. 1919, Toronto Board of Trade to Drury.

34 UTA FF, A1972-0025/144, 9 June 1924, Howe to R.S. Hosmer; /142, 11 June 1924, Howe to Marshall.

35 UTA OTP, A1967-0007/071, 3 Dec. 1921 and 20 Mar. 1922, Howe to Falconer, and 21 Mar. 1922, Falconer to Howe.

36 UTA PF, B1983-0022/008. UTA FF, A1972-0025/186.

37 UTA FF, A2004-0017/032 (W.E.H. Munro).

38 Ibid., relevant student files.

39 Ibid., /001 (R.D.K. Acheson), Student Record; /036 (J.M. Robinson); /040 (G.J. Thomson).

40 Ibid., /033 (H.H. Parsons): H.H. Parsons, 'Aerial Timber Sketching Memoirs,' ca. 1977 (author's copy); /021 (F.T. Jenkins). AO, RG1-E-10, 217, F.T. Jenkins, 31 Dec. 1951, F.T. Jenkins.

41 UTA FF, A2004-0017/045 (W.G. Wright); /028 (J.B. Matthews); /006 (A.H. Burk); /017 (D.W. Gray).

42 Ibid., /021 (J.C.W. Irwin).

43 Ibid., /017 (G.C. Grant); /038 (L.R. Seheult); /001 (J.L. Alexander); /020 (R.C. Hosie). R.L. Schmidt, *The History of Cowichan Lake Research Station* (Victoria: British Columbia Ministry of Forestry, 1992), 1–3. D. G. Bryant, ed., *The Fiftieth Anniversary of the Faculty of Forestry at the University of New Brunswick* (Fredericton: Unipress, 1958), 107–8.

44 UTA FF, A2004-0017/019 (C.C. Heimburger).

45 Ibid., /038 (J.F. Sharpe); /027 (F.A. MacDougall).

46 Ibid., /018 (W.E.D. Halliday).

47 Ibid., /016 (A.W. Goodfellow). Governor General's website, Order of Canada, 'Allan W. Goodfellow.'

48 UTA FF, A2004-0017/034 (A.W. Porter), 17 Oct. 1927, Porter to Howe, and 21 Oct. 1927, Howe to Porter.

49 Ibid., /026 (G.M. Linton); /017 (W.B. Greenwood); /001 (P. Addison). Lambert and Pross, 481–3.

50 UTA FF, A2004-0017/024 (H.H. Krug), 'Draft' biography of 'Howard Krug, BScF,' available at www.ontarionature.org/enviroandcons/reserves/res_ kinghurst.html.
51 UTA FF, A2004-0017/025 (G.R. Lane), 25 Sept. 1930, Howe to Lane.

4 'Forestry's Darkest Hour,' 1930–1941

1   UTA OTP, A1968-0006/001 (P.Z. Caverhill), 13 Feb. 1933, Caverhill to H.J. Cody; UTA FF, A2004-0017/017 (G.C. Grant), 2 May 1931, C.D. Howe to Grant.
2   Kuhlberg diss., ch. 11. Johnstone, chs. 13 and 14. UTA FF, A2004-0017/013 (L.M. Ellis), 18 Feb. 1932, Howe to Ellis; A1972-0025/149, 2 May 1931, Howe to E. Wilson. MacMillan Papers, Box 1, File 4, 1 June 1931, Howe to MacMillan.
3   W.B. Greeley, *Forests and Men: A Veteran Forest Leader Tells the Story of The Last Fifty Years of American Forestry* (Garden City, NJ: Doubleday, 1951), ch. 8. T.D. Clark, *The Greening of the South: The Recovery of Land and Forest* (Kentucky: University of Kentucky Press, 1984), chs. 6 and 7. UTA FF, A1972-0025/143, 27 Feb. 1934, Howe to W.W. Pearse.
4   UTA FF, A2004-0017/007 (H.E. Capp), 9 Dec. 1936, Howe to F.A. Moure; /017 (G.C. Grant), 2 May 1931, Howe to Grant.
5   Ibid., /027 (J.A. Macdonald).
6   Ibid., /009 (C. Cooper); /038 (R.E. Cooper); /021 (C.H. Irwin); /040 (A.C. Thrupp).
7   Ibid., /021 (F.G. Jackson).
8   UTA OTP, A1967-0007/062, 24 Dec. 1919, Memo for the President Re: Permanent Practice Area and Camp for the Faculty of Forestry, C.D. Howe.
9   Howe arranged for some of the province's leading lumbermen to lobby the government for a ranger school, which the latter also sought, but to no avail. UTA FF, A1972-0025/151, correspondence between J.S. Gillies and Howe from late 1923 and early 1924.
10 Ibid., /149, 16 July 1920, E.J. Zavitz to Howe; /146, 17 Sept. 1920, Howe to W.N. Millar; /065, 21 June 1920, Howe to Falconer, encl. is 21 June 1920, Howe to Zavitz; /038 (J.F. Sharpe), 25 Oct. 1920, Howe to Sharpe.
11 *Varsity*, 4 Feb. 1921.
12 UTA FF, A1972-0025/146, 7 May 1921, Millar to Howe.
13 Ibid., /127, 12 Jan. 1923, Memo Re Recommended Location of Camp at Bronte, Ontario.
14 Ibid.: 14 Mar. 1923, Millar to Howe, encl. Memo of Transactions at Faculty Staff Min.; 23 Feb. and 28 Mar. 1923, Howe to Millar.

15 Ibid., A2004-0017/041 (J.L. Van Camp), 3 Sept. 1924, Howe to Van Camp.
16 This and the following few paragraphs are drawn from Kuhlberg 2002b, 358–67.
17 UTA OTP, A1967-0007/127a, 20 June 1931, Millar to Falconer.
18 *Pulp and Paper Magazine of Canada*, May 1932, 117, and June 1932, 158.
19 UTA OTP, A1967-0007/131a, 1 June 1932, E.H. Finlayson to Falconer. Friedland, ch. 26.
20 UTA OTP, A1967-0007/132a, ca. May–June 1932, 'Comments on Professor Millar's Article on Self Education,' Howe. Kuhlberg 2002b, 366–67.
21 UTA BOG, A1970-0024/013, 12 May and 15 June 1932, Min. of BOG Mtg., incl. 15 June 1932, Falconer to BOG. UTA OTP, A1967-0007/132a, ca. 6 and 8 June 1932, Millar to Falconer, and 10 June 1932, President's Secretary to Millar.
22 UTA OTP, A1968-0006/010, 27 July 1933, Howe to Cody.
23 Ibid., /003, 6 Feb. 1933, Howe to Falconer; /010, 3 and 10, Oct. 1933, Howe to Cody, and ca. Oct. 1933, Suggested Draft to Pres. Carnegie Foundation, Dr Henry Suzzalo. UTA BOG, A1970-0024/013, 13 and 26 Oct. 1933, Min. of BOG Mtg.
24 McKillop, ch. 14. UTA FF, A1972-0025/141, 26 Sept. 1933, Howe to Cody, and 25 June 1934, 'Report of the Dean of the FF.' UTA OTP, A1968-0006/010, ca. 3 Oct. 1933, Suggested Draft to Pres. Carnegie Foundation, Howe.
25 UTA FF, A2004-0017/010 (G.G. Cosens). Howe had first discussed the job with Benjamin F. Avery, Abitibi's Chief Forester, but Avery had declined the offer in the hope that he would realize his 'dream' of practising sustained yield management on all the company's timberlands; ibid., A1972-0025/186, 16 Mar. 1934, Howe to Cody.
26 Ibid., /141: 7 Feb. and 5 May 1936, 14 Feb. and 29 Nov. 1938, Howe to Cody; Salaries for 1938–39. UTA OTP, A1968-0006/032, 7 Mar. 1938, Howe to Cody.
27 J.T. Saywell, *'Just Call Me Mitch': The Life of Mitchell F. Hepburn* (Toronto: University of Toronto Press, 1991), chs. 7–8. UTA FF, A1972-0025/149, 7 Nov. 1934, Howe to Wilson; A2004-0017/0025 (G.R. Lane), 7 Nov. 1934, Howe to Lane.
28 UTA FF, A1972-0025/142, 13 June 1935, Howe to C.D. Orchard; /149, 7 Nov. 1934 and 12 June 1935, Howe to Wilson; ibid., A2004-0017/025 (G.R. Lane), 7 Nov. and 20 Dec. 1934, Howe to Lane; /039 (K.A. Stewart), 12 Dec. 1934, Howe to Stewart.
29 Ibid., A2004-0017/027 (F.A. MacDougall), 7 Jan. 1935, Howe to MacDougall; /034 (F.M. Plahte), 6 Dec. 1934, Howe to Plahte.

30 Ibid., A1972-0025/141, 9 May 1935, Howe to Cody, encl. copy of Howe's draft letter to the Minister of Lands and Forests. A2004-0017/027, 7 Jan. 1935, Howe to MacDougall; /039, 12 Dec. 1934, Howe to Stewart. AO, RG3-9 (DLF, General 1935), 13 May 1935, Howe to M.F. Hepburn.
31 UTA FF, A1972-0025/149, 26 and 27 May 1935, Howe to Wilson.
32 Ibid., /144, 3 Dec. 1935, Howe to A.B. Recknagel.
33 Ibid., A2004-0017/008 (H.R. Christie): 24 Oct. 1932, Christie to Howe, encl. 15 Sept. 1932, 'Report of the Department of Forestry for the Year Ending August 31st, 1932,' H.R. Christie; 2 Nov. 1932, Howe to Christie. UTA OTP, A1967-0007/094, 7 Dec. 1925, Howe to Falconer.
34 UTA PF, B1983-0022/006, 28 Nov. 1931, Howe to Dear Sir. UTA FF, A1972-0025/138, 8 Dec. 1931, M. Ardenne to Howe; /151, Fernow Letters.
35 UTA FF, A2004-0025/014 (FF Calendars, 1930s). A1972-0025/141, 17 June 1935, 'Report of the Dean of the FF,' and 19 June 1936, 'Report of the Dean of the FF.' Sisam, 105–6.
36 UTA FF, A2004-0017/027 (F.A. MacDougall), 7 Jan. 1935, Howe to MacDougall. A1972-0025/141, 17 June 1935, 'Report of the Dean.'
37 Ibid., A2004-0017/031 (P.M. Morley).
38 Ibid., relevant student files.
39 Ibid., /026 (S.T.B. Losee). St Mary's Paper Archives (SMPA), documents from 1946 to the early 1970s.
40 UTA FF, A2004-0017/010 (D.I. Crossley); /004 (E. Bonner); /030 (J.B. Millar). R. Bott and P. Murphy, *Living Legacy: Sustainable Forest Development at Hinton, Alberta* (Hinton: Weldwood of Canada, 1997), passim.
41 UTA FF, A2004-0017/006 (A.F. Buell); /030 (G.D. Millson).
42 Ibid., /006 (A.W.A. Brown). *ESO Newsletter,* June 2005.
43 UTA FF, A2004-0017/027 (D.A. MacLulich). E-mail correspondence with Dr Mary Ann Fieldes, 21 Oct. and 2 Nov. 2006.
44 UTA FF, A2004-0017/004 (J.E. Bier).
45 Ibid., /017 (W.R. Grinnell); /019 (Q. Hess); /008 (C.H.D. Clarke).
46 Ibid., /005 (A.S. Bray).
47 Ibid., /003 (G.H. Bayly).
48 Ibid., /028 (K.M. Mayall); /044 (M.R. Wilson).
49 Ibid., A1972-0025/149, 24 Feb. 1939, Howe to R. Zon; A2004-0017/007 (A.T. Catto), 5 May 1937, Howe to Catto.
50 Ibid., A1972-0025/144, 20 July 1936, Howe to C. Schott; A2004-0017/026 (A.P. Leslie), ca. late Mar. 1937, Leslie to R.C. Hosie.
51 Ibid., A1972-0025/144, 30 Oct. 1931, Howe to A.O. Shedd. B.W. Hodgins and J. Benedickson, *The Temagami Experience: Recreation, Resources, and Aboriginal Rights in the Northern Ontario Wilderness* (Toronto: University of Toronto

Press, 1989), passim. G. Killan, *Protected Places: A History of Ontario's Provincial Park System* (Toronto: Queen's Printer in association with Dundurn Press, 1993), passim.

52 UTA FF, A2004-0017/010 (D.I. Crossley), 29 June 1936, Howe to Crossley.

53 Ibid., A1972-0025/149, 4 Mar. 1938, Howe to Wilson. UTA OTP, A1968-0006/032, 12 Nov. 1937, Howe to Cody.

54 UTA OTP, A1968-0006/032, 7 Mar. 1938, Howe to Cody; /046, 9 Oct. 1940, Howe to Cody, encl. copy of 7 Mar. 1938, Howe to Cody, on which Howe recorded his comments.

55 Ibid., /045, 28 Mar. 1941, Cody to Cosens, and 7 Apr. 1941, Cosens to Cody; /037, 18 Jan. 1939, Howe to Cody. UTA FF, A1972-0025/141, 9 Oct. 1940, and 13 Jan. and 17 Feb. 1941, Howe to Cody.

56 UTA OTP, A1968-0006/055, 10 June 1943, Howe to Cody (2), and 11 June 1943, Cody to Howe. UTA FF, A1972-0025/146, Dean C.D. Howe, all clippings.

5: 'The Present Pressure for Registration in Forestry Is Temporary,' 1941–1947

1 UTA FF, A1972-0025/154 (OFIA), 17 Dec. 1945, OFIA Circular No. 45-33, W.A. Delahey. UTA OTP, A1968-0007/003 (FF, 1946–47), all documents, with the citation taken from 29 Jan. 1946, W.J. LeClair to S.E. Smith.

2 Kuhlberg diss., Section IV. AO, RG1-122, TB1F232, A-35; RG1-E-10, 167–8 (Kimberly-Clark). UTA OTP, A1968-0006/054, 20 Jan. 1943, G.G. Cosens to J.H. Cody.

3 *Globe and Mail,* 23 Oct. 1939 and 1 June 1953.

4 AO, RG18-125 (Public Hearings Held … Part VII), 760; Howard Kennedy was the chairman of this Royal Commission.

5 UTA FF, A1972-0025/187, Dec. 1946, 'Brief from FF to Royal Commission on Forestry' (hereafter RCF).

6 Spruce Falls Inc. Archives (SFIA), Woodlands Manager Annual Reports File I, 30 Apr. 1936, 'Forestry 1936,' J.B. Millar.

7 UTA OTP, A1968-0006/063, 24 Feb. 1945, Cosens to Cody. UTA FF, A2004-0017/010 (G.G. Cosens), correspondence from 1943 to 1945 dealing with the Bothwell award; /022 (H.J. Johnstone), 22 Apr. 1943, Cosens to W. McNutt. *Globe and Mail,* 8 July 1943.

8 UTA OTP, A1968-0006/054, 4 Dec. 1942, Cosens to Cody. UTA FF, A2004-0017/041 (G. Tunstell), Record of Employment; /011 (B.G. Day), correspondence dealing with Day's war service.

9 UTA FF, A1972-0025/155 (Employment, Graduates and Students …), all correspondence; A2004-0017/035 (J.E.C. Pringle).

10 *Toronto Evening Telegram*, 1 Mar. 1946. *Varsity*, 25 Feb. 1946.

11 UTA FF, A2004-0017/011 (M.M. Dixon), 28 Dec. 1944, Cosens to Dixon.

12 Ibid., /001 (M.A. Adamson), 27 Nov. 1942, Cosens et al. to Adamson.

13 Ibid., /038 (F.E. Sider), 8 May 1942, Cosens to Sider.

14 UTA OTP, A1968-0006/054, 4 Dec. 1942, Cosens to Cody; /059, 1 Dec. 1943, Cosens to Cody. UTA FF, A1972-0025/155, all correspondence; /152, 23 Mar. 1944, Cosens to G. Cameron.

15 UTA, A2004-0017/043 (F.N. Wiley), all documents, with the citations taken from: 27 Feb. 1944, Wiley to Cosens, and 6 and 24 Mar. 1944, Cosens to Wiley; /004 (E. Bonner); /001 (W.W. Adams).

16 Friedland, 360–3. R. Bothwell, I. Drummond, and J. English, *Canada since 1945* (Toronto: University of Toronto Press, 1989), 109–11.

17 UTA OTP, A1968-0006/063, 26 Mar. and 12 Apr. 1945, Cosens to Cody, and 19 May 1945, Col. Lepans to President.

18 Ibid., 11 May 1945 (Enquiries about Entering, 1944–5 FF); /003, 10 Sept. 1945, Cosens to S.E. Smith.

19 Ibid., 10 and 27 Sept. 1945, Cosens to Smith.

20 Ibid., 27 Sept. 1945, Cosens to Smith.

21 Ibid., 11 Dec. 1945 and 28 Jan. 1946, Cosens to Smith; UTA FF, A2004-0017/013 (J.L. Farrar); /016 (J.H. Godden), 31 Jan. 1946, Cosens to Godden, and 6 Feb. 1946, Godden to Cosens, and CV, John Herbert Gooden; /030 (A.S. Michell).

22 Ibid., A1972-0025/187, Dec. 1946, 'Brief from FF to RCF.'

23 UTA OTP, A1968-0007/003, 9 Feb. and 11 Mar. 1946, Cosens to Smith, and 28 Feb. 1946, Smith to BOG; /019: 1 Mar. 1946, Admission 1946–47; 20 Sept. and 23 Nov. 1946, Smith to Cosens; 11 Sept. 1946, UofT FF, Report of Cte on Admissions Session 1946–7; 15 Oct. 1946, Cosens to Smith; 21 Nov. 1946, W.G. Thompson to Smith; 5 Dec. 1946, Smith to H.R. Scott. UTA FF, A1972-0025/187, Dec. 1946, 'Brief from FF to RCF.'

24 P.A. Baskerville, *Sites of Power: A Concise History of Ontario* (Toronto: Oxford University Press, 2005), 209–11. Bothwell et al., ch. 19.

25 AO, F1014, MU1785 (DLF, 1940–46), 'What the CCF Will Do,' E.B. Jolliffe. UTA FF, A1972-0025/155, 22 Apr. 1944, Cosens to A. Koroleff.

26 *Globe and Mail*, 9 July 1943. Forestry commissions were in use in Scandinavia, and it appears that this is from where the idea had originated.

27 UTA FF, A1972-0025/154 (Ontario RCF), 20 Mar. 1944, Mem (GGC). *Globe and Mail*, 4 Apr. 1944; AO, F1014, MU1786 (RCF 1943–44), all notes.

28 *Globe and Mail*, 4 Apr. 1944. AO, F1014, MU1786 (RCF), F.A. MacDougall's notes; MU1787 (Howard Kennedy and the RCF 1946), all notes. AO, RG3-23, 136 (Ontario RCF): 1 Nov. 1943, G.A. Drew to H. Kennedy; 14 Nov. 1943 and 29 Dec. 1945, Kennedy to Drew; 28 Nov. 1945, Memo. AO, RG3-17,

454, Progressive Conservative Association, 2-2 Point Program, 13 Dec. 1944, Transcript of Speech by Premier George A. Drew over CBL and Ontario Network.

29 *Globe and Mail*, 13 Feb. 1943.

30 AO, RG3-23/81(John. C.W. Irwin): 13 Feb. 1943, Cosens to C.G. McCullagh; 15 Feb. 1943, Cosens to Drew. UTA FF, A2004-0017/021 (J.C.W. Irwin), 13 Feb. 1943, Cosens to Irwin. *Globe and Mail*, 16 Feb. 1943.

31 UTA FF, A1972-0025/155, 22 Apr. 1944, Cosens to Koroleff.

32 Ibid., /152 (Abitibi Power and Paper), 28 Jan. 1947, Cosens to C.B. Davis.

33 Ibid., /211, 25 Sept. 1946, Min. of First Mtg. of Advisory Cte on Forestry Research of the Ontario Research Commission.

34 Ibid., /141, 29 Nov. 1938, Howe to Cody; A2004-0017/034 (G.W. Phipps), 21 Feb. 1944, R.C. Hosie to Phipps.

35 SFIA, 1930–1980 (SFIA), Misc., Forestry Reports, Sept. 1945, 'Report of Regeneration Studies on the Limits of Spruce Falls Power and Paper Co., Ltd., Kapuskasing, Ontario,' Hosie. UTA FF, A1972-0025/024 (all files); /020 (R.C. Hosie). R.C. Hosie, *Forest Regeneration in Ontario* (Toronto: University of Toronto Press and the Research Council of Ontario, 1953).

36 'Forester's Tribute – F.A. MacDougall 1896–1966,' A. Herridge (author's copy).

37 UTA OTP, A1968-0006/032, 19 Dec. 1939 and 29 Apr. 1940 (encl. 27 Apr. 1940, Royal Bank to C.D. Howe), Howe to Cody.

38 Ibid., /042, 25 Jan. 1940, Howe to Cody; /006 (A.W.A. Brown), 13 Mar. 1944, Howe to Brown.

39 Ibid., /046, 24 Apr. 1940, Howe to Cody; /032, 29 Apr. 1940, Howe to Cody; /127: 1 and 24 Apr. 1940, Howe to President; 12 Feb. 1941, Howe to W.J. Dunlop; 27 Feb. 1941, Cody to P. Heenan; 28 Nov. 1951, T.W. Dwight to J.W.B. Sisam. UTA BOG, A2002-0003, 8 and 22 Feb., 13 June, 12 and 26 Sept. 1941, Min. of BOG Mtgs.

40 UTA FF, A2004-0017/001 (G.S. Andrews), 14 Feb. 1944, Cosens to Andrews; A1972-0025/127, 1 and 24 Apr. 1940, Howe to President. UTA OTP, A1968-0007/003, 26 Oct. and 8 Dec. 1945, S.E. Smith to Cosens, and 15 May 1946, Cosens to A. Gaine; A1968-0006/063, 'Proposed Addition to the University Forest,' Cosens and T.W. Dwight, May 1945.

41 AO, RG3-17/443 (DLF, Forest Ranger Schools), all correspondence. UTA FF, A2004-0017/029 (P. McEwen), all correspondence; /001 (M.A. Adamson), 27 Nov. 1942, Cosens to Adamson. UTA OTP, A1968-0007/076, 27 June 1944, (Copy of) 'Agreement made this 27th day of June A.D. 1944, between … the University of Toronto and His Majesty the King.'

42 UTA OTP, A1968-0007/003, 4 July 1945, F.A. MacDougall to S.E. Smith, and

12 July 1945, Smith to MacDougall. /019: 26 Sept. 1946, W.G. Thompson to Smith; 27 Sept. 1946, Smith to Cosens (2); 28 Sept. 1946, Cosens to Smith; 19 June 1947, H.R. Scott to Smith; 27 June 1947, Smith to Scott.

43 *Globe and Mail*, 14 Nov. 1942. UTA OTP, A1968-0006/059, 27 May 1944, Cody to W.J. LeClair; /063, 14 Feb. 1945, Cosens to Cody; A1968-0007/003, 28 Nov. 1945, Cosens to Smith; /019, 21 Jan. 1947, 'Proposed Agreement on Establishment of Fellowships at the University of Toronto,' and 21 April 1947, 'Announcement.'

44 UTA FF, A1972-0025/152, 21 Dec. 1944 and 3 Jan. 1945, Cosens to A. Bedard, and 28 Dec. 1944, Bedard to Cosens.

45 Ibid., A2004-0017/037 (R.A. Shand); /022 (M.A. Kasturik); /006 (W.G.E. Brown).

46 Ibid., /002 (D.D. Avery); /008 (D.C.E. Clark); /005 (J.G. Boultbee); /025 (H.C. Larsson).

47 Ibid., /035 (G.L. Puttock); /022 (W.J. Johnston); /032 (D. Naysmith); /004 (J.R.F. Blais); /041 (K.B. Turner).

48 Harvey Robbins, phone interview, 24 Oct. 2005. UTA FF, A2004-0017/015 (H.D. Graham); /019 (K.W. Hearnden), all documents, with the citations taken from 22 June 1943, Hearnden to Cosens, and 3 July 1943, Cosens to Hearnden.

49 UTA FF, A2004-0017/037 (R. Schafer).

50 Ibid., /014 (W.T. Foster).

51 Ibid., /064 (A.H. Richardson). A.H. Richardson, *Conservation by the People: The History of the Conservation Movement in Ontario to 1970* (Toronto: University of Toronto Press, 1974).

52 UTA FF, A1972-0025/152, 6 Feb. 1946, Cosens to C.M. Cushnie.

53 Ibid., A2004-0017/043 (F.N. Wiley), 6 Mar. 1944, Cosens to Wiley. UTA OTP, A1968-0007/019: 20 Feb. 1947, Cosens to Smith, encl. 21 Feb. 1947, C.H. Sage to Cosens; 13 Mar. 1947, Smith to BOG; 14 Mar. 1947, Smith to Cosens.

54 UTA FF, A2004-0017/041 (J.L. Van Camp), 15 Mar. 1947, Van Camp to Cosens, and 27 Mar. 1947, Cosens to Van Camp.

55 UTA OTP, A1968-0006/063, 21 Nov. 1944, Cosens to Cody, and CV, John William Bernard Sisam. A1968-0007/003 (all from 1945): 20 Aug., S. McAree to S.E. Smith; 30 Aug., Cosens to Sisam; 30 Aug., Smith to Cosens; 3 Sept., F. Hudd to Smith; 10 Sept., Sisam to Smith; 1 Nov., Smith to V. Massey; 5 Nov., Acting High Commissioner to Smith; 15 Nov., Sisam to Cosens.

56 AO, F1014, MU1788, Memo to Minister of Lands and Forests 1947, 10 and 15 Mar. 1947, MacDougall to Minister. UTA OTP, A1968-0007/019, Smith to Sisam.

57 The Cosens' generosity to the University of Toronto was not confined to the
   Faculty of Forestry. In September 1947 his first wife bequeathed over $3,000
   to establish an 'open scholarship for women in University College.' UTA
   OTP, A1968-0007/035, 30 Apr. 1948, Smith to Cosens.
58 UTA OTP, A1968-0007/019: 20 Feb. 1947, Cosens to Smith; 24 Feb. 1950,
   A.E. Wicks to Sisam; 3 Mar. 1950, Smith to Wicks. /076: 20 Sept. 1950,
   Cosens to F.K. Morrow; 6 Dec. 1950, Cosens to Smith; 8 Dec. 1950, Smith
   to Cosens. /097, 31 July 1952, Cosens to Smith, and 18 Aug. 1952, Smith
   to Cosens. /110, 22 Mar. 1954, Smith to Sisam, and 25 Mar. 1954, Sisam to
   Smith. /133: 3 Nov. 1955, Cosens to J.T. Phillips and to Smith; 6 Apr. 1956,
   Smith to Sisam; 9 Apr. and 4 June 1956, Sisam to Smith. A1971-0011/014,
   25 Mar. 1958, M. St A. Woodside to Cosens. A1978-0028/017, 22 July 1971,
   Sisam to J.H. Sword.

6: 'Today It is Not always Ranked Professionally as First,' 1947–1957

1 UTA OTP, A1968-0006/063, 21 Nov. 1944, G.G. Cosens to H.J. Cody, and
   CV, J.W.B. Sisam; A1968-0007/019, 28 Mar. 1947, S.E. Smith to Sisam.
2 Friedland, ch. 29. UTA, Dept. of Graduate Records, A1973-0026, 2, R.N.
   Aro, 18 June 1951, D.V. Love to Aro.
3 Gillis and Roach, 242–8. AO, RG1-E-10/56 (Timber Management Mtgs.,
   TMM), vol. 1.
4 Pross and Lambert, chs. 18–19. AO, RG1-335 contains the reams of documentation that were generated by the province's nascent forest management planning activities. *Report of the Royal Commission on Forestry 1947*
   (Toronto: Baptist John, Printer to the King's Most Excellent Majesty, 1947).
5 AO, RG1-E-10/56 (TMM), vols. 1 and 2. *Windsor Star*, 2 Sept. 1949.
6 Hosie, 1953. UTA FF, A1972-0025/211(ORC); /210, 16 July 1953, Sisam to
   J.O. Wilhelm. SMPA, Regeneration Files 1950/09, Project RC-17.
7 AO, RG1-A-I-10, all boxes.
8 UTA FF, A1972-0025/196. AO, RG1-BB1/8 (Leslie Frost), 30 Nov. 1965, Interview with Frost, P. Pross and R. Lambert.
9 AO, F150, Box 1 (Advisory Committee [ACte] 1950–59), all correspondence. RG1-A-I-10/1 (ACte Min.), 18 Feb. 1950, Sisam to H.R. Scott, and 6
   Sept. 1950, Min. of ACte Mtg.; /6, all files; /7, 16 July 1951, Sisam to Minister of Lands and Forests (MLF), and Nov. 1951 – Jan. 1952, Mgmt (Timber). UTA FF, A72-0025/196, 12 Feb. 1950, Sisam to (all ACte members).
10 Ibid., White Paper 1954, all correspondence. Lambert and Pross, ch. 19.
   AO, RG1-A-I-10/1, ACte Min.; /8 Cte Administration; RG1-49-200.
11 This raised a major issue between industry and government. The former

argued that, if it paid for even part of the regeneration costs, then it gained
equity in the future crop, a contention the latter disputed. The fear that
industry's case rested on firm legal ground, however, caused the Ontario
government to take back responsibility for forest management in 1962.

12 AO, RG1-A-I-10/7, 4 Feb. 1952, Sisam to H.R. Scott; /1, ACte Min., all ger-
mane Mtgs. but esp. those in late 1953 and throughout 1954 that deal with
Bill 89; ACte Min., all Mtgs. from 1955 to 1956; 16 Dec. 1955, Min. of Cte
Mtg.

13 AO, RG3-23/91 (DLF), Minister's File, 9 Jan. 1957, C.E. Mapledoram to
L.M. Frost; /90, DLF, ACte to MLF, 7 Oct. 1958, R.A. Farrell to J.W. Spooner, and 12 Jan. 1961, Frost to Spooner. RG1-A-I-10/1, ACte Min. 1956, and
11 July 1958, Sisam to Frost. UTA PF, B1989-0100/001 (J.W.B. Sisam), Transcript of Interview with Sisam.

14 *Report of the Ontario Royal Commission on Forestry, 1947* (Kennedy Report)
(Toronto: Baptist John, Printer to the King's Most Excellent Majesty, 1947),
179. AO, RG1-A-I-10/1, ACte Min. of Special Mtg.

15 UTA OTP, A1968-0007/076, Sisam to Smith: 21 Oct., 29 Dec. 1947; 28 Jan.,
11 May, 10 Nov. 1948; 10 Mar. 1949; Smith to Sisam: 11 May, 5 Nov. 1948; 24
Jan., 8 Mar., 24 June 1949; /133, 18 Jan. 1956, Sisam to Smith, and 20 Jan.
1956, Smith to Sisam.

16 Ibid., /10: '1952 – Interim Report on "Balfron,"' A.D. Hall; 8 Sept. 1952,
S.E. Smith to Sisam; 5 Jan. 1953, A.G. Rankin to Sisam; 9 Jan. 1953, Sisam to
Rankin. /110, 'Report on the UofT Forest for 1954'; /121, 1 Dec. 1954 and
2 Mar. 1955, Sisam to Smith; /133: 15 Mar. 1956, F.R. Stone to Smith; 'Road
Construction in the University Forest, 1956–1958'; 'Report on the UofT
Forest, Jan. 1 1955 to Mar. 31 1956.'

17 Ibid., /101, 13 Mar. 1953, Memo for the President, A.G. Rankin; /064, all
correspondence relating to Forest Ranger School charges; /101, 7 July
1952, Sisam to K.F. Tupper, and 13 Mar. 1953, Rankin to S.E. Smith.

18 White, ch. 4. Friedland, 373–81.

19 UTA OTP, A1968-0007/019, 3 Apr. 1947, Cosens to Smith. /035: 10 Jan.
1948, Smith to Sisam; 30 June 1947, Sisam to J.A. MacFarlane, encl. 30 June
1947, 'UofT FF, Note on Plans for Future Development in the FF,' and to
Smith; 29 Dec. 1947, Sisam to Smith.

20 Ibid., /089, 15 Oct. 1951, Smith to Sisam, and 18 Oct. 1951, Sisam to Smith.

21 Ibid., /050, 22 Sept. 1948, Smith to Sisam, and 4 Oct. 1948, Sisam to Smith;
/110, 12 Oct. 1949, Sisam to Smith, and 5 Dec. 1949, R.E. Spence to Smith.

22 Ibid., /110, 5 Dec. 1949, Spence to Smith, on which Smith recorded his views.

23 Bliss, 337, 346, and 348. 'Glory Days,' John Court, *Profiles: The York University
Magazine for Alumni and Friends*, Sept. 1997.

24 UTA OTP, A1968-0007/110, 8 Oct. 1942, Codicil to the last will and testament of Agnes Euphemia Wood of Toronto.

25 Ibid.: 12 June 1950, H.S. Jackson to Smith; 14 June 1950, G.H. Duff to Smith, encl. 'Creation of a Joint Research Centre at the Wood Estate'; 23 Oct. 1950, E.R. Arthur to Smith; 24 Nov. 1950, Smith to W.D. Patterson; 12 Dec. 1950, K.F. Tupper to Smith; /133, 10 Jan. 1956, Memo Re: Radioactive Disposal.

26 Ibid., /110: 14 June 1950, Duff to Smith, encl. 'Creation of a Joint Research Centre at the Wood Estate – Memo Embodying the Views of Representatives of the University and Other Interested Bodies'; 21 June 1950, R.L. Charles to Smith; 'Memo in Relation to the Use of Glendon Hall Property. Conference at 1 p.m. at the York Club, Mon., Oct. 2nd'; Feb. 1951, 'Proposal for the Development at Glendon Park of Permanent Facilities for a Scientific Garden, Arboretum, Experimental Grounds and Facilities for Research and Teaching' (hereafter 'Proposal'); 22 Aug. 1951, Duff to Comptroller. UTA BOG, A2002-0003, 25 May and 23 Nov. 1950, Min. of BOG Mtg.

27 UTA OTP, A1968-0007/110: 11 Nov. 1950, Duff to Smith; 23 Nov. 1950, Smith to BOG; 16 Feb. 1951, Duff to Smith, encl. Feb. 1951, 'Proposal.'

28 Ibid., /076, 4 Jan. 1951, Sisam to Smith.

29 Ibid., also 18 June 1951, Sisam to Smith, encl. (draft) 'UofT Forestry Series'; /089, 28 Aug. 1951, Sisam to Smith.

30 Ibid., /089: 22 Feb. 1952, Sisam to Smith, encl. 23 Jan. 1952, 'Note on the Contribution the FF Might Make to Silvicultural Research and a Way in which This Can be Encouraged and Supported by Industry,' Sisam; 26 Feb. 1952, Smith to Sisam. /076, 18 June 1951, Smith to Sisam. UTA FF, A1985-0036/048, 19 Feb. 1952, Min. of Staff Mtg.

31 UTA OTP, A1968-0007/110: 17 Jan. 1951, Smith to H.A. Innis; 19 Dec. 1952, Dunlop to Smith; 27 Aug. 1952, Cassell to Phillips.

32 Ibid., 17 July 1952, Duff to Smith.

33 Ibid., /089, 18 Feb. 1952, Sisam to Smith.

34 Ibid., /110: 14 Apr. 1953, Sisam to Smith; 15 June 1953, Duff to Smith; 17 June 1953, Smith to Duff; all documents germane to negotiations between the Ontario government and the University of Toronto regarding the new normal school.

35 Ibid.: 14 Apr. 1953, Sisam to Smith. /101: 15 June 1953, Smith to Sisam; 17 Dec. 1952, and 22 June 1953, Sisam to Smith.

36 Ibid., /121, 26 Jan. 1955, Sisam to Frost.

37 Ibid.: /133: 7 Nov. 1955, Smith's notes; 7 Nov. 1955, Smith to Sisam; 18 Nov. 1955, G.L. Court to Smith.

38 Ibid., 16 Nov. 1955, Sisam to Phillips and Sisam to Smith.

39 Bliss, 467–70, 479, 499, and 547.
40 UTA OTP, A1968-0007/133, 26 Jan. 1956, Smith to C.E. Mapledoram, and
    17 Feb. 1956, Smith to M.W. McCutcheon.
41 Ibid., Memo of discussion: Mr D.W. Ambridge, W.E. Phillips, S.E. Smith –
    York Club, 1.30 p.m., Wed., Nov. 23rd, 1955, and Smith's notes.
42 Ibid., 16 Nov. 1955, Sisam to Phillips, and Sisam to Smith.
43 Ibid.: 4 Jan. 1956, Sisam to Smith; 13 Jan. 1956, Smith to Phillips; 19 Jan.
    1956, Sisam to Smith, encl. 'Proposed Programme – Forest Economics – FF';
    17 Feb. 1956, Smith to McCutcheon; 20 Feb. 1956, McCutcheon to Smith.
44 Ibid. (all 1956): 4 June, Sisam to Smith; 5 June, Smith to Sisam; 15 June,
    Mapledoram to Smith; 19 June, Smith to Mapledoram. UTA BOG, A2002-
    0003, 21 June 1956, Min. of BOG Mtg. UTA FF, A1985-0036/004 (OPC),
    all correspondence from the mid- to late 1950s between FF and OPC. UTA
    OTP, A1971-0011/004, 25 Sept. and 13 Nov. 1956, Sisam to Smith, and 26
    Sept. 1956, Smith to Sisam.
45 UTA OTP, A1968-0007/133 (all 1956): 16 Feb. and 2 Apr., Sisam to D.W.
    Ambridge; 18 Apr. 1956, Smith to Ambridge; 15 May 1956, Sisam to Smith.
    UTA BOG, A2002-0003, 3 May, 21 June, and 5 October 1956, Min. of BOG
    Mtg. I.C.M. Place, *75 Years of Research in the Woods: A History of Petawawa Forest
    Experiment Station and Petawawa National Forestry Institute* (Bunstown: General
    Store Publishing House, 2002), 79–80.
46 UTA FF, A2004-0017, germane boxes and files.
47 Ibid., /060 (Chung-Li Huang); /066 (Tung-Kung Yuan).
48 UTA, A2004-0017, 1041 (A.R. Turner); /019 (J.H. Hewetson); /011 (R.M.
    Dixon); /020 (T.G. Honer); /030 (V.G. Merritt); /038 (D.A. Skeates).
49 Ibid., /001 (J.R.T. Andrews); /002 (E.B. Ayer); /006 (T.A. Buell); /009
    (W.G.L Cleaveley); /012 (D.P. Drysdale); /021 (J.D. Irwin); /029 (T.P. McEl-
    hanney); /030 (B.F. Merwin); /031 (R.W. Morison); /033 (D.G. Parsons);
    /038 (J.T. Somerville); /041 (N.J. Turnbull).
50 Ibid., /008 (M. Chubb); /018 (L.S. Hamilton) and (D.G.E. Harris); /028
    (P.G. Masterson).
51 Ibid., /026 (O.L. Loucks), esp. regular correspondence, 18 July and 14 Aug.
    2006.
52 UTA, A2004-0017, germane files.
53 J. Zucchi, *Italians in Toronto: Development of a National Identity, 1875–1935*
    (Toronto: University of Toronto Press, 1988), ch. 2.
54 UTA FF, A2004-0017/003 (E.J.K. Bagg); /013 (R.Y. Edwards); /030 (J.L.
    Mennill). Regular correspondence with Rachel Mills, July 2006.
55 Ibid., /044 (G.C. Wilkes). *Capital Xtra*, 13 Oct. 2000. G.C. Wilkes, correspon-
    dence and phone interviews, July 2006.

56 UTA OTP, A2004-0017/006 (T.A. Buell); /018 (W.G. Harris); /034 (D. Penna); /011 (A.E. Davis); /005 (J.E. Bothwell).
57 Ibid., /019 (A.J. Herridge); /044 (D.R. Wilson); /012 (D. Drysdale); /022 (J.W. Keenan); /015 (J.W. Giles).
58 Ibid., /007 (R.D. Carman).
59 *Toronto Daily Star*, 19 Oct. 1957.
60 Killam, passim. Lambert and Pross, ch. 22.
61 UTA OTP, A2004-0017/020 (S.G. Holmes), 21 June 1954, Holmes to Sisam.
62 UTA Senate, A2002-0004, 9 Apr. 1936, Min. of Senate Mtg.
63 Bryant, 69–153.
64 UTA OTP, A1968-0007/076, 20 Sept. 1950, and 3 Jan. and 13 Feb. 1951, Sisam to Smith; /064, 23 May 1950, Sisam to Smith; /101, 25 Feb. 1953, Sisam to Members of the Council, FF, and 26 Feb. 1953, Sisam to Smith. UTA Senate, A2002-0004, 9 Feb. 1951, Min. of Senate Mtg.
65 UTA OTP, A1968-0007/133, 27 Apr. 1956, Smith to Sisam.

7: 'Forestry Has Suffered Its Share of Frustrations,' 1957–1971

1 Friedland, ch. 30.
2 In 1958, Ontario's Department of Lands and Forests (DLF), with the minister's support, was moving towards fully funding a meaningful silvicultural program ('Project Regeneration'), but Treasury Board vetoed the initiative: the details can be traced in AO, RG1-E-10/74 (Timber Management, TM), 'Regeneration Policy,' vol. 3.
3 UTA FF, A1972-0025/168, 14 Oct. 1966, A.D. Hall to Sisam.
4 Friedland, ch. 31.
5 UTA OTP, A1971-0011/014, 3 Mar. 1958, Sisam to M. St A. Woodside, encl. 3 Mar. 1958, Memo with Respect to the Possible Re-location …, Sisam.
6 Ibid., 11 Mar. and 12 May 1958, Woodside to Sisam. UTA FF, A1985-0036/048: 21 Jan. 1958, Memo Re FF Expansion; 23 Jan. 1958, Memo of Mtg.; 30 June 1960, Min. of Mtg.; /007, correspondence from the early to mid-1960s regarding problems with the FF building, esp. 30 Mar. 1965, Sisam to F.J. Hastie. *Annual Report of the President of the University of Toronto, 1958–9*, 50–1 (hereafter *President's Report*).
7 UTA FF, A1985-0036/048, 15 Dec. 1960, Min. of Mtg.
8 Ibid., 25 Feb. 1960, Min. of Mtg. Bothwell et al., 216. Gillis and Roach, 249–52. Sisam, 54–61.
9 Available at http://www.snre.umich.edu/about-snre/hundred-history.php; http://www.dnr.cornell.edu/.
10 UTA OTP, A1975-0021/093: 2 Oct. 1964, Sisam to J.H. Sword; 2 Oct. 1964,

C.T. Bissell to Members of ACte to the FF. UTA FF, A1985-0036/048, 26 June 1961, Min. of Mtg; A2004-0017/030 (J.W. McNutt), 27 Mar. and 26 June 1962, Sisam to McNutt. *President's Report, 1959–1960*, 53, and *1961–1962*, 56.

11 *President's Report, 1959–1960*, 53. UTA OPV, A1977-0020/030, all documents.
12 UTA FF, A1985-0036/048: 25 Nov. 1966, Min. of Mtg.; Undergraduate Program; 8 Feb. and 26 June 1961, Min. of Mtg.; /002, 5 Dec. 1966, Sisam to Members of the Board of Forestry Studies; /007, 13 Dec. 1968, Min. of Mtg. *President's Report, 1960–1*, 54–5.
13 *National Audubon Society Field Guide to North American Trees* (New York: Knopf, 1980), 419. Erik Jorgensen, phone interview, 24 July 2006.
14 UTA OTP, A1971-0011/060, all documents dealing with the Dutch elm disease and establishment of the Shade Tree Research Laboratory (hereafter STR Lab). UTA OTP, A1975-0021/137, 9 Jan. 1963, J.F. Westhead to Sword. *Globe and Mail*, 22 and 24 Jan. 1962. *Canadian Audubon Magazine*, Nov.–Dec. 1962, 'The Dutch Elm Disease – What Can Be Done about It?,' Jorgensen.
15 UTA FF, A1971-0011/060, 7 May 1963, Sword to Bissell, and 11 June 1963, Sword to Bissell and Sisam. UTA BOG, A2002-0003/024 (Oct. 1963), Min. of Mtg.
16 UTA OPV, A1977-0020/030, 25 June 1964, 'The STR Lab,' Jorgensen, and 19 Jan. 1965, Jorgensen to Sword. UTA OTP, A1975-0021/030 (all 1967): 19 Oct., Jorgensen to A.K. Roberts; 20 Oct., Jorgensen to G. de B. Robinson; 26 Sept., Jorgensen to Sword; 10 Oct., Sword to Jorgensen. A1975-0021/093 (1968): 'Urban Forestry: Some Problems and Proposals, Prepared for the 9th Commonwealth Forestry Conference,' Jorgensen; 4 Mar., Jorgensen to H.W. Thomas; 15 Mar., Jorgensen to Sword; 'Observations on Some Commonwealth Forestry Problems Based on a Visit to India, Malaysia and Kenya in Connection with the IX Commonwealth Forestry Conference,' Jorgensen. UTA FF, A1985-0036/004: 22 Dec. 1970 and 30 Apr. 1971, R. Brunelle to Jorgensen; 7 Jan. 1971, Jorgensen to S. Dymond; 21 May 1971, Jorgensen, 'Application for a Special Grant in Support of a Research Associate in Urban Forestry in the STR Lab of the FF, 1971–1972.' Sisam, 110.
17 UTA OPV, A1977-0020/030, 7 June 1965, Sisam to Sword, and 24 June 1964, 'STR Lab,' Jorgensen; /046, 13 July 1966, Jorgensen to Sword. UTA FF, A1977-0019/004, 21 Jan. 1969, Sisam to Bissell. Jorgensen, phone interview.
18 H.S. Braun, *A Northern Vision: The Development of Lakehead University* (Thunder Bay: Lakehead University, 1987), chs. 2–3. G.A. Garratt, *Forestry Education in Canada* (Montreal: CIF, 1971), ch. 10. UTA OTP, A1968-0007/050, 9 Nov. 1948, Sisam to S.E. Smith; UTA FF, A2004-0017/007 (C.J.H. Campbell).

298   Notes to pages 175–80

19 UTA OTP, A1975-0021/024, 13 Apr. 1965, H.S. Braun to Bissell. A.K. Mc-

Dougall, *John P. Robarts: His Life and Government* (Toronto: University of To-

ronto Press, 1986), ch. 6. Braun, chs. 4–10. Garratt, 68–70.

20 Braun, 24.

21 UTA FF, A2004-0017/039 (A.L.K. Switzer), 30 Nov. 1962, and 1 Apr. and 15
May 1963, Switzer to Sisam, and 6 Dec. 1962, Sisam to Switzer. UTA OTP,
A1975-0021/005, ca. June 1964, 'A Degree Course in Forestry at Lakehead
College,' Appendix (J. Haggerty). Braun, 62.

22 UTA OTP, A1975-0021/005, 16 July 1964, A.J. Herridge to Bissell, encl. 'A
Degree Course.' UTA FF, A1985-0036/048, 23 Sept. 1963, Professional Education in Ontario; A2004-0017/042 (P.C. Ward), 6 Aug. and 10 Oct. 1963,
J.R. Nicholson to D.M. Fisher; *Port Arthur News-Chronicle*, 25 Jan. and 13 Mar.
1964. Haggerty also apparently began publishing scathing, and discombob-
ulated, attacks on the University of Toronto in the *Forestry Chronicle* at this
time.

23 Friedland, 437–8 and ch. 32. UTA FF, A2004-0017/039 (A.L.K. Switzer), 15
Feb. 1964, Switzer to Sisam, and 16 Mar. 1964, Sisam to Switzer.

24 UTA FF, A2004-0017/018 (W.G. Harris), 31 Mar. 1964, Sisam to Harris;
/042 (P.C. Ward) (all 1964): 17 Mar., Ward to Sisam; 20 Mar. and 1 May,
Sisam to Ward; 10 Apr., Mtg. on Forestry Education, Lakehead College of
Arts, Science and Technology; 8 May, Min. of Annual Mtg., Canadian In-
stitute of Forestry (CIF), Northwestern Ontario Section. UTA OTP, A1975-
0021/005, 18 June 1964, Sword to Bissell, encl. ca. June 1964, 'Draft.' AO,
RG32-1-1, M373, Lakehead College, 4 Nov. 1964, Herridge to Robarts.
Braun, 72.

25 AO, RG32-1-1, M373, Lakehead College (all 1964): 25 June, A.K. Roberts to
W.G. Davis; 24 July and 23 Dec., Davis to Roberts; 17 Dec., Sisam to Roberts;
(all 1965): 4 Jan., Davis to Roberts; 7 Jan., Roberts to Davis; 4 Jan. and 1 and
17 Mar., G.C. Wardrope to Davis; 28 Jan., 9 Mar., and 3 Apr., Davis to Wardrope; 19 Feb., Wardrope to Davis, encl. 18 and 24 Feb. and 11 Mar. 1965,
K.G. Crowhurst to Wardrope. UTA OTP, A1975-0021/024, 13 Apr. 1965,
Braun to Bissell; /093, 'Report on FF, UofT,' 1965.

26 Ibid., /122, 6 Apr. 1961, Memo for File – Estimates of FF, Sword. UTA FF,
A1985-0036/048, 6 Nov. 1962, Min. of Mtg.

27 UTA OTP, A1975-0021/093, ca. 14 July 1965, Sisam to Bissell, encl. 'Report
on FF, UofT.'

28 Friedland, ch. 34. White, ch. 4.

29 UTA OTP, A1975-0021/030 (all 1965): 15 Nov., R.W. Kennedy to Bissell;
24 Nov., Bissell to Kennedy; 3 Dec., Sisam to Sword. UTA OPV, A1977-
0020/041, 6 Jan. 1966, Sisam to Sword; A1975-0021/093, Feb. 1966, 'Sub-

mission to the Commission to Study Development of Graduate Programs in Ontario,' FF, UofT; /046, Nov. 1966, 'Proposals for the Future Development of Research and Postgraduate Education,' FF, UofT, submitted to the Dept. of Forestry and Rural Development; /095, 11 and 21 Oct. 1966, Sisam to Sword. UTA FF, A1985-0036/048, 17 Feb. 1966, Min. of Mtg.

30 UTA FF, A1985-0036/007, 4 Mar. 1966, Sisam to F.J. Hastie, and 8 Mar. 1966, Hastie to Sisam.

31 Ibid., /048, 19 Apr. 1966 and 30 May 1967, Min. of Mtgs. UTA OTP, A1975-0021/093, 15 Aug. 1967, Sisam to Sword.

32 UTA OTP, A1975-0021/093, 21 Aug. 1967, Woodside to Sword.

33 Ibid., 10 Oct. 1967, Memo of the FF for the President's Advisory Council. Sisam also forwarded a copy to F.A. Ireland, the president's research assistant: ibid., 21 Aug. 1967, Sisam to Ireland.

34 UTA OPV, A1977-0020/095: 3 Jan. 1967, F.R. Stone to Sword; /093, 'Proposal for the Relocation and Expansion of the Botany-Forestry Research Unit Now Situated at Glendon College Campus,' N.P. Badenhuizen and Sisam; 25 Oct. 1967, Sisam to Stone; 26 Oct. 1967, Stone to Sisam.

35 UTA OTP, A1975-0021/093, 14 Dec. 1967, Sisam to Woodside. UTA OPV, A1977-0020/025, 21 May 1968, Sisam to Sword.

36 UTA OTP, A1977-0019/004, 5 Dec. 1968, Jorgensen to Bissell.

37 Ibid., 6 Dec. 1968, Sisam to Bissell.

38 UTA FF, A1985-0036/007, 13 Nov. 1970, Sisam to Hearnden.

39 UTA OTP, A1977-0019/044, 4 May 1970, Sisam to Bissell, and 'Report on the Future of the FF, UofT,' 1970.

40 UTA FF, A1985-0036/048: 18 May 1968, 'Lakehead'; Nov.–Dec. 1968, 'Log Book – The School of Forestry at Lakehead University,' A. Bartholomew; 'Future Plans of the School of Forestry'; 13 Nov. 1968, Min. of Mtg. UTA OPV, A1977-0020/077, Nov. 1968, OPFA, Resolution No. 3.

41 UTA OTP, A1977-0019/004, 2 Apr. 1969, Bissell to Sisam; /025, 21 Oct. 1969, Sword to Bissell, and 21 Jan. 1970, Sword to A.G. Rankin.

42 UTA FF, A1985-0036/004, 8 Jan. 1970, Jorgensen to Sisam; 28 Sept. 1970, Sisam to Brunelle. Gillis and Roach, 255–7. Killan, ch. 5. Beginning in 1968, the OFIA memoranda were replete with discussion of the unprecedented danger posed by this emotionalism.

43 UTA OTP, A1977-0019/044, 19 Feb. and 31 Oct. 1969, Herrige to Bissell. UTA FF, A1985-0036/007 (OFPA), 26 Aug. 1970, 'A Statement Concerning the Association Study of Ontario Professional Forestry Education,' Hearnden, encl. in 26 Aug. 1970, A.R. Fenwick to all OPFA Members.

44 Ibid., 1 May 1970, Hearnden to Bissell, and encl.

45 Immediately after tabling his highly controversial report for the OPFA's

education committee, Herridge had asked Sisam if the dean wished for
Herridge to resign from the faculty's advisory committee. If Sisam had sus-
pected that Herridge had pulled the rug from under his feet, this was the
perfect opportunity to be done with his former student. But Sisam asked
Herridge to stay on. And when Herridge resigned upon hearing that Sisam
was going to be replaced as dean (all members offered to do so in order
to permit the new dean to select his own members for the committee),
Sisam was effusive in his praise for Herridge's contribution. UTA FF, A1985-
0036/007, 8 Apr. 1971, Herridge to Sisam, and 19 Apr. 1971, Sisam to
Herridge.

46 The reaction to Herridge's report by the Committee on University Affairs
reinforces this interpretation. Although Sisam understood Herridge's ul-
terior motive in handling this matter, the dean was profoundly hurt by the
OPFA's actions and was convinced he smelled a rat at work. Sisam, and others, directed their anger at Hearnden, who was now the OPFA's president
and a recent appointment to Lakehead's forestry school. The problem was
exacerbated by the OPFA's decision to handle this matter behind a cloak
of secrecy that could not help but generate an air of suspicion, particularly
around Hearnden. He would categorically deny any role in this affair, and
the evidence bears him out. After all, if Sisam's suspicions were true, the
OPFA would have recommended moving the faculty to Lakehead, but it did
not even include Thunder Bay as a possible site to which the faculty should
be relocated. UTA OTP, A1977-0019/044, 13 July 1970, D. Wright to D.C.
Williams. UTA FF, A1985-0036/048, 21 July 1970, Re: Faculty Development.
/007 (all 1970): 2 July, McNutt to Sisam; 12 Aug., McNutt to Herridge; 24
Aug., Sisam to McNutt; 'A Statement Concerning ... Professional Forestry
Education,' encl. in 26 Aug., Fenwick to all OPFA Members. Carman, phone
interview, 13 July 2006.

47 UTA FF, A1985-0036/036, 'Newsletter and Directory, Forestry Alumni As-
sociation,' 19070. UTA OTP, A1977-0019/044 (all 1970): 5 June, Sisam to
Bissell; 16 June, 'Professional Forestry Education,' OPFA; 18 June, Bissell to
D.T. Wright; Excerpt from Min. of the Cte on University Affairs Held June
23, 1970; 13 July, Wright to D.C. Williams; 24 July, Sword to J.B. Macdonald.
UTA OPV, A1977-0020/077, 5 Aug. 1970, Sisam to Sword.

48 UTA OPV, A1977-0020/077, 23 Sept. 1970, Memo to the President, Sword.

49 *Globe and Mail,* 30 Oct. 1970.

50 UTA FF, A1985-0036/048, 'A Statement to the Cte on University Affairs by
the UofT,' Nov. 1970, 4–5. UTA OTP, A1977-0019/044, 3 Dec. 1970, Sword
to Bissell. *Toronto Telegram,* 11 Dec. 1970.

51 Garratt, ch. 22, esp. 365. AO, RG32-1-1, M383 (UofT Forestry Program), 1

Dec. 1971, G.H. Bayly to E.E. Stewart. UTA FF, A2004-0017/019 (Q.F. Hess), 13 Nov. 1970, T. Daly to Hess, and 7 Jan. 1971, Hess to Daly; /027 (T.E. Mackey), 13 Nov. 1970, Daly to Mackey, and 17 Dec. 1970, Sisam to Mackey. /021 (J.C.W. Irwin) (all 1971): 14 May, Irwin to L.N. Earl; 21 June, Daly to Sisam; 28 June, Sisam to Irwin; 5 July, Sisam to Daly and Sword; 24 Nov., Irwin to V. Nordin. UTA OTP, A1978-0028/017 (all 1971): 14 May, Irwin to Earl; 21 June, Daly to Sisam; 5 July, Sisam to Sword; 22 Nov. and 14 Dec., D.S. Bruce to Sword; 30 Nov., Sword to Bruce.

52 UTA OTP, A1977-0019/044, 7 Jan. 1971, Bissell to R.B. Loughlan, and 8 Jan. 1971, Bissell to J.D. Coats.

53 UTA FF, A1985-0036/048, 11 Dec. 1970, Decanal Search Cte; /007, 25 Jan. 1971, Sisam to L.A. Smithers; A2004-0017/044 (J.R.M. Williams), 8 Jan. 1971, Sisam to Williams. AO, RG32-1-1, M382 (UofT Forestry Program), 21 Dec., Sisam to Stewart.

54 AO, RG32-1-1, M383 (UofT Forestry Program 1971): 14 Jan., D. Reid to Members of the Cte on University Affairs; 14 Jan., Forestry Degree Students to William G. Davis (telegram); 19 Jan., Stewart to Forestry Degree Students; 19 Jan., Stewart to W.G. Tamblyn; 20 Jan., J. Jessiman to Davis; 26 Jan., Stewart to Jessiman.

55 Ibid., 29 Jan., Tamblyn to Stewart, encl. 26 Jan. 1971, W.D. Bohm to Braun.

56 Ibid., 18 Feb., Stewart to Bayly. UTA OTP, A1977-0019/044, 23 Apr. 1971, A. Vidnal to J. White.

57 AO, RG32-1-1, M383, 2 Mar. 1971, Bayly to Stewart. Braun, ch. 15.

58 J.H.G. Smith, *UBC Forestry 1921–1990: An Informal History* (Vancouver: Faculty of Forestry, UBC, 1990), 94–5.

59 UTA FF, A2004-0017/001, (M.S. Allen) and (J.E. Ambrose); /006 (L.O.W. Burrige); /007 (K.P. Campbell); /032 (B.J. Myers); /033 (R.N. O'Reilly).

60 Ibid., /001 (W.D. Addison); /005 (J.D. Brodie); /030 (J.F.K. McNutt); /035 (G.D. Puttock); /042 (J.D. Walker).

61 Ibid., /004 (C.A. Benson); /008 (R.E. Chopowick); /009 (P.A. Cooper); /010 (G.N. Crombie); /012 (J. Dreifelds); /014 (R.J. Fessenden); /015 (D.J. Gerrard); /016 (W.S. Good); /023 (L.T. Kirby); /027 (E.B. MacDougall); /031 (H.P.B. Moens); /032 (P.A. Murtha).

62 Ibid. R.D. Fry, phone interview, 19 Feb. 2007. J.M. Taylor, phone interview and e-mail, 20 Feb. 2007.

63 UTA FF, A2004-0017/013 (J.W. Ebbs); /014 (N.W. Foster); /020 (F.J. Hutcheson); /021 (I.H.H. Jennings); /024 (E.S. Kondo); /025 (L.S. Lambert); /029 (H.J. McGonigal); /031 (R.M. Monzon) and (I.K. Morrison); /040 (W.D. Tieman); /044 (E.G. Wilson).

64 Ibid., /006 (D.E. Buck), esp. *Toronto Star*, 1 Dec. 1988.

65 Ibid., /044 (J.R.M. Williams), 22 June 1972, Williams to M. Grasley; /040 (W.D. Tieman), 21 July 1959, Tieman to M. Harman, and 13 Aug. 1959, Harman to Tieman.

66 Ibid., /035 (R.M. Rauter). K.A. Armson, phone interviews, 17 Aug. and 16 Sept. 2006.

67 *Toronto Star*, 2 June 1971. UTA FF, A2004-0017/005 (M.K. Boyd).

68 One of the faculty's own drove this point home to Sisam at a dinner the dean attended in 1968. Peter Addison (2T9), chief of the Ontario government's Parks Branch, lamented how he was unable to find any university graduates in Canada who were qualified for positions on his staff in the area of outdoor recreation. All Sisam could do to respond was point out that, although there was 'a good deal of discussion at the Canadian forestry schools' regarding this matter, there was no movement from his faculty to satisfy this need. UTA FF, A2004-0017/001 (P. Addison), 2 May 1968, Sisam to Addison.

69 UTA FF, A1985-0036/004, 20 Jan. 1971, Memo, W.C. Harrison; /044, 16 Mar. 1971, Sword to President.

70 Ibid., /007, 23 Apr. 1971, L.A. Smithers to Members of the ACte to the FF, and 4 May 1971, Smithers to J. Hamilton.

8: 'Rebuilding a Neglected and Deplorably Weak Faculty,' 1971–1985

1 AO, RG1-41/28 (FF, UofT), 12 Dec. 1973, 'Biography of Vidar Nordin.' UTA OTP, A1978-0028/017, 'Biography of Vidar J. Nordin.'

2 UTA OTP, A1978-0028/045, 7 Nov. 1972, D.F. Forster to J.R. Evans.

3 Ibid., A1979-0051/008, 13 May 1974, Nordin to Forster. UTA FF, A2004-0025/009, 30 Sept. 1971, Nordin to M.L. Prebble, and 5 Nov. 1973, Nordin to M.D. Seeley.

4 UTA OTP, A1978-0028/017: ca. 6 Apr. 1972, J.W. Simmons to J.H. Sword; 26 Aug. 1971, Forster to Nordin; 14 Oct. 1971, Nordin to Forster; 24 Jan. 1972, Forster to M.F. Murrill.

5 Ibid.: 14 July and 4 Aug. 1971, 8 and 12 Apr. 1972, Nordin to Sword; 9 Aug. and 29 Oct. 1971, Forster to Nordin; 13 Aug. 1971, Nordin to J.D. Hamilton; 23 Sept. 1971, Faculty Mtg. (notes); 4 Nov. 1971, K.S. Gregory to Forster; 5 Nov. 1971, Forster to Gregory. UTA FF, A2004-0025/009, 2 Nov. 1971, Nordin to Forster; 29 Nov. 1971, Nordin to F. Hastie.

6 UTA FF, A2004-0025/009, 6 Oct. 1971, Nordin to Staff. *President's Report*, 1971–2 (The Dean). UTA FF, A2004-0025/009, 27 Apr. 1973, Nordin to L. Bertin. UTA OTP, A1978-0028/045, 18 Apr. 1973, Nordin to Forster, encl. 9 Apr. 1973, M. Wayman to W.F. Graydon.

7 Ibid., /017, 21 Mar. 1972, Nordin to Forster, encl. 'Dean's Message' for *Annual Ring 1972*. A2004-0025/009: 27 June 1972, Nordin to Forster; 3–5 Nov. 1971, EMR Contract Proposals by Nordin; 11 Nov. 1971, Nordin to R.D.T. Birchall; 23 Nov. 1971, File Note; 10 Dec. 1971, Nordin to B.A. Gingras; 19 Dec. 1971, Nordin to F.A. Zuana; 25 Jan. 1972, Nordin to R. Brunelle; 27 Jan. 1972, 'Explanatory Notes, Budget Estimates 1972–1973'; 15 Mar. 1972, Nordin to W.Q. Macnee. A1985-0036/002, 22 Nov. 1971 and 14 Feb. 1972, Nordin to Forster.

8 UTA FF, A2004-0025/031, corr. Sept.–Dec. 1972 germane to Act. /009: 11 Dec. 1972, Nordin to J. Crispo; 14 Dec. 1972, G.H. Bayly to Nordin; 20 Nov. 1973, Nordin to J. Bene; 12 Feb. 1974, Nordin to A.G. Rankin; 14 Oct. 1977, Nordin to Members of the Faculty Advisory Board, and encl.; 8 Mar. 1974, Nordin to members of new Advisory Board.

9 UTA OTP, A1979-0051/008, 3 June 1974, A.H. Zimmerman to Nordin, and 18 June 1974, Evans to Nordin.

10 Ibid., 1984-0026/025, 1 Nov. 1976 and 21 Jan. 1977, Nordin to D.A. Chant. UTA FF, A2004-0025/009: 13 Oct. 1971, Nordin to Bene; 9 Nov. 1971, Nordin to A. Barsan; 11 Mar. 1974, Nordin to R.Z. Callahan; 30 Apr. 1974, article for *University of Toronto Bulletin*; 30 Apr. 1974, File Note. A1985-0036/043, 23 Sept. 1976, Min of (ExCte) Mtg.

11 UTA OTP, A1978-0028/017, 23 Aug. 1971, Nordin to Sword. UTA FF, A2004-0025/009, 9 Feb. 1972, Nordin to J. Davis; 23 May 1975, Nordin to Deans of Forestry Schools in Canada; /020 (Association of University Forestry Schools of Canada, Min. and Constitution). UTA OTP, A1979-0051/008: 11 July 1973, Nordin to Evans et al.; Oct. 1973, 'A National Statement by the Schools of Forestry at Canadian Universities'; ca. 1974, 'Forestry Education in Canada: The Future Is Now,' Nordin.

12 E.R. Townsend, *Algonquin Forestry Authority ... A 20 Year History* (Hunstville: Algonquin Forestry Authority, 1995), 7–11. UTA OTP, A1979-0051/074, 15 Oct. 1974, Nordin to Evans, and 16 Oct. 1974, Evans to Nordin. UTA FF, A2004-0025/009, 31 Oct. 1974, Nordin to Herridge, and 10 Jan. 1975, Nordin to Evans. UTA OTP, A1984-0026/080, 2 Dec. 1977, Nordin to Evans, and encl.

13 UTA FF, A2004-0025/009, 1 May 1973, Nordin to Academic Staff, and 13 Feb. 1975, Nordin to Evans. UTA OTP, A1971-0011/004, 14 Jan. 1957, D.C.F. Fayle to S.E. Smith.

14 UTA FF, A2004-0025/008, 25 Sept. 1972, 'The FF at the UofT'; /009, 30 Jan. 1974, Nordin to G. Smith. UTA OTP, A1978-0028/045, 2 Nov. 1972, Nordin to Forster, Evans, and Sword.

15 UTA FF, A2004-0025/009, 26 June 1972, Nordin to Forster. UTA OTP,

A1978-0028/017, 14 June 1972, Nordin to Forster; 11 July 1972, Nordin to Jorgensen. UTA FF, A2004-0025/009, 14 Mar. 1973, File Note.

16 UTA OTP, A1978-0028/045, 7 Nov. 1972, Forster to Evans; /017, 2 May 1972, Nordin to G. de Robinson, and 2 May 1972, Nordin to Forster. UTA FF, A2004-0025/009, 11 Nov. 1972, Nordin to Forster, and 17 Jan. 1972, Nordin to Sword. Nordin took the same approach in dealing with the Department of Landscape Architecture, which enjoyed a rocky marriage with – and quick divorce from – the faculty as a result.

17 UTA OTP, A1978-0028/045, 29 May 1973, Jorgensen to Evans, and 13 June 1973, Nordin to Jorgensen. UTA FF, A2004-0025/009, 1 Mar. 1974, Nordin to H. Van Dyke; A1985-0036/043, 25 May 1976, Min. of Mtg. UTA OTP, A1987-0020/005, 27 Jan. 1981, H.C. Eastman to Planning Subcommittee, encl. 26 Jan. 1981, 'Draft Administrative Response to the Plans and Priorities of the Faculty of Forestry.'

18 K.J. Rea, *The Prosperous Years: The Economic History of Canada, 1939–1975* (Toronto: University of Toronto Press, 1985), 254. Friedland, chs. 38–9.

19 UTA OTP, A1979-0051/008: 23 Aug. and 2 Oct. 1973, and 4 Feb. 1975, Nordin to Forster; 23 Nov. 1973, Forster to Nordin; /074 (Forestry): 3 Oct. 1974, Nordin to Evans; 11 Nov. 1974, Evans to Nordin; 16 Dec. 1974, Nordin to P.P.M. Meincke. UTA FF, A2004-0025/009, 25 Nov. 1974, Nordin to Forster.

20 UTA FF, A2004-0025/009: 4 Feb. 1974, File Note; 4 Feb. 1974, Nordin to Evans; 27 Feb. 1974, Nordin to P.D. McTaggart-Cowan; 25 Mar. 1974, Nordin to K. Hearnden; 2 Oct. 1974, File Note. UTA OTP, A1979-0051/008: 7 Jan. 1974, Nordin to Evans; 14 Jan. 1974, Evans to Nordin; 30 Jan. 1974, L. Collins to Evans; 4 Feb. 1974, Mtg. with Ontario Provincial Government Representatives; 4 Feb. 1974, Nordin to A.B.R. Lawrence.

21 UTA FF, A2004-0025/031, 25 Feb. 1974, Nordin to Evans, and encl. Friedland, 563.

22 UTA FF, A2004-0025/009, 16 Dec. 1974, Nordin to Burchall. UTA OTP, A1979-0051/074: 4 Dec. 1974, G.P. Hiebert to Evans et al.; 9 Dec. 1974, Memo for Discussion with Nordin and Robinson; 23 June 1975, Meincke to M. Israel.

23 UTA FF, A2004-0025/015: 7 Jan. 1975, J.W. Anderson to D.V. Love; 9 Jan. 1975, K.A. Armson to Love; 2 June 9175, Nordin to Love, encl. 29 May 1975, File Note; /009, Nordin to Evans.

24 Ibid., A1985-0036/048, 20 Nov. 1975, Ref. Faculty ExCte Mtg. of Nov. 11, 1975.

25 Ibid., A2004-0025/031, Notes of June 11th (1975) Mtg., 2.30 p.m. , Forestry, For File; 20 June 1975, Confidential – FF.

26 AO, RG1-41, 181 (FF, UofT), 23 May 1980, J.A.C. Auld to K.G. Laver. UTA

FF, A1985-0036/043, 11 Nov. 1975, Min. of Mtg.; A2004-0025/009, 25 Oct.
1976, Nordin to Evans; /008 (Faculty Development, General, Report of
Progress 1977), Nordin; /008 (President, General), 13 Dec. 1979, File Note.
UTA OTP, A1987-0020/005, 23 May 1980, Reynolds to Chant.

27 UTA OTP, A1984-0026/080, 20 Apr. 1977, to Nordin (Draft), encl. in 21
Apr. 1977, D.W. Lang to Evans et al.

28 Ibid., 4 Nov. 1977, Nordin to Evans, and 8 Nov. 1977, Evans to Nordin.

29 Ibid.: 6 Dec. 1977, A. Zimmerman to Evans; 31 Jan. 1978, Nordin to Evans;
2 Feb. 1978, Evans to Nordin.

30 Ibid., 17 Feb. 1978, Evans to Nordin. UTA FF, A2004-0025/031, 10 Feb.
1978, Reynolds to Nordin.

31 UTA OTP, A1978-0028/045: 9 Jan. 1973, B.M. Levitt to Forster; 10 Jan.
1973, Forster to Evans; 11 Jan. 1973, Sword to Evans; 18 Jan. 1973, Forster to
Nordin. UTA FF, A2004-0025/031: 16 Feb. and 17 May 1978, Nordin to Eastman; 24 Feb. 1978, Nordin to Zimmerman; 23 June 1978, File Note. /008:
19 Oct. 1977, Nordin to Evans; 17 Feb. 1978, H.S.B. Jones to R.W. Missen;
20 Apr. 1978, Nordin to Eastman; 10 May 1979, Eastman to Nordin.

32 UTA FF, A2004-0025/031, 14 Oct. 1977, Nordin to Members of the Faculty
Advisory Board, encl. Report of the South-West Campus Redevelopment
Task Force. UTA OTP, A1992-0024/002, 25 Oct. 1977, D.W. Strangway to
Eastman.

33 UTA FF, A2004-0017/037 (M.D. Sandoe), 25 Oct. 1978, Nordin to Sandoe.
UTA OTP, A1985-0016/005, 18 Sept. 1978, Nordin to P.E. Trudeau (2), and
25 Sept. 1978, Trudeau to Nordin.

34 R.J. Fessenden, phone interview, 15 Aug. 2006, and K.A. Armson, phone
interview, 17 Aug. and 16 Sept. 2006. UTA OTP, A1985-0016/005: 30 June
1978, Armson to Nordin; 8 Sept. 1978, Chant to J.M. Ham; 31 Oct. 1978,
Missen to Nordin. UTA FF, A2004-0025/008 (Provost, Budget), 28 Sept.
1978, Nordin to Chant.

35 UTA FF, A2004-0025/031, 20 Mar. 1979, Nordin to Zimmerman. UTA OTP,
A1985-0016/005, 21 Dec. 1978, (Research and Planning Officers) to D.W.
Lang, and 1 Feb. 1979, Eastman to Ham.

36 UTA OTP, A1986-0021/004, 27 June 1979, Ham to B. Stephenson.

37 Ibid., 4 July 1979, Memo to File, original emphasis.

38 Ibid., A1985-0016/005, 16 Jan. 1979, Nordin to Ham, encl. Phase III Planning Document. UTA FF, A2004-0025/008, Mar. 1978, 'Phase III: Planning
and Priorities, Faculty of Forestry and Landscape Architecture.' UTA OTP,
A1986-0021/004, 10 July 1979, Nordin to Ham.

39 UTA OTP, A1985-0016/005, 14 Feb. and 6 Mar. 1979, W.R. Clark to Ham,
and 2 Mar. 1979, Ham to Clark.

40 Ibid., A1987-0020/005, Aug. 1979, 'Report from the Task Force on the Fea-
    sibility of Locating the Faculty of Forestry at Scarborough College.' UTA FF,
    A2004-0025/031, 4 Oct. 1979, Ham to Nordin and J. Foley, and (Draft) Min.
    of the 173rd Council Mtg., Oct. 12, 1979. UTA OTP, A1986-0021/004, 12
    Oct. 1979, Min. of Council Mtg., FF.
41 *Varsity*, 28 Nov. 1979. *Newspaper*, 17 Oct. and 28 Nov. 1979. UTA FF, A2004-
    0025/031, Oct. 1979, Instructions [to Questionnaire] and Questionnaire;
    A1985-0036/043, 'Report of Ad Hoc Committee on Task Force Report on
    the Feasibility of Locating the Faculty of Forestry on Scarborough Campus,'
    and 'Response of the Council of the FF,' vol. II.
42 UTA FF, A1985-0036/043, 30 Oct. 1979, V.G. Smith to Love, in 'Response of
    the Council of the FF,' vol. II.
43 UTA OTP, A1986-0021/004, 13 Nov. 1977, Foley to Ham, encl. 12 Nov.
    1979, 'Response of Scarborough College to ... Locating the Faculty on the
    Scarborough Campus.'
44 *Toronto Star*, 4 Dec. 1979.
45 UTA OTP, A1986-0021/004, Dec. 1979, FF, UofT, 'Response to Move to
    Scarborough College,' original emphasis. UTA FF, A2004-0025/006, 28 Dec.
    1979, Nordin to Ham.
46 UTA OTP, A1986-0021/004, 11 Feb. 1980, Foley to Ham, original emphasis.
47 UTA FF, A2004-0025/008 (President, General), 13 Dec. 1979, File Note;
    /031, 6 June 1980, Nordin to Zimmerman. UTA OTP, A1987-0020/005, 14
    Jan. 1981, Love to Ham.
48 UTA OTP, A1987-0020/005: 18 Nov. 1980, Missen to Nordin; 18 Nov. 1980,
    Lang to Ham et al., and encl.; 19 Dec. 1980, Ham to Boswell.
49 Ibid., 27 Jan. 1981, Eastman to Planning Subcommittee, encl. 26 Jan. 1981,
    Draft 'Administrative Response to the Plans and Priorities of the FF.'
50 UTA FF, A2004-0025/006: 25 May 1981, Strangway to Nordin; 2 Sept. 1981
    and 5 Nov. 1982, Ham to W.G. Davis; 19 Feb. 1982, Davis to Nordin; *Bulletin*,
    26 Apr. 1982; 17 Sept. 1982, Nordin to Academic Staff; /044, 25 Mar. 1983,
    Nordin to R.N. Wolff. UTA OTP, A1989-0016/003, 9 Nov. 1981, Nordin to
    Ham, and 1 Sept. 1982, Nordin to Zimmerman; A1990-0021/022, 2 Aug.
    1983, Nordin to J. Roberts.
51 UTA FF, A2004-0025/006, 25 Mar. 1982, Notes, J.M. Ham; /008, 11 Mar.
    1982, White Paper Task Force. UTA OTP, A1992-0024/002, 22 Nov. 1984,
    Wolff to G.E. Connell.
52 UTA FF, A2004-0017/009 (R.E. Cooper); /015 (G. Gignac); /016 (D.W.
    Gilmore); /018 (R.K. Harris); /023 (M.O. Kemmsies); /027 (P.J. Martins);
    /037 (W. Sarafyn); /040 (C.L. Tan); /044 (J.S. Williams). M.O. Kemmsies,
    e-mail correspondence, 9 Apr. 2007.

53 UTA FF, A2004-0017/002 (H.M. Armleders) and (W.R. Armleders); /012 (D.C. Drysdale); /021 (G.I. Jameson); /035 (E.M. Raitanen) and (W.E. Raitanen); /036 (K.V. and R.R. Robinson); /037 (E.E. Salo) and (J.A. Salo); /039 (J.C. Stewart); /041 (A.B. Tworzyanski) and (T.J. Tworzyanski); /043 (H.A. Wiecek) and (S.A. Wiecek).
54 Ibid., /005 (G.R. Bartolotti); /009 (S.J. Colombo); /012 (J.M. Duncanson); /029 (D.A. McGorman); /030 (L.M. Miller) and (D.J. Milton); /031 (A.J. Mosseler); /032 (B.D. Nicks); /037 (R.A. Rosebrugh); /064 (S.M. Smith); /040 (R.G. Tomchik); /041 (R.M. Ubbens).
55 Ibid., /060 (S.L. Hummel); /025 (Min. of Governing Council), 11 Mar. 1993, Submission to Governing Council, incl. 19 Feb. 1993, M. Hummel to R.J. McGavin.
56 UTA OTP, A1989-0016/032: 15 Sept. 1982, Wolff to FF Staff; 16 Nov. 1982, Wolff to Ham; 7 Dec. 1982, Wolff to E. Wilson. A1990-0021/022: 20 July 1983, Wolff to FF Staff; 18 Oct. 1983, Wolff to Forster; 24 Feb. 1984, Strangway to M.J. Marques; 9 Mar. 1984, Marques to Strangway; 22 Mar. 1984, (Message for) 'Roger' (R.N. Wolff); 14 June 1984, Strangway to Love; A1992-0024/002, 9 Nov. 1984, Wolff to F. Iacobucci.
57 Ibid., A1992-0024/002, 25 Nov. 1983, Report of 8th Mtg. of the Faculty Advisory Board.

9: 'Forestry at U. of T. Is Not Dead Yet,' 1985–2005

1 UTA FF, A2004-0017/022 (S.G. Johnston), all correspondence; /028 (A.J. Mayes). UTA OTP, A1992-0024/002, 6 Nov. 1984, Employment, F. Keenan.
2 Friedland, chs. 39–40.
3 UTA OPV, A2001-0018/054, 19 Nov. 1985, J.R. Carrow to G.E. Connell.
4 Friedland, chs. 39–40.
5 Unless otherwise indicated, this chapter is based upon: UTA FF, A2004-0017/007 (J.R. Carrow), all documents, and the following phone interviews (all but first Carrow interview in 2007): R. Bryan, 17 May; J.R. Carrow, 11 Sept. 2006 and 4 May; L. Eckel, 17 May; M. Innes, 18 May; P.C. Schleifenbaum, 28 May; S. Smith, 29 May; T. Smith, 25 May; V.R. Timmer, 24 May.
6 UTA OTP, A1992-0024/002 (NRC), M. Hubbes; A1994-0013/002, 18 Oct. 1985, Nordin to Connell. Phone interviews: Carrow, 11 Sept. 2006 and 4 May 2007, and A. Zimmerman, 18 Aug. and 6 Sept. 2006.
7 Carrow, phone interviews, 11 Sept. 2006 and 4 May 2007. UTA FF, A2004-0025/044, Notes from Mtg. Held on Mar. 27, 1987, with Dean Carrow and Prof. V.G. Smith, FF.
8 UTA OTP, A1994-0013/002 (all 1986): 14 Apr., Carrow to Connell; 17 Apr.,

J.J. Balatinecz to R.G. Rosehart; 23 Apr., Zimmerman to V.G. Kerrio; 24 Apr., Rosehart to Balatinecz; 7 May, Connell to Rosehart; 12 May, Kerrio to Zimmerman.

9 UTA FF, A2004-0025/044: 3 Apr. 1985, R.N. Wolff to D.V. Love; 11 Apr. 1985, Love to Wolff; 22 and 25 Feb. and 21 Apr. 1988, Carrow to J. Keffer; 14 Mar. 1988, Keffer to Carrow.

10 UTA OTP, A1998-0008/022 (all 1988): 6 Sept., Zimmerman to Connell; 16 Sept., Connell to J.E. Foley (note); 4 Oct., Connell to Zimmerman.

11 Ibid.: 21 Oct., D.W. Lang to Keffer and M. Johnson; 27 Oct., Foley to Lang; 8 Dec., Carrow to Foley (on which Connell's notes are written and dated 31 Dec.); 13 Dec., Foley to Carrow.

12 Ibid.: 25 Oct., Lang to Johnson, and 8 Dec., Lang to Connell.

13 UTA BOG, A2005-0120/025, Apr. 1992, 'Report of the Working Group on the Future of the FF.' UTA OPV, A2001-0018/029 (all 1989): 24 Aug., Carrow to D. Pringle, encl. 'Towards 2000 – Synopsis'; 18 Sept., Lang to Pringle; 21 Sept., V.G. Smith to Lang; 16 Oct., Lang to Planning and Priorities.

14 UTA BOG, A2005-0120/025, 18 Jan. 1993, 'A Comparison of the Statement by Vice-President and Provost Joan E. Foley to the FF, 7 Jan. 1993 and to the Planning and Priorities Cte, 12 Jan. 1993.'

15 UTA OPV, A2001-0018/038, 13 Mar. 1990, A.H. Melcher to Carrow.

16 UTA BOG, A2005-0120/025, 18 Jan. 1993, 'A Comparison of the Statement.'

17 UTA OTP, A1998-0008/097, 19 Dec. 1991, Carrow to Foley, and 20 Dec. 1991, Balatinecz to Foley.

18 UTA FF, A2004-0025/025, 4 Feb. 1992, Foley to all FF Members. UTA OTP, A1998-0008/097, 28 Feb. 1992, J.E. Till to Foley, and 2 Mar. 1992, Foley to Till.

19 UTA BOG, A2005-0120/025, Apr. 1992, 'Report of the Working Group on the Future of the FF.'

20 UTA FF, A2004-0025/025, 13 July 1992, Academic and Non-academic Staff to Foley.

21 Ibid., 5 Aug. 1992, Foley to M. Candy.

22 Ibid., 19 Oct. 1992, D. Balsillie to Foley, in 'FF UofT, Submission to Governing Council for Thursday, Mar. 11, 1993' (hereafter 'Submission Mar. 1993'), and 14 Oct. 1992, J.K. Naysmith to Foley and W.J. Roll to Foley.

23 Ibid., 14 Oct. 1992, Zimmerman to Foley.

24 UTA OTP, A1998-0008/107, 8 Dec. 1992, S. Prichard to R. Allen. E-mail correspondence (in 2007) with: B. Shapiro, 14 May, and R. Allen, 16 May.

25 UTA BOG, A2005-0120/025, 12 Jan. 1993, P.L. Aird to R.J. McGavin, and 18 Jan. 1993, 'A Comparison of the Statement.' *Globe and Mail*, 7 Jan. 1993.

26 UTA BOG, A2005-0120/025, 'Report No. 47 of the Academic Board – Jan. 7th 1993.'

27 Ibid., cs. 14 Jan. 1993, Carrow to Dear Colleagues.

28 *Varsity*, 11 Jan. 1993. *University of Toronto Bulletin*, 25 Jan. 1993. *Newspaper*, 10 Feb. 1993.

29 UTA BOG, A2005-0120/025, 12 Jan. 1993, Foley to Members of the Planning and Priorities Cte, and 'Report No. 40 of the Planning and Priorities Cte – Jan. 19 and 26 1993,' and all attachments. *University of Toronto Bulletin*, 25 Jan. and 8 Feb. 1993.

30 UTA FF, A2004-0025/025, 28 Jan. 1993, Zimmerman to McGavin, in 'Submission Mar. 1993,' and 1993/01 – Feb. 11 1993 (01),' Supplementary Documentation, particularly Address to Mtg. of Academic Board, D.V. Love.

31 UTA BOG, A2005-0120/025, 13 Jan. 199[3], James Dat to J.F. Burke, original emphasis, and 13 Jan. 1993, C. Hutchinson to Burke.

32 *Globe and Mail*, 9 Feb. 1993. *Varsity*, 22 Feb. 1992.

33 UTA FF, A2004-0025/025, 19 Feb. 1993, M. Hummel to McGavin, in 'Submission Mar. 1993.'

34 Ibid., 9 Mar. 1993, Carrow to McGavin in 'Submission Mar. 1993.'

35 Ibid., 29 Mar. 1993, 'Report No. 48 of the Academic Board – Feb. 11th and 25th 1993.' *Toronto Star*, 26 Feb. 1993. *Varsity*, 1 Mar. 1993.

36 UTA FF, A2004-0025/025, 8 Mar. 1993, J.M. Duncanson to McGavin, in 'Submission Mar. 1993,' and 8 Mar. 1993, T. Buell to the Governing Council. *University of Toronto Faculty Association Newsletter – Special Report # 2*, 8 Mar. 1993.

37 UTA FF, A2004-0025/025, 9 Mar. 1993, Carrow to McGavin, in 'Submission Mar. 1993.'

38 Ibid., 29 Mar. 1993, Min. of the Governing Council Mtg held on Thursday, Mar. 11th, 1993 at 4:30 p.m. in the Council Chamber, Simcoe Hall.

39 Ibid., 16 Mar. 1993, J.E. Foley to Carrow.

40 *Varsity*, 16 Mar. 1993.

41 UTA FF, A2004-0017/027 (I.B. Mackenzie), 17 Mar. 1993, Mackenzie to Breakthrough Campaign; /037 (M.R. Rosen), 10 Feb. 1994, Rosen to R. Frankle.

42 *University of Toronto Bulletin*, 26 Apr. 1993. *Varsity*, 29 Mar. 1993. UTA OTP, A1998-0008/107, 8 Mar. and 5 Apr. 1993, D. Wang to Pritchard.

43 UTA FF, A2004-0017/016 (J.E. Glatt).

44 Carrow, phone interviews, 11 Sept. 2006 and 4 May 2007. UTA FF, A2004-0017/054 (N.A. Kliewer), 28 July 1993, Kliewer to Carrow.

45 UTA OPV, A2001-0018/061, 18 June 1993, Carrow to Faculty Staff et al., and encl., and 23 June 1993, Carrow to P.W. Gooch. UTA BOG, A2005-

0120/1994-02, 28 Apr. 1994, A.S. Sedra to J.F. Burke, encl. Dec. 1993 (Rev. 22 Mar. 1994), 'Towards 2001: Academic Plan, FF UofT, 1994 to 2000.'

46 UTA BOG, A2005-0120/025, 28 Apr. 1994, Sedra to Burke.

47 *Forestscope* 1(2), 1994.

48 *Academic Review 1999*, FF UofT (Dean's Copy).

49 June 1998, 'Graduate Studies in Forestry at the UofT: Document prepared for submission to the Ontario Council of Graduate Studies for period appraisal of the Forestry graduate degree programmes in the School of Graduate Studies, Section 1: The Programme' (Dean's Office, FF). *2004 Academic Plan, FF – University of Toronto* (Dean's Copy) (hereafter *2004 Academic Plan*).

50 *2004 Academic Plan.*

51 'Establishment of a Collaborative Research Programme between the FF, UofT, and Haliburton Forest and Wildlife, Inc., to Promote Sustainable Forest Management' (Dean's Copy), ca. 1994. *University of Toronto Magazine,* Autumn 1996. Research Report, FF UofT, 2000 (Dean's Office, FF). Schleifenbaum, phone interview, 28 May 2007.

52 *2004 Academic Plan.*

53 Ibid.

54 Ibid.

55 Dean's Copy, Dean's Office, FF, 'UofT – Brief for the Periodic Appraisal of the M.Sc.F., Ph.D., and M.F.C. Programs in the FF – Submitted to the OCGS, August 2005, vol. I: The Program' (hereafter 'Brief, August 2005').

56 UTA FF, A204-0017/046 (M.R. Coyea), (B. Farquhar) and (D.C. Krahn); /047 (J.P. Brandt), (K. Dewar), (D.C. Krahn), (T. Scarr), (S. Spezzaferro), (M. Streit), (S. Strobl), (A.M. Taylor), (S. Wetzel) and (S. Zuberec); /048 (R. Schuetz); /049 (M. Buchanan) and (G.D. Nigh); /050 (K. Farr) and (D. Kneeshaw); /052 (P. Fotiadis). 'Brief, August 2005,' passim. Phone interviews (2007): G.D. Nigh, 23 May, and K. Dewar, 15 May.

57 UTA FF, A2004-017/034 (R.J. Carvey); /046 (M.B. Karsh). *2004 Academic Plan.*

Conclusion

1 UTA FF, A1972-0025/190, 1908 List of Forest Engineers in Canada; /210, 'Forest Resources and Problems of Canada, B.E. Fernow, Delivered to the Society of American Foresters, 28 Dec. 1911,' 144; /187, 'Report of Senate Cte on FF, 1906–7.' *Globe and Mail,* 30 June 2007.

2 L.F. Riley, phone interview, 28 May 2007.

3 *Toronto Star,* 20 May 2007.

4 T. Smith, phone interview, 25 May 2007.

References

Primary Sources

Archives of Ontario (AO)

PRIVATE COLLECTIONS
F5, J.P. Whitney
F6, W.H. Hearst
F150, Gillies Lumber Company
F1014, F.A. MacDougall
F1095, A.H.D. Ross

GOVERNMENT COLLECTIONS
RG1, Department of Lands and Forests/Ministry of Natural Resources: Series
   A-I-10, BB1, and E-10; Series 41, 49, 122, and 335
RG2, Department of Education: Series 29 and 42
RG3, Office of the Premier: Series 2, 3, 4, 5, 6, 7, 8, 9, 10, 17, and 23
RG4, Department of the Attorney-General: Series 32
RG6, Department of the Treasury
RG8, Department of the Provincial Secretary: Series 5
RG18, Royal Commissions: Series 79, 125
RG32, Department of Colleges and Universities: Series 1

The Biltmore Company Archives (TBCA)

Forestry Department Manager's Records (FDMR): Series C, Box 7, File 45
   (Judson F. Clark), and Box 13, File 13 (Judson F. Clark); Series S, Box 34, File
   7 (Thomas Southworth), and Box 36, File 43 (C.D. Howe)

Cornell University Archives (CU)

Private Collections: 20-1-561, Bernhard Eduard Fernow, 1885–1930

Library and Archives Canada (LAC)

GOVERNMENT COLLECTIONS
RG32, Public Service Commission
RG39, Dominion Forest Service
Private Collections: MG30-A59, Gordon M. Dallyn

Queen's University Archives (QUA)

W.L. Goodwin Fonds

St Mary's Paper Archives (SMPA) – All fonds

Spruce Falls Inc. Archives (SFIA) – All fonds

University of British Columbia Archives (UBCA)
Private Collections: H.R. MacMillan Personal Papers

University of Toronto: Faculty of Forestry, Dean's Papers

University of Toronto Archives (UTA)

ADMINISTRATIVE RECORDS
Board of Governors/Governing Council (BOG): A1970-0024, A1973-0025,
    A2002-0003, and A2005-0120
Department of Admissions: A1969-0008
Department of Graduate Records: A1973-0026
Faculty of Forestry (FF): A1972-0018, A1972-0025, A1976-0006, A1979-0015,
    A1985-0036, A1990-0017, A2004-0017, and A2004-0025
Office of the President/Executive Assistant to the President (OTP): A1967-
    0007, A1968-0006, A1968-0007, A1971-0011, A1973-0029, A1975-0005,
    A1975-0019, A1975-0021, A1976-0020, A1977-0019, A1978-0028, A1979-0030,
    A1979-0042, A1979-0051, A1979-0057, A1982-0021, A1984-0026, A1985-0016,
    A1986-0002, A1986-0021, A1987-0020, A1989-0016, A1990-0021, A1992-0024,
    A1994-0013, and A1998-0008

Office of the Provost/Vice-Provost (OPV): A1977-0009, A1977-0020, and A2001-0018
Registrar: A1973-0051
Senate: A1970-0005 and A2002-0004

PRIVATE FONDS (PF)
B1972-0031, J. Loudon
B1979-0034, J.K. Harkness
B1974-0047, J.W.B. Sisam
B1989-0010, J.W.B. Sisam
B1983-0022, J.H. White

Government Documents

Ontario Sessional Papers
Statutes of Ontario

Newspapers/Periodicals

B.C. Lumberman
Canada Lumberman
Canadian Audubon Magazine
Canadian Forestry Journal
Capital Xtra
Educational Monthly
Forestry Chronicle
Forestry Quarterly
Globe and Mail
Newspaper
Port Arthur News Chronicle
Profiles: The York University Magazine for Alumni and Friends
Pulp and Paper Magazine of Canada
St Thomas Evening Journal
Toronto Daily Star
Toronto Evening Telegram
Toronto News
Toronto Star
Toronto World
Torontonensis
University of Toronto Bulletin

University of Toronto Magazine
Varsity
Windsor Star

Interviews (by Phone, E-mail, or Regular Correspondence)

Allen, M., e-mail, 3 April 2007.
Allen, R., e-mail, 16 May 2007.
Armson, K.A., phone, 17 Aug. and 16 Sept. 2006.
Bell, G., phone, 7 Oct. 2005.
Bryan, R., phone, 17 May 2007.
Carman, R.D., phone, 13 July 2006.
Carrow, J.R., phone, 11 Sept. 2006 and 4 May 2007.
Connell, A., phone, 28 Oct. 2005.
Daigle, N., phone, 28 May 2007.
Dewar, K., phone, 15 May 2007.
Eckel, L., phone, 17 May 2007.
Fayle, D., e-mail, 5 Nov. and 8 Dec. 2006 and June 2007.
Fernandes, D., phone and e-mail, 30 May and 3 June 2007.
Fessenden, R.J., phone, 15 Aug. 2006.
Fieldes, M.A., e-mail, 21 Oct. and 2 Nov. 2006.
Fotiadis, P., phone, 15 May 2007.
Fry, R.D., phone, 19 Feb. 2007.
Gignac, G., e-mail, 9 April 2007.
Innes, M.., phone, 18 May 2007.
Jennings, I.H., phone, 13 June 2007.
Jorgensen, E., phone, 24 July 2006.
Kemmsies, M., e-mail, 9 April 2007.
Loucks, O.L., e-mail and regular mail, 18 July and 14 Aug. 2006.
Miller, G., phone, 4 June 2007.
Mills, Rachel, e-mail, 19 July 2006.
Nigh, G.D., phone, 23 May 2007.
Nordin, V.J., e-mail, 7 Aug. 2006.
Pulkki, R., phone, 28 May 2007.
Riley, L., phone, 28 May 2007.
Robbins, Harvey, phone, 24 Oct. 2005.
Roche, M., e-mail and regular mail, 17, 19, and 20 Oct. 2005.
Schleifenbaum, Peter C., phone, 28 May 2007.
Shapiro, B., e-mail, 14 May 2007.
Smith, S., phone, 29 May 2007.

Smith, T., phone, 25 May 2007.
Smith, V.G., phone, 14 Aug. 2006.
Taylor, J.M., phone, 20 Feb. 2007.
Timmer, V.R., phone, 24 May 2007.
Veneziano, Amalia, phone and e-mail, 15 June 2007.
Watt, L., phone and e-mail, 4 June 2007.
Wilkes, George C., regular correspondence and phone, July 2006.
Zimmerman, Adam, phone, 18 Aug. and 6 Sept. 2006.

Secondary

Published

Avery, Donald. *Reluctant Host: Canada's Response to Immigrant Workers, 1890–1930*. Toronto: McClelland and Stewart, 1979.
Baskerville, Peter A. *Sites of Power: A Concise History of Ontario*. Toronto: Oxford University Press, 2005.
Bissell, Claude. *Halfway Up Parnassus: A Personal Account of the University of Toronto, 1932–1971*. Toronto: University of Toronto Press, 1974.
Bliss, Michael. *Northern Enterprise: Five Centuries of Canadian Business*. Toronto: McClelland and Stewart, 1987.
Bothwell, Robert, Ian Drummond, and John English. *Canada Since 1945*. Toronto: University of Toronto Press, 1989.
Bott, Robert, and Peter Murphy. *Living Legacy: Sustainable Forest Development at Hinton, Alberta*. Hinton: Weldwood of Canada, 1997.
Braun, Harold S. *A Northern Vision: The Development of Lakehead University*. Thunder Bay: Lakehead University, 1987.
Bryant, David G., ed. *The Fiftieth Anniversary of the Faculty of Forestry at the University of New Brunswick*. Fredericton: Unipress, 1958.
Calvin, D.D. *Queen's University at Kingston: The First Century of a Scottish-Canadian Foundation, 1841–1941*. Kingston: Trustees of the University, 1941.
Canada, Committee on Forests. *Forests of British Columbia*. Ottawa: Commission of Conservation, 1918.
Canada, *Report of the Canadian Forestry Convention Held at Ottawa, January 10, 11 and 12*. Ottawa: Government Printing Bureau, 1906.
Clark, Thomas D. *The Greening of the South: The Recovery of Land and Forest*. Kentucky: University of Kentucky Press, 1984.
Fernow, B.E. *Forest Conditions of Nova Scotia*. Ottawa: Commission of Conservation, 1912.
Freese, Frank. *A Collection of Log Rules*. Washington, DC: USDA, Forest Service General Technical Report, FPL 1: ca. – General Technical Report 1.

Friedland, Martin. *The University of Toronto: A History*. Toronto: University of Toronto Press, 2002.

Garratt, George A. *Forestry Education in Canada*. Montreal: Canadian Institute of Forestry, 1971.

Gillis, R. Peter, and Thomas R. Roach. *Lost Initiatives: Canada's Forest Industries, Forest Policy and Forest Conservation*. New York: Greenwood Press, 1986.

Girard, Michel. *L'écologisme retrouvé: Essor et déclin de la Commission de la Conservation du Canada*. Ottawa: Les Presses de l'Université d'Ottawa, 1994.

Graves, H.S., and E.A. Ziegler, *The Woodsman's Handbook*. Washington, DC: USDA, Forest Service, 1910.

Greenlee, James G., *Sir Robert Falconer: A Biography*. Toronto: University of Toronto Press, 1988.

Greeley, William B. *Forests and Men: A Veteran Forest Leader Tells the Story of The Last Fifty Years of American Forestry*. Garden City, NJ: Doubleday, 1951.

Hodgins, Bruce W., and Jamie Benedickson. *The Temagami Experience: Recreation, Resources, and Aboriginal Rights in the Northern Ontario Wilderness*. Toronto: University of Toronto Press, 1989.

Hosie, R.C. *Forest Regeneration in Ontario*. Toronto: University of Toronto Press and the Research Council of Ontario, 1953.

Howe, C.D. *Forest Regeneration on Certain Cut-Over Pulpwood Lands in Quebec*. Ottawa: Commission of Conservation, 1918.

Howe, C.D., and J.H. White. *Trent Watershed Survey*. Ottawa: Commission of Conservation, 1913.

Humphries, Charles W. *'Honest Enough to be Bold': The Life and Times of Sir James Pliny Whitney*. Toronto: University of Toronto Press, 1985.

Johnston, Charles M. *E.C. Drury: Agrarian Idealist*. Toronto: University of Toronto Press, 1986.

Johnstone, Kenneth. *Timber and Trauma: 75 Years with the Federal Forestry Service, 1899–1974*. Ottawa: Supply and Services Canada, 1991.

Killan, Gerald. *Protected Places: A History of Ontario's Provincial Park System* Toronto: Queen's Printer in association with Dundurn Press, 1993.

Kuhlberg, M. 'Ontario's Nascent Environmentalists: Seeing the Foresters for the Trees in Southern Ontario, 1919–1929.' *Ontario History* 88(2), 1996, 119–43.

– '"We have 'Sold' Forestry to the Management of the Company": Abitibi's Forestry Initiatives in Ontario, 1919–1929.' *Journal of Canadian Studies* 34(3), 1999, 187–210.

– '"We Are the Pioneers in This Business": Spanish River's Forestry Initiatives after the Great War.' *Ontario History* 93(2), 2001, 150–78.

– 'A Failed Attempt to Circumvent the Limits on Academic Freedom: C.D.

Howe, the Forestry Board, and "Window Dressing" Forestry in Ontario in the Late 1920s.' *History of Intellectual Culture* 2, 2002a, 1–23.

– '"By Just What Procedure Am I to Be Guillotined?": Academic Freedom in the Toronto Forestry Faculty between the Wars.' *History of Education* 31(4), 2002b, 351–70.

Lambert, Richard S., and Paul Pross. *Renewing Nature's Wealth: A Centennial History of the Public Management of Lands, Forests and Wildlife in Ontario, 1763–1967.* Toronto: Ontario Department of Lands and Forests, 1967.

MacKay, Donald. *Empire of Wood: The MacMillan Bloedel Story.* Vancouver: Douglas and McIntyre, 1982.

McDougall, A.K. *John P. Robarts: His Life and Government.* Toronto: University of Toronto Press, 1986.

McKillop, A.B. *Matters of Mind: The University in Ontario, 1791–1951.* Toronto: University of Toronto Press, 1994.

Murray, David R. *Hatching the Cowbird's Egg: The Creation of the University of Guelph.* Guelph: University of Guelph Press, 1989.

National Audubon Society. *National Audubon Society Field Guide to North American Trees.* New York: Knopf, 1980.

Neatby, Hilda. '*And Not to Yield': Queen's University*, vol. 1, *1841–1917.* Kingston and Montreal: McGill-Queen's University Press, 1978.

Nelles, H.Vivian. *The Politics of Development: Forests, Mines and Hydro-Electric Power in Ontario, 1849–1941.* Hamden, Conn.: Archon Books, 1974.

Norrie, Kenneth, and Douglas Owram. *A History of the Canadian Economy.* Toronto: Harcourt, Brace Jovanovich, 1991.

Oliver, Peter. *Public and Private Persons: The Ontario Political Culture, 1914–1934.* Toronto: Clarke, Irwin, 1975.

– *G. Howard Ferguson: Ontario Tory.* Toronto: University of Toronto Press, 1977.

Place, I.C.M. *75 Years of Research in the Woods: A History of Petawawa Forest Experiment Station and Petawawa National Forestry Institute.* Bunstown:General Store Publishing, 2002.

Rea, K.J. *The Prosperous Years: The Economic History of Canada, 1939–1975.* Toronto: University of Toronto Press, 1985.

Report of the Ontario Royal Commission on Forestry 1947. Toronto: Baptist John, Printer to the King's Most Excellent Majesty, 1947.

Richardson, A.H. *Conservation by the People: The History of the Conservation Movement in Ontario to 1970.* Toronto: University of Toronto Press, 1974.

Rodgers, Andrew Denny. *Bernhard Eduard Fernow: A Story of North American Forestry.* Princeton: Princeton University Press, 1951.

Saywell, J.T. *'Just Call Me Mitch': The Life of Mitchell F. Hepburn.* Toronto: University of Toronto Press, 1991.

Schmidt, R.L. *The History of Cowichan Lake Research Station*. Victoria: B.C. Ministry of Forestry, 1992.

Sisam, J.W.B. *Forestry and Forestry Education in a Developing Country: A Canadian Dilemma*. Toronto: University of Toronto Press, 1982.

Sissons, C.B. *A History of Victoria University*. Toronto: University of Toronto Press, 1952.

Smith, J.H.G. *UBC Forestry 1921–1990: An Informal History*. Vancouver: Faculty of Forestry, University of British Columbia, 1990.

Townsend, E. Ray. *Algonquin Forestry Authority … A 20 Year History*. Hunstville: Algonquin Forestry Authority, 1995.

Tripp-Knowles, Peggy. 'The Feminine Face of Forestry in Canada,' in *Challenging Professions: Historical and Contemporary Perspectives on Women's Professional Work*. Toronto: University of Toronto Press, 1999, 194–212.

White, Richard. *The Skule Story: The University of Toronto Faculty of Applied Science and Engineering, 1873–2000*. Toronto: Faculty of Applied Science and Engineering, 2000.

Zavitz, Edmund John. *Recollections, 1875–1964*. Toronto: Department of Lands and Forests, 1964.

Zucchi, John. *Italians in Toronto: Development of a National Identity, 1875–1935*. Toronto: University of Toronto Press, 1988.

UNPUBLISHED

Ayre, David John. 'Universities and the Legislature: Political Aspects of the Ontario University Question, 1868–1906.' Doctoral dissertation, University of Toronto, 1981.

Kuhlberg, Mark. '"In the power of the Government": The Rise and Fall of Newsprint in Ontario, 1894–1932.' Doctoral dissertation, York University, 2002.

Parsons, H.H. 'Aerial Timber Sketching Memoirs,' ca. 1977 (author's copy).

Index

Association of University Forestry
   Schools of Canada, 204, 209
Auld, J., 219
Australia, 55
Australian National University, 193
Austria, 261
Averill, H., x
Avery, B.F., 74, 134
Avery, D.D., 134
Ayer, E.B., 159

Bagg, J.K., 162
Balatinecz, J.J., 193
Balm, P., 8
Balsillie, D., xi
Barbados, 134
Barnum, F.J.B., 97–8
Bayly, G.H., 56, 108, 192
Bayly, G.W., 53, 56, 70, 81
BC Lumberman, 20
Beall, H.W., 122
Bedard, A., 133
Bell, T., 210
Benson, C.A., 194
Bentley, A.W., 49, 137
Bevilacqua, E., 229
Bier, J.E., 107
Biltmore Forest School (North Caro-
   lina), 25–6, 40, 66
Bissell, C., 177–9, 183, 185, 187–90,
   197
Black, R., 97–8
Blair, J.H., 161
Blais, J.R.F., 134
Blake, T., 256
Bliss, M., 156
bluebirds, 86
Boeing Company, 91
Bohm, W.D., 190
Bonner, E., 106, 128

Bortolotti, G.R., 229
Boswell, E.F., 226
Bothwell, G.E., 116
Bothwell, J.E., 163
Bothwell, R., 121
Boultbee, J.G., 134
Boultbee, R., 81, 134
Bourchier, R.J., 161
Bowater, 137
Brandt, J.P., 263
Bray, A.S., 108
Brazil, 105, 204, 228
Brebner, J., 70
Breckenridge, J., 244
British Columbia, 26, 58, 60–2,
   107
British Columbia Forest Service, 57,
   60, 83, 263, 267
British Columbia, University of, 60,
   107, 162; forestry school, 77, 103,
   179, 193, 199, 233, 235, 247–8
British Empire, 80
British Isles, 55, 80. See also United
   Kingdom
British West Indies, 158
Brodie, J.A., 75, 81, 102, 144
Brodie, J.D., 193–4
Bronte, Ontario, 94
Brown, A.W.A., 106
Brown, C., 262
Brown, W.G.E., 134
Bruce Peninsula, 86, 128
Bryan, R. See Faculty of Forestry,
   University of Toronto: Bryan's
   deanship
Buchanan, M., 263
Buck, D.E., 195
Buckley, C.E.H., 270
Buell, A.F., 105–6, 117
Buell, T.A., 159, 163, 251